AF361801

Micromachines for Dielectrophoresis

Micromachines for Dielectrophoresis

Editor

Rodrigo Martinez-Duarte

MDPI • Basel • Beijing • Wuhan • Barcelona • Belgrade • Manchester • Tokyo • Cluj • Tianjin

Editor
Rodrigo Martinez-Duarte
Mechanical Engineering
Clemson University
Clemson
United States

Editorial Office
MDPI
St. Alban-Anlage 66
4052 Basel, Switzerland

This is a reprint of articles from the Special Issue published online in the open access journal *Micromachines* (ISSN 2072-666X) (available at: www.mdpi.com/journal/micromachines/special_issues/Micromachines_for_Dielectrophoresis).

For citation purposes, cite each article independently as indicated on the article page online and as indicated below:

LastName, A.A.; LastName, B.B.; LastName, C.C. Article Title. *Journal Name* **Year**, *Volume Number*, Page Range.

ISBN 978-3-0365-4338-3 (Hbk)
ISBN 978-3-0365-4337-6 (PDF)

Contents

About the Editor vii

Preface to "Micromachines for Dielectrophoresis" ix

Rodrigo Martinez-Duarte
Editorial for the Special Issue on Micromachines for Dielectrophoresis
Reprinted from: *Micromachines* **2022**, *13*, 417, doi:10.3390/mi13030417 1

Akshay Kale, Amirreza Malekanfard and Xiangchun Xuan
Analytical Guidelines for Designing Curvature-Induced Dielectrophoretic Particle
Manipulation Systems
Reprinted from: *Micromachines* **2020**, *11*, 707, doi:10.3390/mi11070707 5

**Song-Yu Lu, Amirreza Malekanfard, Shayesteh Beladi-Behbahani, Wuzhou Zu, Akshay Kale
and Tzuen-Rong Tzeng et al.**
Passive Dielectrophoretic Focusing of Particles and Cells in Ratchet Microchannels
Reprinted from: *Micromachines* **2020**, *11*, 451, doi:10.3390/mi11050451 25

Ralph Hölzel and Ronald Pethig
Protein Dielectrophoresis: I. Status of Experiments and an Empirical Theory
Reprinted from: *Micromachines* **2020**, *11*, 533, doi:10.3390/mi11050533 39

Devin Keck, Callie Stuart, Josie Duncan, Emily Gullette and Rodrigo Martinez-Duarte
Highly Localized Enrichment of *Trypanosoma brucei* Parasites Using Dielectrophoresis
Reprinted from: *Micromachines* **2020**, *11*, 625, doi:10.3390/mi11060625 61

Chong Shen, Zhiyu Jiang, Lanfang Li, James F. Gilchrist and H. Daniel Ou-Yang
Frequency Response of Induced-Charge Electrophoretic Metallic Janus Particles
Reprinted from: *Micromachines* **2020**, *11*, 334, doi:10.3390/mi11030334 73

**Pierre-Emmanuel Thiriet, Joern Pezoldt, Gabriele Gambardella, Kevin Keim, Bart
Deplancke and Carlotta Guiducci**
Selective Retrieval of Individual Cells from Microfluidic Arrays Combining Dielectrophoretic
Force and Directed Hydrodynamic Flow
Reprinted from: *Micromachines* **2020**, *11*, 322, doi:10.3390/mi11030322 87

Honeyeh Matbaechi Ettehad, Pouya Soltani Zarrin, Ralph Hölzel and Christian Wenger
Dielectrophoretic Immobilization of Yeast Cells Using CMOS Integrated Microfluidics
Reprinted from: *Micromachines* **2020**, *11*, 501, doi:10.3390/mi11050501 101

Zichuan Yi, Zhenyu Huang, Shufa Lai, Wenyao He, Li Wang and Feng Chi et al.
Driving Waveform Design of Electrowetting Displays Based on an Exponential Function for a
Stable Grayscale and a Short Driving Time
Reprinted from: *Micromachines* **2020**, *11*, 313, doi:10.3390/mi11030313 119

Wenyao He, Zichuan Yi, Shitao Shen, Zhenyu Huang, Linwei Liu and Taiyuan Zhang et al.
Driving Waveform Design of Electrophoretic Display Based on Optimized Particle Activation
for a Rapid Response Speed
Reprinted from: *Micromachines* **2020**, *11*, 498, doi:10.3390/mi11050498 131

Benjamin G. Hawkins, Nelson Lai and David S. Clague
High-Sensitivity in Dielectrophoresis Separations
Reprinted from: *Micromachines* **2020**, *11*, 391, doi:10.3390/mi11040391 147

Jasper Giesler, Georg R. Pesch, Laura Weirauch, Marc-Peter Schmidt, Jorg Thöming and Michael Baune
Polarizability-Dependent Sorting of Microparticles Using Continuous-Flow Dielectrophoretic Chromatography with a Frequency Modulation Method
Reprinted from: *Micromachines* **2019**, *11*, 38, doi:10.3390/mi11010038 161

About the Editor

Rodrigo Martinez-Duarte

Rodrigo Martinez-Duarte is an Associate Professor in the Department of Mechanical Engineering at Clemson University (USA) and Head of the Multiscale Manufacturing Laboratory. His group's expertise lies at the interface between micro/nanofabrication, carbonaceous materials, electrokinetics and microfluidics. Rodrigo is known as the pioneer of carbon-electrode Dielectrophoresis (carbonDEP), a technique for bioparticle manipulation using carbon electrodes and microfluidics devices with application to diagnostics and therapeutics. He is also internationally known for pushing the envelope on the use of renewable materials and non-traditional techniques such as origami and robocasting to manufacture shaped geometries that serve as precursors to architected carbon and carbide structures. At the nanoscale, his group is innovating ways to use microbial factories as nanoweavers of biofibers. A recurrent theme in his Multiscale Manufacturing Laboratory is assessing the effect of processing on the properties of carbonaceous materials and structures at multiple length scales, towards tailoring their performance. At Clemson University he teaches manufacturing processes and their application, as well as fundamentals of micro/nanofabrication. His pedagogical approach emphasizes teamwork, flipped classrooms, and project-based learning.

Besides the US, Rodrigo has lived and worked in Switzerland, Spain, India, Mexico and South Korea and has a track record of service and leadership. He is also a past President of the AES Electrophoresis Society. He is or has chaired several sessions and international meetings on Carbon and/or Electrokinetics within the Electrochemical Society, Society for Hispanic Professional Engineers, AES, and the CarbonMEMS community.

Preface to "Micromachines for Dielectrophoresis"

Dielectrophoresis (DEP) remains an effective technique for the label-free identification and manipulation of targeted particles ranging from inert particles to biomolecules and cells. Applications are numerous, including clinical diagnostics and therapeutics, advanced manufacturing, electronic displays, and colloidal microrobots. This collection integrates novel contributions to the field in 2020 from multiple groups around the world.

Rodrigo Martinez-Duarte
Editor

micromachines

MDPI

Editorial

Editorial for the Special Issue on Micromachines for Dielectrophoresis

Rodrigo Martinez-Duarte

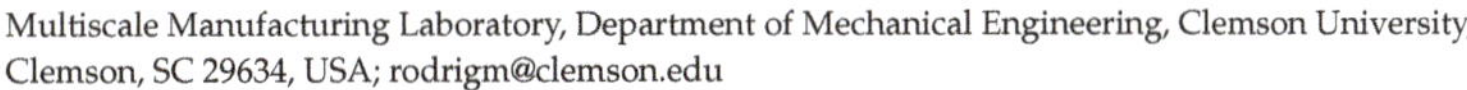

Multiscale Manufacturing Laboratory, Department of Mechanical Engineering, Clemson University, Clemson, SC 29634, USA; rodrigm@clemson.edu

---

Dielectrophoresis (DEP) remains an effective technique for the label-free identification and manipulation of targeted particles ranging from inert particles to biomolecules and cells. Applications are numerous, including clinical diagnostics and therapeutics, advanced manufacturing, electronic displays, and colloidal microrobots. This Special Issue includes 11 novel contributions to the field in multiple aspects from theory to application.

Kale et al. in "Analytical Guidelines for Designing Curvature-Induced Dielectrophoretic Particle Manipulation Systems" [1] present a novel mathematical framework to analyze particle dynamics inside a circular arc microchannel using computational modeling. Their analysis reveals that the design of such devices can be synthesized to three dimensionless parameters and provide validated equations to facilitate the design of curvature-induced DEP systems. The Xuan group at Clemson University also contributes "Passive Dielectrophoretic Focusing of Particles and Cells in Ratchet Microchannels" [2] by Lu et al., which is a fundamental study of the passive focusing of particles in ratchet microchannels using direct current DEP. Via computational modeling and experimentation, they demonstrate how particles were better focused using symmetric ratchet microchannels instead of asymmetric ones and postulate an equation to determine the particle focusing ratio depending on the particle's DEP and electrokinetic mobilities, channel width, electric field in the constriction, and the number and shape of ratchets.

Hölzel and Pethig contribute "Protein Dielectrophoresis: I. Status of Experiments and an Empirical Theory" [3], where their analysis of the DEP data for 22 different globular proteins revealed that 19 of such works reported protein DEP behavior at an electric field gradient much smaller than the $\sim 4 \times 10^{21}$ V^2/m^3 required to overcome the dispersive forces associated with Brownian motion, according to current DEP theory. They note that current DEP theory neglects the contribution of the permanent dipole moment of proteins to the DEP force and present a novel molecular version of the Clausius–Mossotti factor that was derived empirically and, when considered, brings most of the reported protein DEP above the minimum required to overcome dispersive Brownian thermal effects.

Regarding the application of DEP, different groups present their latest results regarding the use of DEP for the concentration of parasites in the context of global health, the use of Janus particles as colloidal microrobots, an integrated microfluidic system for single-cell isolation and retrieval, and the use of complementary metal–oxide–semiconductor (CMOS) fabrication processes to implement a cell viability assay. Keck et al. present "Highly Localized Enrichment of *Trypanosoma brucei* Parasites Using Dielectrophoresis" [4], where titanium electrodes are used to characterize the DEP response of *T. brucei* and enable its rapid enrichment in specific locations on-chip. This work is a step towards facilitating the direct identification of *T. brucei* when attempting to diagnose human African trypanosomiasis, also known as sleeping sickness. Regarding colloidal microrobots, Shen et al. contribute "Frequency Response of Induced-Charge Electrophoretic Metallic Janus Particles" [5], where they describe how electric and magnetic fields can be used to control the direction and speed of Janus particles by exploiting induced-charge electrophoresis (ICEP). Particle motion was characterized through phoretic force spectroscopy across the

Citation: Martinez-Duarte, R. Editorial for the Special Issue on Micromachines for Dielectrophoresis. *Micromachines* 2022, *13*, 417. https://doi.org/10.3390/mi13030417

Received: 7 March 2022

Accepted: 7 March 2022

Published: 8 March 2022

Publisher's Note: MDPI stays neutral with regard to jurisdictional claims in published maps and institutional affiliations.

range 1 kHz–1 MHz, and the authors report a change in direction at ~30 kHz, where particles transition from moving towards their dielectric side below 30 kHz to towards their metallic side above 30 kHz. In "Selective Retrieval of Individual Cells from Microfluidic Arrays Combining Dielectrophoretic Force and Directed Hydrodynamic Flow" [6], Thiriet and co-authors introduce a device for the isolation, retrieval, and off-chip recovery of single cells. Their design uses 3D electrodes embedded in a microfluidic channel to allow for the selective trapping of cells in specific sites through hydrodynamics, and their selective release using a negative DEP force. Cells were then recovered and analyzed off-chip with transcriptional analysis, revealing only a marginal alteration of their molecular profile. In "Dielectrophoretic Immobilization of Yeast Cells Using CMOS Integrated Microfluidics" [7], Ettehad and co-workers validated the use of CMOS-integrated microfluidic devices for the separation and purification of live yeast cells from dead ones using dielectrophoretic forces. This is an important contribution as it further demonstrates the feasibility of using well-established CMOS processes to fabricate DEP devices.

Towards improving the performance of electrowetting (EWD) and electrophoretic (EPD) electronic displays, the industry–academia collaboration between the University of Electronic Science and Technology, South China Academy of Advanced Optoelectronics, and the Shenzhen Guohua Optoelectronics Technology Co. presents "Driving waveform design of electrowetting displays based on an exponential function for a stable grayscale and a short driving time" [8] by Yi et al., and "Driving waveform design of electrophoretic display based on optimized particle activation for a rapid response speed" [9] by He et al. Yi and co-workers postulate an exponential function to drive EWD after studying the impact of the function time constant to reduce flicker and improve the static display performance of EWDs; meanwhile, He and co-workers' experimental results show that their postulated waveform leads to an improved display quality and a reduction in the flicker intensity in EPDs, when compared to a conventional waveform.

Lastly, important contributions to improving DEP separations are presented. Hawkins et al. in "High sensitivity in Dielectrophoresis separations" [10] critically review multiple ways to improve the sensitivity of DEP-based particle separations. These include combinations of 2D and 3D electrode structures, single or multiple field magnitudes and/or frequencies, and variations in the media suspending the particles. Giesler et al. in "Polarizability-Dependent Sorting of Microparticles Using Continuous-Flow Dielectrophoretic Chromatography with a Frequency Modulation Method" [11] present an improvement in the dielectrophoretic particle chromatography (DPC) of latex particles by exploiting differences in both their DEP mobility and crossover frequencies. To this end, they modulate the frequency of the electric field to induce periodic transitions from positive to negative movement and achieve multiple cycles of particle trap and release.

I would like to thank all the authors for contributing to this first installment of "Micromachines for Dielectrophoresis" as well as all the reviewers whose insightful feedback helped improve the impact of these contributions.

Conflicts of Interest: The author declares no conflict of interest.

References

1. Kale, A.; Malekanfard, A.; Xuan, X. Analytical Guidelines for Designing Curvature-induced dielectrophoretic particle manipulation systems. *Micromachines* **2020**, *11*, 707. [CrossRef] [PubMed]
2. Lu, S.Y.; Malekanfard, A.; Beladi-Behbahani, S.; Zu, W.; Kale, A.; Tzeng, T.R.; Wang, Y.N.; Xuan, X. Passive dielectrophoretic focusing of particles and cells in ratchet microchannels. *Micromachines* **2020**, *11*, 451. [CrossRef] [PubMed]
3. Hölzel, R.; Pethig, R. Protein dielectrophoresis: I. status of experiments and an empirical theory. *Micromachines* **2020**, *11*, 533. [CrossRef] [PubMed]
4. Keck, D.; Stuart, C.; Duncan, J.; Gullette, E.; Martinez-Duarte, R. Highly localized enrichment of *Trypanosoma brucei* parasites using dielectrophoresis. *Micromachines* **2020**, *11*, 625. [CrossRef] [PubMed]
5. Shen, C.; Jiang, Z.; Li, L.; Gilchrist, J.F.; Ou-Yang, H.D. Frequency response of induced-charge electrophoretic metallic janus particles. *Micromachines* **2020**, *11*, 334. [CrossRef] [PubMed]

6. Thiriet, P.E.; Pezoldt, J.; Gambardella, G.; Keim, K.; Deplancke, B.; Guiducci, C. Selective retrieval of individual cells from microfluidic arrays combining dielectrophoretic force and directed hydrodynamic flow. *Micromachines* **2020**, *11*, 322. [CrossRef] [PubMed]
7. Ettehad, H.M.; Zarrin, P.S.; Hölzel, R.; Wenger, C. Dielectrophoretic immobilization of yeast cells using CMOS integrated microfluidics. *Micromachines* **2020**, *11*, 501. [CrossRef] [PubMed]
8. Yi, Z.; Huang, Z.; Lai, S.; He, W.; Wang, L.; Chi, F.; Zhang, C.; Shui, L.; Zhou, G. Driving waveform design of Electrowetting Displays Based on an exponential function for a stable grayscale and a short driving time. *Micromachines* **2020**, *11*, 313. [CrossRef] [PubMed]
9. He, W.; Yi, Z.; Shen, S.; Huang, Z.; Liu, L.; Zhang, T.; Li, W.; Wang, L.; Shui, L.; Zhang, C.; et al. Driving waveform design of electrophoretic display based on optimized particle activation for a rapid response speed. *Micromachines* **2020**, *11*, 498. [CrossRef] [PubMed]
10. Hawkins, B.G.; Lai, N.; Clague, D.S. High-sensitivity in dielectrophoresis separations. *Micromachines* **2020**, *11*, 391. [CrossRef] [PubMed]
11. Giesler, J.; Pesch, G.R.; Weirauch, L.; Schmidt, M.P.; Thöming, J.; Baune, M. Polarizability-dependent sorting of microparticles using continuous-flow dielectrophoretic chromatography with a frequency modulation method. *Micromachines* **2020**, *11*, 38. [CrossRef] [PubMed]

 *micromachines*

Article

Analytical Guidelines for Designing Curvature-Induced Dielectrophoretic Particle Manipulation Systems

Akshay Kale [1,*], **Amirreza Malekanfard** [2] and **Xiangchun Xuan** [2]

1   Electrical Engineering Division, CAPE Building, Department of Engineering, University of Cambridge, Cambridge CB3 0FA, UK
2   Department of Mechanical Engineering, Clemson University, Clemson, SC 29634, USA; amaleka@g.clemson.edu (A.M.); xcxuan@clemson.edu (X.X.)
*   Correspondence: ak2115@cam.ac.uk

Received: 24 June 2020; Accepted: 20 July 2020; Published: 21 July 2020

Abstract: Curvature-induced dielectrophoresis (C-iDEP) is an established method of applying electrical energy gradients across curved microchannels to obtain a label-free manipulation of particles and cells. This method offers several advantages over the other DEP-based methods, such as increased chip area utilisation, simple fabrication, reduced susceptibility to Joule heating and reduced risk of electrolysis in the active region. Although C-iDEP systems have been extensively demonstrated to achieve focusing and separation of particles, a detailed mathematical analysis of the particle dynamics has not been reported yet. This work computationally confirms a fully analytical dimensionless study of the electric field-induced particle motion inside a circular arc microchannel, the simplest design of a C-iDEP system. Specifically, the analysis reveals that the design of a circular arc microchannel geometry for manipulating particles using an applied voltage is fully determined by three dimensionless parameters. Simple equations are established and numerically confirmed to predict the mutual relationships of the parameters for a comprehensive range of their practically relevant values, while ensuring design for safety. This work aims to serve as a starting point for microfluidics engineers and researchers to have a simple calculator-based guideline to develop C-iDEP particle manipulation systems specific to their applications.

Keywords: microfluidics; dielectrophoresis; curvature-induced; electrokinetic; particle focusing

---

1. Introduction

Obtaining a pre-concentrated sample of particles is an important step for its use in subsequent operations in micro total analysis systems (μTAS) [1–4]. The particle concentration process in these microfluidics-based systems is commonly brought about by manipulation of particle motion inside the "active regions" located in microfluidic flows with the help of an external energy field-induced force in conjunction with the hydrodynamic drag force. Such manipulation leads to a deflection of the particle motion from its unperturbed direction, which is induced by the component of the external force orthogonal to the flow direction. The resulting motion of the particles causes them to flow through a more confined volume within the microfluidic channel, thereby focusing them and increasing their local concentration [5]. This concentrated sample can then be extracted for further downstream processing by designing an outlet junction to the primary channel and exploiting the laminar nature of microfluidic flows. The particle concentration forces induced by the energy fields are functions of the intrinsic properties of the particles relative to those of the fluid. Hence, no physical labels need to be attached to the particles for tagging them. Since these tags can affect the particle phenotype, especially in the case of biological particles, the concentration process are entirely non-invasive to the particles [6]. This non-invasive nature of the concentration process makes it particularly suitable to biomedical research where retaining the phenotype of the bio-particle is extremely crucial. Depending upon the type of external energy field applied, the concentration processes can be induced acoustically [7,8], magnetically [9,10], optically [11,12], hydrodynamically [13,14], or dielectrophoretically [15]. Along with the advantages of a non-invasive nature and a cubic scaling of the concentration force with particle size [16], which are offered by each of the aforementioned methods, dielectrophoretic particle concentration offers additional benefits: (A) easier integration with the subsequent downstream electrically driven detection systems [17] and (B) possibility to exploit the plug-like velocity profile of the bulk electroosmotic fluid flow for transporting the particles, which avoids dispersion issues prevalent in pressure driven flows [18].

The gradients of the external energy field in dielectrophoresis (DEP) are of an electrical nature, and hence, the dielectrophoretic concentration of particles occurs due to the differences in the electrical properties of the particles relative to the fluid [19–21]. The dielectrophoretic force acting on a polarisable sphere with a diameter d, an electrical conductivity σ_p and a dielectric constant ε_p inside a fluid with an electrical conductivity σ_f and a dielectric constant ε_f can be expressed as follows:

$$\mathbf{F}_{\mathrm{DEP}} = \frac{\pi}{4}d^3\varepsilon_f Real(f_{CM})\nabla(\mathbf{E}{\cdot}\mathbf{E}) \text{ where } f_{CM} = \left[\frac{\left(\varepsilon_p - \varepsilon_f\right) + \mathrm{j}\frac{(\sigma_p - \sigma_f)}{2\pi f}}{\left(\varepsilon_p + 2\varepsilon_f\right) + \mathrm{j}\frac{(\sigma_p + 2\sigma_f)}{2\pi f}}\right]; \mathrm{j} = \sqrt{-1} \tag{1}$$

where $\mathbf{E}$ is the externally applied electric field, f is the frequency of the applied voltage and f_{CM} is the Clausius-Mossotti factor of the particle-fluid system. Based on the method of inducing electrical energy gradients, i.e., $\nabla(\mathbf{E}{\cdot}\mathbf{E})$, dielectrophoretic concentration process can be broadly classified as electrode-based (i.e., eDEP, where the gradients are generated by a set of patterned [22–28] or virtual electrodes [29]), insulator-based (i.e., iDEP, where the gradients are generated by non-uniform cross sections within the microfluidic circuit [30–35]), or curvature-induced (i.e., C-iDEP, where the curvature of the microfluidic channel produces unequal electric field intensities across a channel cross section [36]). Simply making the channel curved is sufficient to generate DEP, and hence, the cross section of the microfluidic channel need not be reduced, thereby rendering C-iDEP systems much less susceptible to localised Joule heating effects that are more prevalent in their insulator-based counterparts [37–40]. Similarly, the typical fluid-reservoir-based electrode insertion outside the flow channel not only renders the fabrication simple, but also keeps these systems relatively safe from electrolysis compared to the metallic microelectrode-based eDEP. Additionally, the curvature facilitates the use of a longer channel length within a given area compared to a straight channel, and this greatly increases the area utilisation of lab-on-a-chip systems implementing C-iDEP [41].

While several reports in the existing literature have demonstrated the use of C-iDEP for manipulating particles and cells [42–47], a rigorous mathematical treatment of the dielectrophoretic particle dynamics for these systems is still lacking. Such a treatment will prove to be extremely helpful in providing guidelines to design C-iDEP particle manipulation systems for lab-on-a-chip applications. This work demonstrates a fully analytical treatment of the underlying physics of C-iDEP. Briefly, a circular arc microchannel, which represents the most basic design for C-iDEP particle manipulation systems, is considered for the analysis. Elegant, fully dimensionless equations of the pathline of a particle undergoing DEP in this channel geometry are established from first principles. It is shown that the particle dynamics and the design of circular arc microchannels can be fully controlled by three dimensionless parameters. Two-dimensional finite element simulations are used to support the analysis and are shown to agree with the theoretical equations with great accuracy. This work establishes a simple calculator-based approach to enable microfluidics engineers and scientists to design C-iDEP systems with a factor of safety.

2. Theory and Analysis

2.1. Dielectrophoretic Particle Dynamics in a Circular Arc Microchannel

Figure 1 describes a generalised geometry for the theoretical analysis of particle physics in C-iDEP systems. Let V_0 be a DC voltage drop biased with an AC voltage characterised by an RMS AC to DC ratio α applied across a circular arc microfluidic channel. The channel has a uniform width W, a uniform height H and a mean radius of curvature R_c. The arc subtends an angle β at the centre of the curvature of the channel. Assuming (a) a thin electric double layer limit, (b) negligible displacement currents and (c) uniform liquid properties, the applied DC voltage drop generates an electric field $\mathbf{E}_{DC}$ inside the microchannel, which is governed by the Laplace equation. In cylindrical coordinates, this can be expressed as

$$\frac{1}{r}\frac{\partial}{\partial r}\left(r\frac{\partial V_{DC}}{\partial r}\right) + \frac{1}{r^2}\frac{\partial^2 V_{DC}}{\partial \theta^2} + \frac{\partial^2 V_{DC}}{\partial z^2} = 0 \tag{2}$$

$$\mathbf{E}_{DC} = -\nabla V_{DC} = -\frac{\partial V_{DC}}{\partial r}\hat{\mathbf{r}} - \frac{\partial V_{DC}}{r\partial \theta}\hat{\boldsymbol{\theta}} - \frac{\partial V_{DC}}{\partial z}\hat{\mathbf{z}} \tag{3}$$

where V_{DC} is the DC potential field. $\hat{\mathbf{r}}$, $\hat{\boldsymbol{\theta}}$ and $\hat{\mathbf{z}}$ represent the unit vectors in the radially outward direction, the counter-clockwise angular direction and the upward direction along the channel height respectively. Recognising that the voltage drop is applied across the entire cross-section of the channel represented by the r-z plane, the above equations simplify to an ordinary differential equation expressed as

$$\frac{d^2 V_{DC}}{d\theta^2} = 0; \; \mathbf{E}_{DC} = -\frac{1}{r}\frac{dV_{DC}}{d\theta}\hat{\boldsymbol{\theta}}; \; \text{subject to } V_{DC}(\theta = 0) = V_0; \; V_{DC}(\theta = \beta) = 0 \tag{4}$$

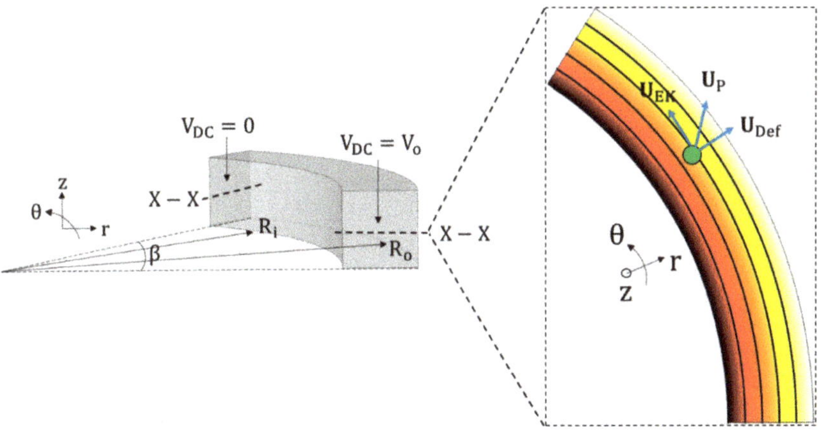

Figure 1. Schematic of a circular arc microchannel explaining the dimensions and the r-θ-z coordinate system, and framework used for a theoretical analysis of the underlying physics of curvature-induced dielectrophoresis (C-iDEP). The 2D projection of the r-θ section plane shows the components of the particle velocity, $\mathbf{u}_P$, inside the microchannel, consisting of the stream-wise electrokinetic ($\mathbf{u}_{EK}$) component and the cross-stream dielectrophoretic deflection ($\mathbf{u}_{Def}$) component. The electric field contour (the darker colour the larger magnitude) and lines (equivalent to the fluid streamlines [48]) are superimposed to explain the orientation of the velocity components relative to the electric field. A situation of negative DEP is shown, where the particle will move from the inner channel wall to the outer wall. For a positive DEP, the resultant velocity and the deflection component will reverse their directions. Note that the underlying physics of C-iDEP are two dimensional because of the application of the voltage over the entire r-z plane.

The solution of Equation (4) is given as

$$V_{DC} = V_0\left(1 - \frac{\theta}{\beta}\right); \quad \mathbf{E}_{DC} = \frac{V_0}{\beta r}\hat{\theta} \tag{5}$$

Due to this electric field, a spherical particle inside the microchannel containing a liquid of dynamic viscosity η_f experiences an electrokinetic velocity $\mathbf{u}_{EK}$, which can be written as

$$\mathbf{u}_{EK} = \frac{\varepsilon_f\left(\zeta_p - \zeta_w\right)\mathbf{E}_{DC}}{\eta_f} = \frac{\varepsilon_f\left(\zeta_p - \zeta_w\right)V_0}{r\beta\eta_f}\hat{\theta} \tag{6}$$

where ζ_p and ζ_w represent the particle and wall zeta potentials respectively. For negative wall and particle zeta potentials (which is usually valid for microfluidic particle handling systems [49]), the electrokinetic motion occurs in the positive (counter-clockwise) direction along the arc length if the wall zeta potential is lower than the particle zeta potential (i.e., $\zeta_w < \zeta_p$) and in the negative (clockwise) direction if vice versa. Note that the AC voltage does not contribute to the electrokinetic motion of the particles because of the linear dependence of the electrokinetic velocity with the electric field (which causes the time average of the AC electric field over a cycle to vanish).

It is also observed from Equation (5) that the electric field inside the microchannel is a function of the radial co-ordinate, thereby making it non-uniform. The resulting electric field gradients inside the channel can be expressed using the expression for DC electric field (Equation (5)) and the definition of the gradient vector in cylindrical co-ordinates (see Equation (3)).

$$\nabla(\mathbf{E}\cdot\mathbf{E}) = \left[\frac{\partial(\mathbf{E}_{DC}\cdot\mathbf{E}_{DC})}{\partial r}\hat{\mathbf{r}} + \frac{\partial(\mathbf{E}_{DC}\cdot\mathbf{E}_{DC})}{r\partial\theta}\hat{\theta} + \frac{\partial(\mathbf{E}_{DC}\cdot\mathbf{E}_{DC})}{\partial z}\hat{\mathbf{z}}\right]\left(1 + \alpha^2\right) = \frac{-2V_0^2\left(1 + \alpha^2\right)}{\beta^2 r^3}\hat{\mathbf{r}} \tag{7}$$

Note that the θ and z components of the electric field gradients vanish due to the electric field being purely a function of the radial coordinate. In addition, note the contribution of the AC voltage through α stemming from the non-zero time-averaged nature of the square of the electric field. A spherical particle of diameter d flowing electrokinetically along an electric field line inside the microchannel (according to Equation (6)) experiences a dielectrophoretic force due to this electric field gradient. Using Equations (1) and (7), the force can be expressed as

$$\mathbf{F}_{DEP} = \frac{-\pi}{2}d^3\varepsilon_f Real(f_{CM})\frac{V_0^2\left(1 + \alpha^2\right)}{\beta^2 r^3}\hat{\mathbf{r}} \tag{8}$$

Equation (8) shows that the negative DEP force is always directed along the positive radial direction (As $Real(f_{CM}) < 0$, $\mathbf{F}_{DEP} > 0$), and hence such a particle moves from the inner channel wall to the outer channel wall. Similarly, the positive DEP force is always directed along the negative radial direction (As $Real(f_{CM}) > 0$, $\mathbf{F}_{DEP} < 0$), and hence, such a particle moves from the outer channel wall to the inner channel wall. It is also clear from Equations (6) and (8) that the DEP force always acts on the particle in a direction orthogonal to its electrokinetic motion and hence cannot directly oppose the same. Hence, the curved arc microchannel is unable to immobilise i.e., trap particles dielectrophoretically, but can only allow a continuous radial deflection of the particles as they move inside the microchannel along the circumferential direction of the arc. In addition, a particle close to the microchannel walls, locally perturbs the electric field between itself and the wall, resulting in a wall repulsion force that always acts on the particle in a direction **n** normal to the wall (i.e., the radial direction in this case). Neglecting the surface conduction effects, the force $\mathbf{F}_w$ acting on a particle sufficiently smaller than the channel curvature can be estimated up to a first order approximation as [50]

$$\mathbf{F_w} \approx \varepsilon_f \frac{d^2}{4}\left(\mathbf{E_{DC}}^W\right)^2\left(1+\alpha^2\right)f\left(\frac{d}{h}\right)\hat{\mathbf{n}} \text{ where } f = \frac{3\pi}{256}\left(\frac{d}{h}\right)^4 \tag{9}$$

where $\mathbf{E_{DC}}^W$ represents the scale of electric field near the wall, and h is the distance between the particle centre and the wall along the normal direction n pointing away from the wall. Similar to dielectrophoresis, the AC field contributes to the DC field in generating the repulsion effect because of the quadratic dependence on the electric field. Using this relation at the inner and outer walls of the circular arc microchannel, the wall repulsion forces at the inner (represented by the superscript "i", i.e., $\mathbf{F_w}^i$) and outer (represented by the superscript "o", i.e., $\mathbf{F_w}^o$) walls may be respectively approximated as [50]

$$\mathbf{F_w}^i \approx \frac{3\pi\varepsilon_f d^6\left(\mathbf{E_{DC}}^i\right)^2\left(1+\alpha^2\right)}{1024(r-R_i)^4}\hat{\mathbf{r}}; \; \mathbf{F_w}^o \approx -\frac{3\pi\varepsilon_f d^6\left(\mathbf{E_{DC}}^o\right)^2\left(1+\alpha^2\right)}{1024(R_o-r)^4}\hat{\mathbf{r}} \tag{10}$$

where $\mathbf{E_{DC}}^i$ and $\mathbf{E_{DC}}^o$ respectively indicate the electric field scales near the inner (i) and outer (o) walls of the microchannel and will be interpreted in more detail after a few steps. Note that the terms "$r-R_i$" and "R_o-r" represent the distance h between the particle centre-line and the inner and outer channel walls respectively, in a direction normal to the walls (i.e., the radial direction in this case). In addition, note that the negative sign for $\mathbf{F_w}^o$ indicates that the action of the force is opposite to the sign convention of a positive radially outward unit vector $\hat{\mathbf{r}}$.

The effective radial deflection force on the particle is obtained by the sum of Equations (8) and (10), which, after balancing against the Stokes drag force ($3\pi\eta_f d\mathbf{u_{Def}}$) for a spherical particle, gives an expression for the terminal deflection velocity $\mathbf{u_{Def}}$ of the particle:

$$\mathbf{u_{Def}} = \left[\frac{-d^2\varepsilon_f Real(f_{CM})V_0^2\left(1+\alpha^2\right)}{6\beta^2 r^3\eta_f} + \frac{\varepsilon_f d^5\left(1+\alpha^2\right)}{1024\eta_f}\left\{\frac{\left(\mathbf{E_{DC}}^i\right)^2}{(r-R_i)^4} - \frac{\left(\mathbf{E_{DC}}^o\right)^2}{(R_o-r)^4}\right\}\right]\hat{\mathbf{r}} \tag{11}$$

Note that Equation (11) assumes a mass-less particle analysis, which holds for miniaturised systems owing to the negligible inertia of the micron-sized particles [18]. The time dependent position vector $\mathbf{r_p}$ of the particle moving inside the microchannel can now be related to the particle velocity $\mathbf{u_p}$ as

$$\mathbf{u_p} = \mathbf{u_{Def}} + \mathbf{u_{EK}} = \frac{d\mathbf{r_p}}{dt} = \frac{dr}{dt}\hat{\mathbf{r}} + r\frac{d\theta}{dt}\hat{\theta} + \frac{dz}{dt}\hat{z} \tag{12}$$

Recognising that the circumferential component of $\mathbf{u_p}$ is represented entirely by the electrokinetic velocity $\mathbf{u_{EK}}$ (Equation (6)) and the radial component by the deflection velocity $\mathbf{u_{Def}}$ (Equation (11)), we can eliminate the vector notations and write

$$\frac{dr}{dt} = \frac{-d^2\varepsilon_f Real(f_{CM})V_0^2\left(1+\alpha^2\right)}{6\beta^2 r^3\eta_f} + \frac{\varepsilon_f d^5\left(1+\alpha^2\right)}{1024\eta_f}\left\{\frac{\left(\mathbf{E_{DC}}^i\right)^2}{(r-R_i)^4} - \frac{\left(\mathbf{E_{DC}}^o\right)^2}{(R_o-r)^4}\right\}; \; r\frac{d\theta}{dt} = \frac{\varepsilon_f\left(\zeta_p-\zeta_w\right)V_0}{\beta\eta_f r} \tag{13}$$

which, after eliminating the time coordinate, gives

$$\frac{dr}{d\theta} = \frac{-d^2 Real(f_{CM})V_0\left(1+\alpha^2\right)}{6\beta r\left(\zeta_p-\zeta_w\right)} + \frac{\beta d^5\left(1+\alpha^2\right)}{1024\left(\zeta_p-\zeta_w\right)V_0}\left\{\frac{\left(r\mathbf{E_{DC}}^i\right)^2}{(r-R_i)^4} - \frac{\left(r\mathbf{E_{DC}}^o\right)^2}{(R_o-r)^4}\right\} \tag{14}$$

Note that the absence of electric field and its gradients in the z direction (i.e., along the channel depth) converts the 3D problem into a 2D problem in the r-θ plane. Equation (14) is an ordinary differential equation that represents the rate of change of the radial coordinate of the particle with respect to its circumferential coordinate and is therefore a direct indication of the dielectrophoretic deflection performance of the C-iDEP system. Note that the term r^2 arising after simplifying Equation (13) is taken inside the brackets with $\mathbf{E_{DC}}^i$ and $\mathbf{E_{DC}}^o$. For further analysis, it is assumed that the particle is sufficiently smaller than the geometry of the channel for the electric field scale around a particle at one wall to be unaffected by the other wall (i.e., for a semi-infinite domain approach to be valid). This assumption is reasonably good in practical microfluidic systems. Under this assumption, the local scales of the quantities $r\mathbf{E}_{DC}{}^{i}$ and $r\mathbf{E}_{DC}{}^{o}$ around the particle can be estimated using Equation (5) as $r\mathbf{E}_{DC}{}^{i} \sim R_i\left(\frac{V_0}{\beta R_i}\right) \sim \frac{V_0}{\beta}$ and $r\mathbf{E}_{DC}{}^{o} \sim R_o\left(\frac{V_0}{\beta R_o}\right) \sim \frac{V_0}{\beta}$. Using these expressions, Equation (14) becomes

$$\frac{dr}{d\theta} = \frac{d^2 V_0\left(1+\alpha^2\right)}{6\beta\left(\zeta_p - \zeta_w\right)}\left[\frac{-Real(f_{CM})}{r} + \frac{3d^3}{512}\left\{\frac{1}{(r-R_i)^4} - \frac{1}{(R_o-r)^4}\right\}\right] \tag{15}$$

Equation (15) can be converted into a fully dimensionless equation by defining a dimensionless radial coordinate r^* and a dimensionless angular co-ordinate θ^* as $r^* = (r - R_i)/W$ and $\theta^* = \theta/\beta$ where $W = R_o - R_i$ is the channel width. In addition, the mean radius of curvature R_C of the channel is introduced by using the relation $R_C = R_i + (W/2)$. With these substitutions, the non-dimensional form of Equation (15) can be written as

$$\frac{dr^*}{d\theta^*} = V_{App}{}^* d^{*2}\left[\frac{-Real(f_{CM})}{R_C{}^* + r^* - 0.5} + \frac{3}{512}d^{*3}\left\{\frac{1}{r^{*4}} - \frac{1}{(1-r^*)^4}\right\}\right] \tag{16}$$

where

$$V_{App}{}^* = \frac{V_0\left(1+\alpha^2\right)}{6\left(\zeta_p - \zeta_w\right)}, R_C{}^* = \frac{R_C}{W}, d^* = \frac{d}{W} \tag{17}$$

Equation (16) shows that the dielectrophoretic motion of a particle in a circular arc microchannel can be completely characterised using three dimensionless numbers, which are defined in Equation (17). $V_{App}{}^*$ is a measure of the strength of the applied voltage relative to the electrical double layer (EDL) potential, which the liquid induces at the surface of the particle and the channel wall. $V_{App}{}^*$ is positive for a counter-clockwise electrokinetic motion of particles and negative for a clockwise one. $R_C{}^*$ and d^* respectively indicate the dimensionless curvature ratio of the microchannel and the dimensionless particle blockage ratio. It is also important to note that the particle motion is completely independent of the electrical permittivity and dynamic viscosity of the fluid because of the linear dependence of the individual particle velocity components on the quantity ε_f/η_f and the resulting cancellation of this quantity while deriving the differential equation of the particle motion.

2.2. A Simplified Exact Solution

Equation (16) is not possible to be solved analytically using straightforward integration methods. However, it can be simplified to an elegant form if the influence of repulsion forces near the channel walls becomes negligible. This happens when the particle is treated as an infinitesimally small point, an assumption which holds if its size is much smaller compared to the characteristic dimensions of the channel geometry (in other words, $d^* \ll 1$ or d << W,R_C). In this scenario, Equation (16) becomes

$$\frac{dr^*}{d\theta^*} = \frac{-Real(f_{CM})V_{App}{}^* d^{*2}}{R_C{}^* + r^* - 0.5} \tag{18}$$

The solution of Equation (18) between any two locations $(r_1{}^*, \theta_1{}^*)$ and $(r_2{}^*, \theta_2{}^*)$ in the channel can be derived as

$$\frac{(r_2{}^* + R_C{}^* - 0.5)^2 - (r_1{}^* + R_C{}^* - 0.5)^2}{2d^{*2}Real(f_{CM})} = V_{App}{}^*(\theta_1{}^* - \theta_2{}^*) \tag{19}$$

Equation (19) holds provided the particles are much smaller than the characteristic dimensions of the channel. This equation may also be viewed as an ideal solution for the particle motion inside the curved microchannel as it ignores the inevitable existence of wall repulsion forces in a special case.

Of particular importance for microfluidics engineers is the use of Equation (19) for quantifying the dimensionless design parameters in order to achieve a full focusing of an incoming particle suspension at the outlet of the channel. For example, in the case of a negative DEP occurring for a particle suspension moving electrokinetically in a positive angular (counter-clockwise) direction, full focusing can be realised by setting $r_1^* = 0$, $r_2^* = 1$ and $\theta_1^* = 0$, $\theta_2^* = 1$. Upon making these substitutions, one can simplify Equation (19) as

$$\frac{R_C^*}{d^{*2}Real(f_{CM})} = -V_{App}^* \tag{20}$$

Note that even if the right hand side (RHS) of Equation (20) is negative because of $\theta_1^* < \theta_2^*$, the negative value of $Real(f_{CM})$ maintains the positivity of the equation. It can be confirmed that Equation (20) retains its mathematical form for all other combinations of the directions of electrokinetic and DEP motion: (a) positive electrokinetic motion, positive DEP, (b) negative electrokinetic motion, negative DEP and (c) negative electrokinetic motion, positive DEP. The only change that occurs for each of these situations is the sign of the equation. Recognising this and taking the magnitude of f_{CM} and $(\theta_1^* - \theta_2^*)$ into account, the relationship between the dimensionless parameters for full focusing can be absorbed into an elegant expression as follows:

$$\frac{\left|V_{App}^*\right|d^{*2}\left|Real(f_{CM})\right|}{R_C^*} = 1 \tag{21}$$

Equation (21) can be rearranged in several forms as shown below in order to determine the threshold value of a dimensionless design parameter for achieving a full focusing from the channel inlet to the channel outlet when the other parameters are known.

$$\left|V_{App}^*\right|_{Min} = \frac{R_C^*}{d^{*2}\left|Real(f_{CM})\right|}; \ d^*_{Min} = \sqrt{\frac{R_C^*}{\left|V_{App}^*\right|\left|Real(f_{CM})\right|}}; \ R_{C\ Max}^* = \left|V_{App}^*\right|d^{*2}\left|Real(f_{CM})\right| \tag{22}$$

Equation (22), in combination with Equation (17), can be used to determine: (a) the minimum voltage that must be applied for fully focusing a given particle size inside a given channel geometry, (b) the smallest particle size that can be fully focused inside a given channel geometry due to a given applied voltage and (c) the maximum curvature radius which one can provide to the channel for fully focusing a given particle size due to a given applied voltage.

2.3. The Full Solution

Realistically, the particle size being finite and non-zero, the particle path-lines deviate from the ideal solution and the wall repulsion forces inevitably contribute to the dielectrophoretic focusing effects. Hence, the full solution of Equation (16) needs to be considered and critically evaluated for characterising the C-iDEP microchannel design. However, the integration methods for solving the equation analytically are not straightforward, and hence, it must be solved using a 1-D numerical integration. Considering the example of negative DEP and a positive electrokinetic motion as before, Equation (16) is integrated within the limits r_1^*, r_2^* and θ_1^*, θ_2^* for the radial and angular co-ordinates respectively. However, the finite size of the particle is now taken into account by setting the initial radial position, i.e., r_1^* as $d^*/2$ instead of 0 (corresponding to the radial position $R_i + \frac{d}{2}$). The angular coordinate limits are taken as 0 and 1 as before. Using these substitutions and rearranging Equation (16), we obtain

$$\frac{1}{d^{*2}} \int_{\frac{d^*}{2}}^{r_2^*} \frac{dr^*}{\left[\frac{-Real(f_{CM})}{R_C^*+r^*-0.5} + \frac{3}{512}d^{*3}\left\{\frac{1}{r^{*4}} - \frac{1}{(1-r^*)^4}\right\}\right]} = -V_{App}^* \tag{23}$$

It must now be recognised that the full focusing of the particles will occur before they reach the outer channel wall, at an equilibrium radial co-ordinate r_{Eq}^* where the DEP force is balanced by the wall repulsion force. The equilibrium co-ordinate can be obtained from Equation (16) by setting $dr^*/d\theta^* = 0$ so that we get

$$\frac{-Real(f_{CM})}{R_C^* + r_{Eq}^* - 0.5} + \frac{3}{512}d^{*3}\left\{\frac{1}{r_{Eq}^{*4}} - \frac{1}{\left(1 - r_{Eq}^*\right)^4}\right\} = 0 \tag{24}$$

The solution of Equation (24) is then used as the integration limit for r_2^* in Equation (23). It is important to note that a similar procedure is followed for all the other combinations of the directions of the DEP and electrokinetic motions, with the only difference being that the integration limit for r_1^* in the case of positive DEP is defined as $1 - \left(\frac{d^*}{2}\right)$ instead of 0 (corresponding to the radial position $R_o - \frac{d}{2}$).

2.4. Data Analysis

Since the integration of the full solution is specific for a given combination of $Real(f_{CM})$, R_C^* and d^*, Equations (23) and (24) are numerically solved in MATLAB using the in-built integration functions for the following ranges of working parameters: (a) 15 values of R_C^* ranging from 1 to 15 in intervals of 1, (b) 27 values of d^*, from 0.001 to 0.01 in steps of 0.001, from 0.01 to 0.02 in steps of 0.0025 and 0.02 to 0.15 in steps of 0.01 and (c) 10 values of $Real(f_{CM})$ ranging from -0.05 to -0.5 in steps of -0.05 for negative DEP and from 0.1 to 1 in steps of 0.1 for positive DEP. This returns a total of 4050 V_{App}^* values for each DEP direction and hence a total of 9100 data points. These parameter ranges are chosen in order to encompass the practical values of these parameters used in the published literature of curvature radii and the particle and cell sizes. A few example calculations of these are included in the Supplementary Materials S1.

For numerical integration purposes, the integration limit for the final position r_2^* in Equation (23) is chosen to be $0.99r_{Eq}^*$ for negative DEP and $1.01r_{Eq}^*$ for positive DEP. This offset of 1% has to be chosen because Equation (24), which is also the denominator of the integrand in Equation (23), vanishes at $r_2^* = r_{Eq}^*$, thereby returning an indeterminate solution at that limit. As a result, the integration approaches the RHS of Equation (23) asymptotically (this is also confirmed from a trial and error study of the integration limit), and the offset is determined sufficient for tending to the solution it is theoretically expected to reach.

It is found that the direction of the electrokinetic motion only changes the sign of the integration solution for both directions of DEP, which is expected as the deviation from the ideal solution, and the factors responsible for it exist in the radial direction alone. Hence, the magnitudes of V_{App}^* and f_{CM} are taken into account for the analysis, and only two sets of data points are generated, one for each direction of DEP.

A visual inspection of the data points reveals that the value of $\left|V_{App}^*\right|$ varies linearly with R_C^* and inversely with d^{*2} and $\left|Real(f_{CM})\right|$ (which is also consistent with Equation (21)). Hence, curve fitting techniques are employed using the "Solver" function in Microsoft Excel to obtain a relationship between these parameters over the defined ranges of these parameters. As the particle blockage ratio d^* is the dimensionless parameter solely responsible for the generation of the wall repulsion force and the resulting deviation from the ideal solution (Equation (21)), the generated data is rearranged to express a combination of $\left|V_{App}^*\right|$, R_C^* and $\left|Real(f_{CM})\right|$ in terms of d^*. First, Equation (21) is rewritten as $\left[\frac{\left|V_{App}^*\right|\left|Real(f_{CM})\right|}{R_C^*}\right]_{Exact} = \frac{1}{d^{*2}}$. Because the data is mathematically similar to this equation, it is processed to generate a three-parameter curve fit expressed by the following equations

$$\left[\frac{\left|V_{App}^*\right|\left|Real(f_{CM})\right|}{R_C^*}\right]_{pDEP,\ Empirical} = \frac{1}{d^{*2}} - \frac{1.53}{d^*} + 3.36 \tag{25}$$

$$\left[\frac{\left|V_{App}^*\right|\left|Real(f_{CM})\right|}{R_C^*}\right]_{nDEP,\ Empirical} = \frac{1}{d^{*2}} - \frac{2.49}{d^*} + 6.75 \tag{26}$$

where the curve fitting parameters are determined from the Solver function in Microsoft Excel by minimising the sum of % errors between the individual MATLAB data points and the corresponding curve fitting data points. The standard sum of least square minimisation technique is not used because the ultimate aim of this work is to analyse the deviation of the DEP particle dynamics from the simplified exact solution due to the inevitable existence of the wall repulsion force and to provide an equation to the scientific community which would be able to reasonably predict the particle dynamics for all the values of particle blockage ratios d^* over the entire chosen parameter range. The sum of least squares method was attempted and was discarded because it was determined to significantly deviate from the MATLAB data for large values of d^*.

Minimising the sum of % errors generates the positive DEP (pDEP) curve fit equation (Equation (25)) with a mean % error of 1.31% and a standard deviation of 1.62%, with 194 outliers (which is about 5% of the total number of 4050 data points) deviating from the MATLAB data by more than 5%. Similarly, minimising the sum of % errors generates the negative DEP (nDEP) curve fit equation (Equation (26)) with a mean % error of 3.55% and a standard deviation of 3.25%, with 290 outliers (which is about 7% of the total number of 4050 data points) deviating from the MATLAB data by more than 8%. Although the nDEP data is a slightly poorer fit compared to the pDEP data, both the equations can be seen to overall provide a highly reliable prediction of the particle dynamics over the entire range of dimensionless parameters considered for this work. Supplementary Materials S2 and S3 provide the Excel files containing and confirming the aforementioned statistics for both positive and negative DEP. The data also justifies the reason for choosing the % error minimisation method over the sum of least squares method.

The % deviation δ of the above curve fit equations from the simplified exact solution can then be calculated as

$$\delta(\%)_{\text{pDEP}} = 100 \left| \frac{\left(\left[\frac{|V_{App}^*||f_{CM}|}{R_C^*} \right]_{\text{pDEP, Empirical}} - \left[\frac{|V_{App}^*||f_{CM}|}{R_C^*} \right]_{\text{Exact}} \right)}{\left[\frac{|V_{App}^*||f_{CM}|}{R_C^*} \right]_{\text{Exact}}} \right| = 100 \times \left| 3.36 d^{*2} - 1.53 d^* \right| \tag{27}$$

$$\delta(\%)_{\text{nDEP}} = 100 \left| \frac{\left(\left[\frac{|V_{App}^*||f_{CM}|}{R_C^*} \right]_{\text{nDEP, Empirical}} - \left[\frac{|V_{App}^*||f_{CM}|}{R_C^*} \right]_{\text{Exact}} \right)}{\left[\frac{|V_{App}^*||f_{CM}|}{R_C^*} \right]_{\text{Exact}}} \right| = 100 \times \left| 6.75 d^{*2} - 2.49 d^* \right| \tag{28}$$

The variation of δ as a function of d^* for both pDEP and nDEP is shown in Figure 2. Figure 2 as well as Equations (27) and (28) show that as the particle blockage ratio increases, the deviation between the curve fit equation and the exact solution becomes larger, which is consistent with the underlying physics of the origins of the wall repulsion force contributions in d^* and the resulting deviation of the particle path-line from the ideally expected path inside the microchannel. It is observed that for any given combination of design parameters, the data for negative DEP deviates more strongly from the ideal solution than positive DEP. This can be explained through the physics of C-iDEP systems and the wall repulsion forces as follows. Regardless of the direction of DEP forces, the influence of wall repulsion forces causes the particles to be fully focused over a smaller radial distance compared to the exact solution. The focusing of a particle undergoing negative DEP is assisted by the repulsion force at the inner wall and opposed by the repulsion at the outer wall. These repulsion forces interchange their functions for positive DEP. Since the inner wall repulsion force is stronger than the outer wall repulsion force (see Equation (10)), negative DEP focusing would require a smaller applied voltage and $|f_{CM}|$ or can tolerate a larger curvature than positive DEP for a given particle size. Hence, the value of $\frac{|V_{App}^*||Real(f_{CM})|}{R_C^*}$ required for a full focusing of negative DEP would be smaller than positive DEP, leading to a greater deviation from the exact solution.

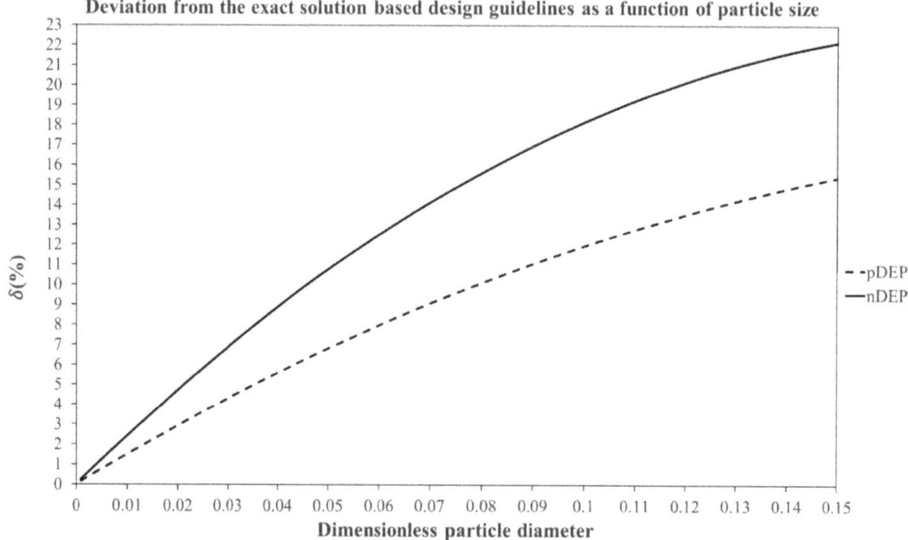

Figure 2. % Deviation of the dimensionless parameter $\left[\frac{|V_{App}^{*}||Real(f_{CM})|}{R_{C}^{*}}\right]$ generated from the curve fitting equations for pDEP (Equation (27)) and nDEP (Equation (28)) from its value predicted by the exact solution (Equation (21)) as a function of the dimensionless particle diameter d^{*}. Note that the curve fitting equations deviate increasingly from the exact solution for both pDEP and nDEP with increasing particle diameters, indicating the increasing influence of the wall repulsion forces on the C-iDEP particle dynamics with particle blockage.

2.5. Numerical Model

COMSOL Multiphysics 5.3a, a commercial finite element code, was used for developing a dimensionless numerical model to validate the analytical solutions. A 2-D set up was sufficient for the analysis considering the absence of electric field along the thickness of the channel. A circular arc microchannel having a unit width, a mean radius of curvature of R_{C}^{*} (to be consistent with the normalisation definition of the mean curvature radius by the channel width) and a unit angle of 1 radian subtended at the centre of curvature was constructed. The dimensionless Laplace equation for electric potential, i.e., $\nabla^{*2}V_{DC}^{*} = 0$, was solved by defining a dimensionless gradient ∇^{*} (defined as $\nabla^{*} = W\nabla$) and a dimensionless DC electric potential V_{DC}^{*} (defined as $V_{DC}^{*} = V/V_{0}$).

Equation (14) is also made non-dimensional through reference velocity scale $\frac{\varepsilon_{f}V_{0}\left(\zeta_{p}-\zeta_{w}\right)}{W\eta_{f}}$ and is expressed as follows upon substitution of the individual terms.

$$\mathbf{u}_{P}^{*} = \mathbf{E}_{DC}^{*} + \frac{V_{App}^{*}d^{*2}}{2}\left[Real(f_{CM})\nabla^{*}\mathbf{E}_{DC}^{*2} + \frac{3}{256}d^{*3}\mathbf{E}_{DC}^{*2}\left\{\frac{1}{r^{*4}} - \frac{1}{(1-r^{*})^{4}}\right\}\hat{\mathbf{r}}\right] \tag{29}$$

where r^{*} is defined as before. The aforementioned unitless Laplace equation and Equation (29) were then solved using the "Electrostatics" interface and the particle tracing function of COMSOL Multiphysics respectively. As boundary conditions for the Laplace equation, Dirichlet boundary conditions for a unit DC voltage drop are applied across the circumferential direction (corresponding to the voltage drop of V_{0} normalised in the dimensionless system), and the other channel walls are assumed electrically insulating. The dimensionless electric field, $\mathbf{E}_{DC}^{*}$, obtained from the finite element solution is then used in Equation (29) to calculate the particle path-lines. The 2D path-line images and the particle position data are generated using the COMSOL post-processing tools and exported for further analysis.

Two test particles that represent a stream of particles occupying a microchannel are chosen for the analysis. The particle whose dynamics are governed by the ideal solution is represented by a red colour in the results, and the particle whose dynamics are governed by the full solution is represented by the blue colour in the results. Considering the directions of the DEP forces, the starting point of the red (ideal) particle is chosen to be at $r^* = 1$, $\theta^* = 0$ for positive DEP and at $r^* = 0$, $\theta^* = 0$ for negative DEP. The starting point of the blue (realistic) particle, however, is chosen to be at the realistic co-ordinates (i.e., $r^* = 1 - d^*/2$, $\theta^* = 0$ for positive DEP and $r^* = d^*/2$, $\theta^* = 0$ for negative DEP) in order to consider the contributions of the repulsion forces. Upon choosing the aforementioned co-ordinates of the test particles, it is ensured that designing the microfluidic system for fully deflecting these particles would automatically ensure the maximum possible concentration of the stream of particles they represent.

A 2-D mapped mesh is used to generate a system of simultaneous finite element equations over the model geometries, and the equations are solved for each simulation in less than 10 s.

3. Results and Discussions

As observed from the data shown in Figure 2, one can divide the working range of the particle blockage ratio into 3 regimes on the basis of the extent of deviation δ from the exact solution. The limits of these regimes were determined from a trial and error analysis of the path-lines of the aforementioned test particles, and the values of these limits were chosen purely based on a reasonable visual distinction between the ideal and the realistic particle path-lines. We acknowledge that this visual interpretation and hence the regime limits could vary slightly. However, we justify through our results that even for the highest particle size regime regardless of its limits, the ideal solution always assures a reliable design of the curved arc microchannel dielectrophoresis with a factor of safety and hence is ultimately proposed for the entire particle size range chosen. We do so using this section to fully analyse the utility of the exact solution and the empirical curve fit equations for each of the three regimes through an example case. Each case considers positive as well as negative DEP and compares the equation-based particle path-line predictions with numerically predicted results. For all the models, a microchannel subtending a unit angle of 1 radian in the centre and a curvature ratio of 5 is chosen. The magnitude of $Real(f_{CM})$ is taken as 0.5 for the DEP forces.

3.1. Regime 1: $\delta \leq 5\%$

In this regime, the deviation between the curve fit equations and the exact solution is very small. Hence, the wall repulsion effect can be deemed negligible enough to affect the particle motion, and thus, the exact solution can be directly used for predicting the DEP particle dynamics in the microfluidic system. For pDEP systems, the data from Figure 2 show that this situation is applicable for particle blockage ratios, i.e., d^* having values up to 0.0354. Similarly for nDEP systems, this situation applies for d^* having values up to 0.0213. Since the exact solution can be applied for designing C-iDEP systems within this regime, it follows from Equation (21) that the value of the dimensionless parameter $|V_{App}{}^*||Real(f_{CM})|/R_C{}^*$ cannot fall below 798 for pDEP and below 2204 for nDEP respectively, if the particle size d^* to be dielectrophoretically focused falls below the aforementioned threshold values

Figure 3 shows an example situation of regime 1 with the help of two-dimensional path-lines of the two test particles (see Section 2.5) numerically predicted by COMSOL for a particle blockage ratio of 0.005 in both positive (Figure 3a) as well as negative DEP (Figure 3b) situations. The dimensionless applied voltage $|V_{App}{}^*|$ across the microchannel is chosen as 400,000 to maintain a consistency with the value of 40,000 for the dimensionless quantity $|V_{App}{}^*||Real(f_{CM})|/R_C{}^*$ as per the exact solution.

As seen from the figures, the microchannel design parameters are able to fully concentrate the red as well as blue particles just at the outlet of the microchannel for both pDEP and nDEP, thereby confirming the applicability of the exact solution. As seen from the inset images, the DEP velocity vectors are directed from the inner channel wall to the outer channel wall for nDEP and vice versa for pDEP, which also confirms well with the theoretical predictions. Their increasing strength (indicated by the size of the arrows) in the region of high electric field also confirms the theoretical dependence of the DEP velocities on the radial position. A careful observation of the particle path-lines shows that the blue particle, as expected, is fully focused slightly before it reaches the destination channel walls and continues to travel along the equilibrium radial co-ordinate. This small discrepancy between the final radial coordinates of the red and blue particles can be explained through the curve fitting equations, which indicate that the value of $|V_{App}{}^*||Real(f_{CM})|/R_C{}^*$ must be 39,697.36 and 39,508.75 for pDEP and nDEP respectively in order to ensure a full focusing of the blue particles at the channel outlet. Since both these values are smaller than the value predicted by the exact solution, the applied voltage is slightly stronger than what is required to fully focus these particles. However, for all practical purposes, it can be seen from the simulation that the % error between the final radial co-ordinates of both test particles is negligibly small, and hence, the exact solution holds in regime 1 for designing the C-iDEP systems.

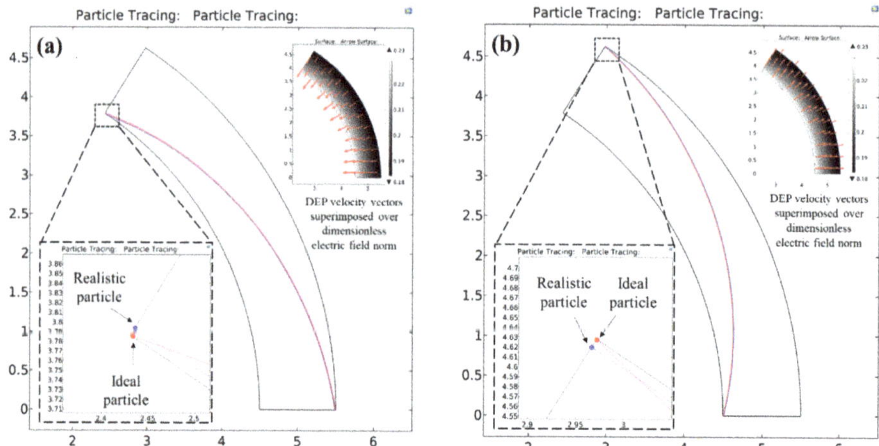

Figure 3. Numerically predicted path-lines for two test particles of a dimensionless diameter $d^* = 0.005$ under the action of (**a**) positive DEP ($Real(f_{CM}) = 0.5$) and (**b**) negative DEP ($Real(f_{CM}) = -0.5$) inside a circular arc microchannel having a curvature ratio $R_C{}^*$ of 5. The dimensionless applied voltage $|V_{App}{}^*|$ for both the DEP directions is 400,000, as calculated from the exact solution. The ideal particle is represented by the red colour and its motion is governed by the exact solution. The realistic particle is represented by the blue colour and its motion is governed by the full solution of particle motion inclusive of wall repulsion effects. This particle size demonstrates a regime 1 behaviour, where the design parameters for the C-iDEP microchannel can be determined reliably by the exact solution. The inset images show the DEP velocity vectors superimposed over the electric field norms, confirming an agreement of the simulated particle dynamics with their theoretical predictions.

One potential limitation of the utility of the analysis in this regime is that the values of d^* in this regime are so small that the required voltages to focus these particles could induce strong Joule heating effects inside the channel which could cause the particle motion to deviate from the exact solution by virtue of change in the fluid properties and electrothermal effects. We expect that given the absence of abrupt and large changes in the electric field in C-iDEP systems, although the overall temperature itself could increase, the increase would be almost uniform in the $r - \theta$ plane (the small non-uniformities stemming from the depth-wise heat dissipation through typical Poly-dimethylsiloxane (PDMS)/glass-based fluidic systems [40]). Hence, the temperature gradients and the resulting electrothermal effects would still be small. Additionally, even if this temperature field would change the electrical permittivity and the dynamic viscosity of the liquid, we have demonstrated that the particle dynamics are independent of these two quantities, and hence, as a conclusion, we expect this analysis to work reliably for regime 1. However, we aim to still confirm our hypothesis experimentally in the near future.

In addition, it is important to note, particularly in context of this regime, that although C-iDEP systems are relatively safer from electrolysis risks compared to the conventional metallic microelectrode-based systems, the risk need not be eliminated. It can still become quite significant in this regime due to strong applied voltages and the subsequent electrokinetic flow of an electrolysis bubble into the microchannel to disrupt the particle dynamics. Microfluidics engineers and scientists can avoid this issue by using an inert electrode element like carbon. The use of this material has been well established in the microelectrode-based eDEP systems [51,52], but can be used for C-iDEP systems too without the loss of generality.

3.2. Regime 2: $5 < \delta \leq 12\%$

In this regime, the deviation between the curve fit equations and the exact solution is slightly too much larger than regime 1 for the effects of wall repulsion not to be neglected. This regime is characterised by particle blockage ratio d^* ranging from 0.0354 to 0.1 for pDEP systems and from 0.0213 to 0.057 for nDEP systems. This corresponds to the quantity $|V_{App}^*||Real(f_{CM})|/R_C^*$ falling within 88.06 and 758.12 for pDEP systems and 270.85 and 2094 for nDEP systems.

Figure 4 shows example situations of regime 2 through a comparison of the numerically predicted path-lines of the two test particles. d^* is taken as 0.08 for pDEP and 0.05 for nDEP. Substituting for these values in the curve fit equations (Equations (27) and (28)) gives the value of $|V_{App}^*|$ to be applied as 1404.85 for pDEP and 3569.5 for nDEP. As seen from Figure 4a,b, applying these respective voltages across the microchannel arc causes the blue particles to achieve a full deflection at the equilibrium radial co-ordinate of just around the microchannel outlet. However, it is observed that the red particles are not able to reach their destination channel wall and achieve a partial focusing for both pDEP and nDEP situations. From Equation (19), the theoretical radial co-ordinates that the red particle would achieve for pDEP and nDEP are 0.1107 and 0.901 respectively, which match very well with their numerically predicted counterparts of 0.111 and 0.9. This behaviour is consistent with the fact that the wall repulsion forces become more significant in this regime and are therefore able to focus the particles fully with the help of a smaller voltage than what is expected from the exact solution.

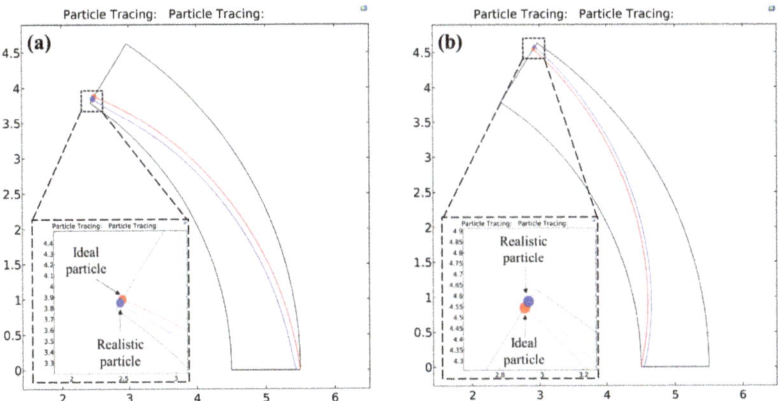

Figure 4. Numerically predicted path-lines for the ideal and realistic test particles under the action of (**a**) positive DEP ($Real(f_{CM}) = 0.5$) and (**b**) negative DEP ($Real(f_{CM}) = -0.5$) inside a circular arc microchannel having a curvature ratio R_C^* of 5. The dimensionless particle diameter d^* is chosen as 0.08 for positive DEP and its motion is driven by a dimensionless voltage of $|V_{App}^*| = 1404.85$. Similarly, the dimensionless particle diameter d^* is chosen as 0.05 for negtive DEP and its motion is driven by a dimensionless voltage of $|V_{App}^*| = 3569.5$. These particle sizes clearly demonstrate a regime 2 behaviour, where the design parameters for the C-iDEP microchannel can be determined reasonably by the exact solution but more reliably by the curve fitting equations.

3.3. Regime 3: $\delta > 12\%$

This regime applies for d^* greater than 0.1 for pDEP and greater than 0.057 for nDEP. In this regime, the large values of d^* result in highly significant contributions of the wall repulsion forces, and hence, the curve fit equations need to be evaluated for tracking the particle dynamics of the C-iDEP microdevice. Using the curve fit equations, it can be calculated that the regime 3 corresponds to a range of 88.06 or lower in the case of pDEP and 270.85 or lower in the case of nDEP for the dimensionless parameter $|V_{App}^*||Real(f_{CM})|/R_C^*$ respectively. At the same time, using the range of d^* for the exact solution returns a value of 100 or lower for pDEP and 307.78 or lower for nDEP. As reflected in the deviation δ, this value is large compared to the ranges defined by the curve fit equations, which implies that for given values of the curvature ratio, R_C^*, and $|Real(f_{CM})|$, the applied voltage $|V_{App}^*|$ to achieve focusing using the exact solution would also be substantially larger than what is required realistically.

This behaviour is also confirmed from the numerically predicted path-lines of the test particles shown in Figure 5. The particle sizes, d^*, chosen are 0.12 and 0.08 for pDEP and nDEP respectively. Substituting these values in the exact solution returns a value of $|V_{App}^*| = 694.44$ and 1562.5 for pDEP and nDEP respectively. As seen in the 2D figure, although this voltage fully deflects the red particle at the microchannel outlet for both pDEP and nDEP situations, the blue particle is seen to travel along the equilibrium radial co-ordinate for a considerable length of the arc. This is because the voltages applied are strong enough to fully focus the blue particles much before they reach the outlet. The inset images, which show the 1-D radial profiles of the dimensionless deflection velocity (DEP + wall repulsion), also support these observations qualitatively and quantitatively. The steep gradient of the velocities close to the wall stems from the strong wall repulsion forces that are dominant only at the walls for both the directions of DEP. The negative sign of the velocity on the profile indicates a motion opposite to the positive radial direction, i.e., from inner channel wall to the outer channel wall, so that the crossover point where the velocity switches sign is identified as the equilibrium co-ordinate. It is 0.1 for pDEP and 0.925 for nDEP, both of which agree very well with the theoretical predictions of 0.098 and 0.9244 respectively as per Equation (19).

All the results indicate that within the practical size ranges of cells and particles published in the existing literature, as d^* increases and one transitions from regime 1 to regime 3, the voltage required for fully focusing the particles becomes increasingly smaller than what is predicted by the exact solution. Similarly, the curvature tolerance for obtaining the focusing also becomes larger. Since the design of experiments is always performed for safety, these results establish that using the exact solution and its corollaries (Equations (21) and (22)) as a guideline for designing the C-iDEP systems would guarantee a realistic focusing of the cells and particles with a factor of safety. Additionally, the actual experimental scenario often possesses uncertainties arising from factors like drift voltage or particle-particle interactions, which inevitably introduce deviations from the design parameter combination to be employed. In such a situation, having an equation which assures focusing before the particles reach the outlet would be a preferred choice for microfluidics engineers. If researchers still wish to have highly precise designs, Figure 2 can be used as a reference chart for back calculating the exact combination of design parameters as its utility has been confirmed through examples.

In addition, one possible limitation of the realistic solutions, especially towards the upper limits of regime 3, is that with particle sizes becoming large, the volume they occupy within the channel can no longer be considered small enough for the point particle definition of DEP used in this work. Hence, aiming to use the curve fit equations for designing the systems in the upper limits of regime 3 or beyond would require the use of computationally expensive boundary element methods for accurate predictions. Using the exact solution as a design guideline would not only eliminate this issue but would also enable engineers to use a simple electronic calculator and design these systems.

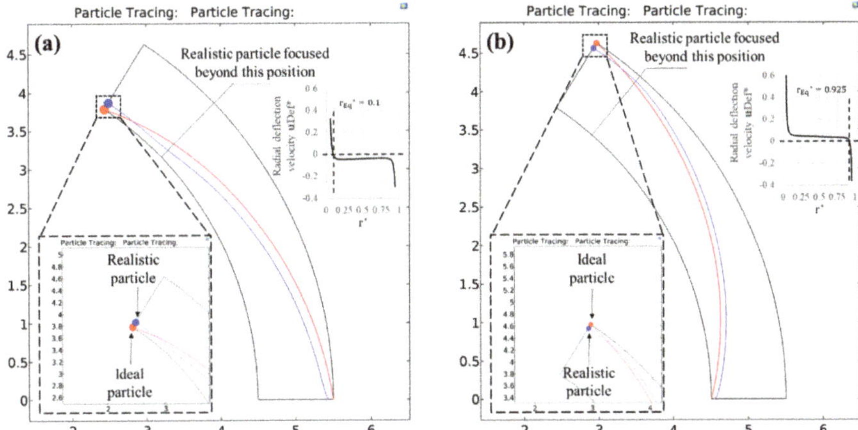

Figure 5. Numerically predicted path-lines for the ideal and realistic test particles under the action of (**a**) positive DEP ($Real(f_{CM}) = 0.5$) and (**b**) negative DEP ($Real(f_{CM}) = -0.5$) inside a circular arc microchannel having a curvature ratio $R_C{}^*$ of 5. The dimensionless particle diameter d^* is chosen as 0.12 for positive DEP and its motion is driven by a dimensionless voltage of $\left|V_{App}{}^*\right| = 694.44$ predicted from the exact solution. Similarly, the dimensionless particle diameter d^* is chosen as 0.08 for negative DEP and its motion is driven by a dimensionless voltage of $\left|V_{App}{}^*\right| = 1562.5$ as predicted by the exact solution. These particle sizes demonstrate a regime 3 behaviour, where the realistic particle motion deviate significantly from the ideal behaviour. This is evident from the fact that the realistic particles in both cases of DEP are focused substantially before they reach the microchannel outlet. The location of full focusing along the arc length is highlighted by the dotted lines, where a kink in the particle path-line is visible, and the particle is seen traveling parallel to the channel wall beyond that point. This regime is characterised by more reduced voltage requirements to focus the particles fully due to the assistance of wall repulsion forces. The inset images represent a 1-D radial profile of the net radial particle velocity component, along with the identification of the equilibrium radial co-ordinates.

3.4. Utility for Multiple Arcs

Since the arc microchannel forms the simplest design for C-iDEP systems, it would be of interest to explore the potential of the analysis of the particle path-line presented in this work for multiple channel turns by treating a single arc as a repeating unit. We consider a two-turn microchannel with identical arc geometries as an example for this work. Such designs are of great use if one needs to work with the same applied voltage as in the single turn and thereby reduce the voltage drop as well electric field strength per unit turn by a factor of two (identical turn geometries maintain a consistency with a single arc geometry and hence a 50% split). This reduction in the field strength would reduce the Joule heating and electrolysis risks associated with a single arc.

Figure 6a shows a COMSOL simulation of the dimensionless voltage drop $V_{DC}{}^*$ for an example opposing two-turn microchannel (each arc being 90° with a curvature ratio of $R_C{}^* = 5$) obtained by solving $\nabla^{*2}V_{DC}{}^* = 0$ (see Section 2.5), which confirms this hypothesis of 50% splitting of the voltage across each arc.

In such a scenario, the analysis of particle position for a single arc can be theoretically used to calculate the final position of the particle at the end of the first turn, which is then used as an initial position for the next turn. We choose the exact solution for exploring this repeating unit approach owing to its established simplicity. Consider an example of a negative DEP ($Real(f_{CM}) = -0.5$) of an ideal particle of a dimensionless diameter d^* of 0.02 (to be within the regime 1 where the exact solution applies) undergoing dielectrophoresis in the two-turn microchannel. In this situation, for a single turn arc, Equation (22) predicts that the voltage needed to fully deflect the particle to the opposite wall is

$\left|V_{App}^{*}\right| = 25,000$. For a two-turn microchannel, this voltage would be split across the two turns by a factor of two, leading to a voltage drop of 12,500 per turn. Setting $\left|V_{App}^{*}\right| = 12,500$ for the initial radial position of a particle as the inner wall of the first arc ($r_1^{*} = 0$) in Equation (19), one can calculate the final particle position at the end of the first turn as $r_2^{*}{}_{\text{Turn 1}} = 0.5249$.

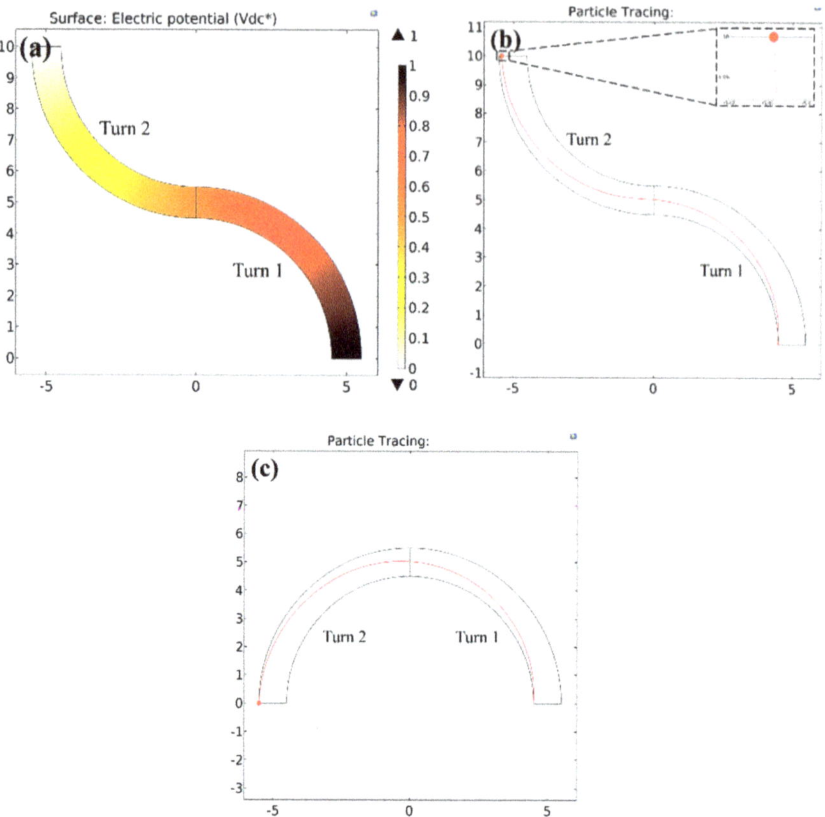

Figure 6. Numerically predicted behaviour of the negative DEP particle dynamics ($Real(f_{CM}) = -0.5$) inside a two-turn microchannel having a curvature ratio of R_C^{*} of 5. (**a**) Numerically predicted dimensionless electric potential plot for a two-turn channel with opposing turns. (**b**) Path-lines of particles of diameter $d^{*} = 0.02$ undergoing a DEP motion for a two-turn channel with opposing turns, under an applied voltage of $\left|V_{App}^{*}\right| = 25,000$, with the inset showing the final position of the particle. (**c**) Path-lines of particles of diameter $d^{*} = 0.02$ undergoing a DEP motion for a two-turn channel with unidirectional turns with other conditions identical to the opposing turn geometry.

Now if the two turns are unidirectional (i.e., in the same direction of curvature as in Figure 6c), this value becomes $r_1^{*}{}_{\text{Turn 2}}$ for the second turn. However, if the turns are opposing (i.e., change of curvature direction), this becomes $(1 - r_1^{*}{}_{\text{Turn 2}})$ for the second turn. The values of R_C^{*}, d^{*} and $\left|V_{App}^{*}\right|$ remain the same. Using Equation (19) again with these conditions for both cases of turns, we obtain $r_2^{*}{}_{\text{Turn 2}} = 1$ for unidirectional turns and $r_2^{*}{}_{\text{Turn 2}} = 0.9545$ for opposing turns. Interestingly, Figure 6b shows the final position of the particle at the end of the second opposing turn to be at $r_2^{*} = 0.908$ (as seen from the inset, it is 5.408, which makes it 0.908 relative to 4.5, which is the radial co-ordinate of the inner wall of the second turn), indicating an offset of 0.0465 from the theoretical prediction. At the same time, Figure 6c shows an exact agreement with the theoretical prediction in unidirectional turns.

This is not surprising because these essentially regenerate a single arc condition and the particle fully reaches the opposite wall.

It is clear from both these observations that the offset produced in the particle position for the opposing turn design, however small relative to the channel dimensions, comes from the change in the direction of the curvature in two-turn channels. Hence, although the analysis of the present work is quite consistent for two-turn designs, this offset points at the existence of an additional parameter or a condition that controls the particle motion in serpentine channels with two or multiple turns and hence can open up new opportunities for an extended analysis. In addition, the present form of the single arc equations in this work hold only if the channel width and curvature is constant. Hence, this research can also open up several directions for an extended analysis in designs where this need not be the case, such as spiral microchannels or curved microchannels with changing widths. Further, these analyses can be made more accurate by studying the wall repulsion force contribution to the particle motion, which would involve an extensive data analysis and numerical integrations. We aim to conduct these studies in the near future through follow-up works. However, we expect that regardless of the channel geometries discussed above, the repulsion forces would still be consistent in their physics and allow particle manipulation over a smaller arc length compared to the exact solution so that the exact solution would remain a preferred design choice.

4. Conclusions and Future Work

In this work, we have provided comprehensive theoretical guidelines for designing curvature-induced dielectrophoresis-based particle concentration systems under the action of both DC and DC biased AC voltages and confirmed them extensively using two-dimensional finite element simulations. An arc microchannel, which forms the most basic design for such systems, is used to provide a detailed mathematical treatment. We have derived an elegant exact solution from first principles which shows that the full focusing of a given size of particles within a given geometry of microchannel is completely governed by three dimensionless parameters and the Clausius-Mossotti factor. These are the applied voltage strength relative to the wall-particle zeta potentials, the curvature ratio and the particle blockage ratio or dimensionless particle diameter.

It is also shown with computational validations that wall repulsion forces which exist at the channel wall are strong functions of the particle blockage ratio alone and perturb the particle dynamics from those predicted by the exact solution. Based on this observation, an extensive numerical integration and test particle analysis is performed over a wide range of practically relevant values of dimensionless parameters to generate curve fitting equations that mutually relate the parameters to each other. Based on the extent of deviation this data exhibits from the exact solutions, three regimes are identified according to a specific range of particle sizes. An example situation from each regime is considered and then the particle path-lines predicted using COMSOL are compared with the theoretically predicted values with quantitative agreement with exact as well as curve fit equations, thereby establishing the reliability of both the equations. The results fully justify the utility of the exact solution to design the arc microchannels with a factor of safety over the entire working range of particles chosen, thereby establishing a calculator-based tool for microfluidics engineers interested in designing this system.

As introduced before in the theory, the present work is applicable for several potential uses in cell research as illustrated ahead. (a) For instance, for cells with well characterised dielectric properties, the analytical model can be used to design a simple arc microchannel, which, either through a positive or a negative DEP, would generate a narrow, highly-concentrated stream of single cells (like the coaches of a train) which can then be directed into further downstream lab-on-a-chip operations such as cell counters, cell culturing chambers and sensors. (b) The analytical equations can also be potentially used for obtaining useful information about the properties of uncharacterised cells in combination with the multi-shell model. (c) The analytical equation can also be used to obtain a size-based separation of cells from a binary mixture by exploiting the wall repulsion forces, which would focus them at different equilibrium coordinates [44].

Apart from a direct use for single arc microchannel design, this work is also shown to open up several new directions for establishing theoretical guidelines to design more complex C-iDEP-based systems involving a change in the strength and/or the direction of curvature. These are ongoing explorations as part of our research and are expected to generate comprehensive theoretical bases for these designs in near future. Establishing experimental protocols to support the dimensionless analysis and to investigate Joule heating effects in these devices is also under consideration.

Supplementary Materials: The following are available online at http://www.mdpi.com/2072-666X/11/7/707/s1. S1: Provides calculations for a few example biological particles and cells suspended in the curved microchannels having the aforementioned range of widths; S2: nDEP_3 parameter fit; S3: pDEP2_3 parameter fit.

Author Contributions: A.K. and X.X. conceived the underlying physical principles of the research. A.K. performed the theoretical analysis, conceptualised the dimensionless computational framework and primarily wrote the paper. A.K. and A.M. developed the computational simulation results. All authors commented on the paper and have read and agreed to the published version of the manuscript.

Funding: This work was supported in part by NSF under grant no. CBET-1704379.

Conflicts of Interest: The authors declare no conflicts of interests.

References

1. Psaltis, D.; Quake, S.R.; Yang, C. Developing optofluidic technology through the fusion of microfluidics and optics. *Nature* **2006**, *442*, 381–386. [CrossRef]
2. Yager, P.; Edwards, T.; Fu, E.; Helton, K.; Nelson, K.; Tam, M.R.; Weigl, B.H. Microfluidic diagnostic technologies for global public health. *Nature* **2006**, *442*, 412–418. [CrossRef] [PubMed]
3. Pratt, E.D.; Huang, C.; Hawkins, B.G.; Gleghorn, J.P.; Kirby, B.J. Rare cell capture in microfluidic devices. *Chem. Eng. Sci.* **2011**, *66*, 1508–1522. [CrossRef]
4. Karimi, A.; Yazai, S.; Ardekani, A.M. Review of cell and particle trapping in microfluidic systems. *Biomicrofluidics* **2013**, *7*, 021501. [CrossRef] [PubMed]
5. Kale, A.; Patel, S.; Xuan, X. Three-dimensional reservoir-based dielectrophoresis (rDEP) for enhanced particle enrichment. *Micromachines* **2018**, *9*, 123. [CrossRef] [PubMed]
6. Gossett, D.R.; Weaver, W.M.; Mach, A.J.; Hur, S.C.; Kwong Tse, H.T.; Lee, W.; Amini, H.; DiCarlo, D. Label-free cell separation and sorting in microfluidic systems. *Anal. Bioanal. Chem.* **2010**, *397*, 3249–3267. [CrossRef]
7. Li, N.; Kale, A.; Stevenson, A.C. Axial acoustic field barrier for fluidic particle manipulation. *Appl. Phys. Lett.* **2019**, *114*, 013702. [CrossRef]
8. Laurell, T.; Petersson, F.; Nilsson, A. Chip integrated strategies for acoustic separation and manipulation of cells and particles. *Chem. Soc. Rev.* **2007**, *36*, 492–506. [CrossRef]
9. Pamme, N. Magnetism and microfluidics. *Lab Chip* **2006**, *6*, 24–38. [CrossRef]
10. Gijs, M.A.; Lacharme, F.; Lehmann, U. Microfluidic applications of magnetic particles for biological analysis and catalysis. *Chem. Rev.* **2010**, *110*, 1518–1563. [CrossRef]
11. Liang, W.; Wang, S.; Dong, Z.; Lee, G.-B.; Li, W.J. Optical spectrum and electric field waveform dependent optically-induced dielectrophoretic (ODEP) micro-manipulation. *Micromachines* **2012**, *3*, 492–508. [CrossRef]
12. Wang, M.M.; Tu, E.; Raymond, D.E.; Yang, J.M.; Zhang, H.; Hagen, N.; Dees, B.; Mercer, E.M.; Forster, A.H.; Kariv, I.; et al. Microfluidic sorting of mammalian cells by optical force switching. *Nat. Biotechnol.* **2005**, *23*, 83–87. [CrossRef]
13. Kale, A.; Song, L.; Lu, X.; Yu, L.; Hu, G.; Xuan, X. Electrothermal enrichment of submicron particles in an insulator-based dielectrophoretic microdevice. *Electrophoresis* **2018**, *39*, 887–896. [CrossRef]
14. Liu, C.; Hu, G. High-throughput particle manipulation based on hydrodynamic effects in microchannels. *Micromachines* **2017**, *8*, 73. [CrossRef]
15. Kale, A.; Lu, X.; Patel, S.; Xuan, X. Continuous flow dielectrophoretic trapping and patterning of colloidal particles in a ratchet microchannel. *J. Micromech. Microeng.* **2014**, *24*, 075007. [CrossRef]
16. Fernandez, R.E.; Rohani, A.; Farmehini, V.; Swami, N.S. Review: Microbial analysis in dielectrophoretic microfluidic systems. *Anal. Chim. Acta* **2017**, *966*, 11–33. [CrossRef] [PubMed]
17. Bazant, M.Z.; Squires, T.M. Induced-charge electrokinetic phenomena. *Curr. Opin. Colloid Interface Sci.* **2010**, *15*, 203–213.

18. Kale, A. Joule Heating Effects in Insulator-Based Dielectrophoresis Microdevices. Ph.D. Thesis, Clemson University, Clemson, SC, USA, 7 August 2015.

19. Lu, C.; Verbridge, S.S. *Microfluidic Methods for Molecular Biology*, 1st ed.; Springer: Manhattan, NY, USA, 2016.

20. Pohl, H.A. *Dielectrophoresis: The Behavior of Neutral Matter in Nonuniform Electric Fields (Cambridge Monographs on Physics)*; Cambridge University Press: Cambridge, UK, 1978.

21. Pohl, H.A. The motion and precipitation of suspensoids in divergent electric fields. *J. Appl. Phys.* **1951**, *22*, 869–871. [CrossRef]

22. Park, B.Y.; Madou, M.J. 3-D electrode designs for flow-through dielectrophoretic systems. *Electrophoresis* **2005**, *26*, 3745–3757. [CrossRef]

23. Park, S.; Beskok, A. Alternating current electrokinetic motion of colloidal particles on interdigitated microelectrodes. *Anal. Chem.* **2008**, *80*, 2832–2841. [CrossRef]

24. Demierre, N.; Braschler, T.; Linderholm, P.; Seger, U.; van Lintel, H.; Renaud, P. Characterization and optimization of liquid electrodes for lateral dielectrophoresis. *Lab Chip* **2007**, *7*, 355–365. [CrossRef] [PubMed]

25. Natu, R.; Martinez-Duarte, R. Numerical model of streaming DEP for stem cell sorting. *Micromachines* **2016**, *7*, 217. [CrossRef] [PubMed]

26. Morgan, H.; Hughes, M.P.; Green, N.G. Separation of submicron bioparticles by dielectrophoresis. *Biophys. J.* **1999**, *77*, 516–525. [CrossRef]

27. Tang, S.-Y.; Zhang, W.; Soffe, R.; Nahavandi, S.; Shukla, R.; Khoshmanesh, K. High resolution scanning electron microscopy of cells using dielectrophoresis. *PLoS ONE* **2014**, *9*, e104109. [CrossRef]

28. Tang, S.-Y.; Zhang, W.; Baratchi, S.; Nasabi, M.; Kalantar-zadeh, K.; Khoshmanesh, K. Modifying dielectrophoretic response of nonviable yeas.t cells by ionic surfactant treatment. *Anal. Chem.* **2013**, *85*, 6364–6371. [CrossRef] [PubMed]

29. Smith, A.J.; O'Rorke, R.D.; Kale, A.; Rimsa, R.; Tomlinson, M.J.; Kirkham, J.; Davies, A.G.; Walti, C.; Wood, C.D. Rapid cell separation with minimal manipulation for autologous cell therapies. *Sci. Rep.* **2017**, *7*, 41872. [CrossRef]

30. Hawkins, B.G.; Smith, A.E.; Syed, Y.A.; Kirby, B.J. Continuous-flow particle separation by 3D insulative dielectrophoresis using coherently shaped, DC-biased, AC electric fields. *Anal. Chem.* **2007**, *79*, 7291–7300. [CrossRef]

31. Romero-Creel, M.F.; Goodrich, E.; Polniak, D.V.; Lapizco-Encinas, B.H. Assessment of sub-micron particles by exploiting charge differences with dielectrophoresis. *Micromachines* **2017**, *8*, 239. [CrossRef]

32. Chen, K.P.; Pacheco, J.R.; Hayes, M.A.; Staton, S.J.R. Insulator-based dielectrophoretic separation of small particles in a sawtooth channel. *Electrophoresis* **2009**, *30*, 1441–1448. [CrossRef] [PubMed]

33. Sanghavi, B.J.; Varhue, W.; Rohani, A.; Liao, K.T.; Bazydlo, L.A.L.; Chou, C.-F.; Swami, N.S. Ultrafast immunoassays by coupling dielectrophoretic biomarker enrichment in nanoslit channel with electrochemical detection on graphene. *Lab Chip* **2015**, *15*, 4563–4570. [CrossRef] [PubMed]

34. Cao, Z.; Zhu, Y.; Liu, Y.; Dong, S.; Chen, X.; Bai, F.; Song, S.; Fu, J. Dielectrophoresis-based protein enrichment for a highly sensitive immunoassay using Ag/SiO2 nanorod arrays. *Small* **2018**, *14*, 1703265. [CrossRef] [PubMed]

35. Zellner, P.; Shake, T.; Hosseini, Y.; Nakidde, D.; Riquelme, L.V.; Sahari, A.; Pruden, A.; Behkam, B.; Agah, M. 3D insulator-based dielectrophoresis using DC-biased, AC electric fields for selective bacterial trapping. *Electrophoresis* **2015**, *36*, 277–283. [CrossRef] [PubMed]

36. Zhu, J.; Xuan, X. Particle electrophoresis and dielectrophoresis in curved microchannels. *J. Colloid Int. Sci.* **2009**, *340*, 285–290. [CrossRef] [PubMed]

37. Xuan, X. Joule heating in electrokinetic flow. *Electrophoresis* **2008**, *29*, 33–43. [CrossRef] [PubMed]

38. Hawkins, B.J.; Kirby, B.J. Electrothermal flow effects in insulating (electrodeless) dielectrophoresis systems. *Electrophoresis* **2010**, *31*, 3622–3633. [CrossRef]

39. Kale, A.; Patel, S.; Qian, S.; Hu, G.; Xuan, X. Joule heating effects on reservoir-based dielectrophoresis. *Electrophoresis* **2014**, *36*, 721–727. [CrossRef] [PubMed]

40. Prabhakaran, R.A.; Zhou, Y.; Patel, S.; Kale, A.; Song, Y.; Hu, G.; Xuan, X. Joule heating effects on electroosmotic entry flow. *Electrophoresis* **2017**, *38*, 572–579. [CrossRef] [PubMed]

41. Xuan, X. Curvature-Induced Dielectrophoresis (C-iDEP) for Microfluidic Particle and Cell Manipulations. In *International Conference on Nanochannels, Microchannels, and Minichannels*; Paper No: ICNMM2012-73042; American Society of Mechanical Engineers: New York, NY, USA, 2012; pp. 681–685.

42. Church, C.; Zhu, J.; Xuan, X. Negative dielectrophoresis-based particle separation by size in a serpentine microchannel. *Electrophoresis* **2011**, *32*, 527–531. [CrossRef] [PubMed]
43. Zhu, J.; Xuan, X. Curvature-induced dielectrophoresis for continuous separation of particles by charge in spiral microchannels. *Biomicrofluidics* **2011**, *5*, 024111. [CrossRef] [PubMed]
44. Dubose, J.; Zhu, J.; Patel, S.; Lu, X.; Tupper, N.; Stonaker, J.M.; Xuan, X. Electrokinetic particle separation in a single-spiral microchannel. *J. Micromech. Microeng.* **2014**, *24*, 115018. [CrossRef]
45. Zhu, J.; Canter, R.C.; Keten, G.; Vedantam, P.; Tzeng, T.R.J.; Xuan, X. Continuous-flow particle and cell separations in a serpentine microchannel via curvature-induced dielectrophoresis. *Microfluid. Nanofluid.* **2011**, *11*, 743–752. [CrossRef]
46. Zhu, J.; Tzeng, T.R.J.; Xuan, X. Continuous dielectrophoretic separation of particles in a spiral microchannel. *Electrophoresis* **2010**, *31*, 1382–1388. [CrossRef]
47. Li, M.; Li, S.; Li, W.; Wen, W.; Alici, G. Continuous manipulation and separation of particles using combined obstacle- and curvature-induced direct current dielectrophoresis. *Electrophoresis* **2013**, *34*, 952–960. [CrossRef] [PubMed]
48. Cummings, E.B.; Griffiths, S.K.; Nilson, R.H.; Paul, P.H. Conditions for Similitude between the Fluid Velocity and Electric Field in Electroosmotic Flow. *Anal. Chem.* **2000**, *72*, 2526–2532. [CrossRef]
49. Kirby, B.J.; Hasselbrink, E.F. Zeta potential of microfluidic substrates: 2. Data for polymers. *Electrophoresis* **2004**, *25*, 203–213. [CrossRef] [PubMed]
50. Yariv, E. "Force-free" electrophoresis? *Phys. Fluids* **2006**, *18*, 031702. [CrossRef]
51. Martinez-Duarte, R. Carbon-Electrode Dielectrophoresis for Bioparticle Manipulation. *ECS Trans.* **2014**, *61*, 11. [CrossRef]
52. Islam, M.; Natu, R.; Larraga-Martinez, M.F.; Marntinez-Duarte, R. Enrichment of diluted cell populations from large sample volumes using 3D carbon-electrode dielectrophoresis. *Biomicrofluidics* **2016**, *10*, 033107. [CrossRef] [PubMed]

 *micromachines*

Article

Passive Dielectrophoretic Focusing of Particles and Cells in Ratchet Microchannels

Song-Yu Lu [1,2,†], Amirreza Malekanfard [1,†], Shayesteh Beladi-Behbahani [3], Wuzhou Zu [1], Akshay Kale [4], Tzuen-Rong Tzeng [3], Yao-Nan Wang [2,*] and Xiangchun Xuan [1,*]

[1] Department of Mechanical Engineering, Clemson University, Clemson, SC 29634-0921, USA; songyul@g.clemson.edu (S.-Y.L.); amaleka@g.clemson.edu (A.M.); wzu@g.clemson.edu (W.Z.)
[2] Department of Vehicle Engineering, National Pingtung University of Science and Technology, Pingtung 912, Taiwan
[3] Department of Biological Sciences, Clemson University, Clemson, SC 29634-0314, USA; sbeladi@g.clemson.edu (S.B.-B.); tzuenrt@clemson.edu (T.-R.T.)
[4] Electrical Engineering Division, CAPE Building, Department of Engineering, University of Cambridge, 9 JJ Thomson Avenue, West Cambridge Site, Cambridge CB3 0FA, UK; ak2115@cam.ac.uk
* Correspondence: yanwang@mail.npust.edu.tw (Y.-N.W.); xcxuan@clemson.edu (X.X.); Tel.: +886-8-770-3202-7456 (Y.-N.W.); +1-864-656-5630 (X.X.)
† These two authors contributed equally to this work.

Received: 26 March 2020; Accepted: 23 April 2020; Published: 25 April 2020

Abstract: Focusing particles into a tight stream is critical for many microfluidic particle-handling devices such as flow cytometers and particle sorters. This work presents a fundamental study of the passive focusing of polystyrene particles in ratchet microchannels via direct current dielectrophoresis (DC DEP). We demonstrate using both experiments and simulation that particles achieve better focusing in a symmetric ratchet microchannel than in an asymmetric one, regardless of the particle movement direction in the latter. The particle focusing ratio, which is defined as the microchannel width over the particle stream width, is found to increase with an increase in particle size or electric field in the symmetric ratchet microchannel. Moreover, it exhibits an almost linear correlation with the number of ratchets, which can be explained by a theoretical formula that is obtained from a scaling analysis. In addition, we have demonstrated a DC dielectrophoretic focusing of yeast cells in the symmetric ratchet microchannel with minimal impact on the cell viability.

Keywords: electrokinetic; dielectrophoresis; particle focusing; microfluidics

---

1. Introduction

Microfluidic devices have been widely used to handle (e.g., focus [1], count [2], trap [3], and sort [4] etc.) various types of particle for biomedical, chemical, and environmental applications. Focusing particles into a tight stream is critical to many of these particle-handling devices such as flow cytometers [5,6] and particle sorters [7–10]. Sheath fluids are often used to confine particles into a well-defined volume, which, however, requires an accurate control of flow rates. This is because sheath-flow focusing acts upon the suspending fluid, not the suspended particles [11]. Therefore, a variety of forces, which may be externally imposed (termed as *active* focusing) or internally induced (termed as *passive* focusing), has been demonstrated to directly manipulate particles for sheath-free focusing [12]. For the *active* focusing of particles, the application of an external acoustic [13], alternating current (AC) electric [14], or magnetic [15] field creates a non-invasive force that drives particles across fluid streamlines. This type of method requires an additional field source other than that pumping the particle suspension, not mentioning the other added difficulties such as the patterning of microelectrodes for acoustic [16] or dielectrophoretic [17] focusing and the magnetic labeling of typically non-magnetic particles [18]. The *passive* focusing of particles relies on a flow- and/or a channel structure-induced transverse force to direct particles towards one or multiple equilibrium positions over the channel cross-section. This type of method requires only one external field source to generate the flow of the particle suspension wherein the particles are automatically focused without any other controls. It is therefore easy to operate and ready to be integrated with a pre- and/or a post-focusing component for lab-on-a-chip systems [12].

Among the flow-induced *passive* particle focusing methods, inertial focusing has been rapidly growing since the seminal work of Di Carlo et al. [19]. It exploits the fluid inertia-induced lift force to focus particles down to multiple or even single streams at high throughput [20–23]. Elastic focusing results from the fluid rheology-induced lift force that is capable of manipulating much smaller particles than inertial focusing [24–27]. The combination of elastic and inertial focusing can further enhance the particle control [28] and extend the working range of flow rates [29]. Among the channel structure-induced *passive* particle focusing methods, hydrophoretic focusing utilizes the anisotropic fluid resistance of slant obstacles to generate transverse flows that carry particles towards the sidewall or channel center [30]. Hydrodynamic filtration-based focusing is based on the split and recombination of fluid flows in multiple loop channels that are symmetrically arranged on both sides of the main microchannel [31]. In addition, a direct current (DC) electric field has been demonstrated to both electrokinetically transport (via fluid electroosmosis and particle electrophoresis) and passively focus particles in a straight uniform microchannel via wall-induced electrical lift [32]. Moreover, its gradient can induce particle dielectrophoresis (DEP) for *passive* focusing in either a straight microchannel with a varying cross-section [33] or a curved microchannel [34]. The so-called insulator-based dielectrophoresis (iDEP) in the former case has been extensively demonstrated to trap [35,36], pattern [37], electroporate [38], and separate [39–43] particles in a continuous electrokinetic flow under either a DC or a DC-biased AC electric field. The effects of insulator structure, electric field, particle properties (e.g., size, charge, and type), and surface treatment have all been investigated [44–46].

However, there has been much less work on particle focusing in iDEP microdevices. A DC-biased AC electric field is necessary for the focusing of particles in a single-constriction microchannel [47], which is an *active* focusing method because the DC component pumps the particle suspension while the AC component supplements particle DEP. The *passive* focusing of particles under a DC electric field has been demonstrated in a single-constriction microchannel only when the size of the constriction closely matches that of the particles [48] or the channel-to-constriction area ratio becomes very large [33]. It can also be realized by the use of an array of ratchets, which, as reported in this work, forms periodic constrictions for a significantly extended working range of DEP. We perform a combined experimental, numerical, and theoretical study of the effects of ratchet structure, electric field, and particle size on the DC dielectrophoretic focusing of particles in ratchet microchannels. We also demonstrate the biological application of this *passive* particle focusing method to yeast cells.

2. Experiment

2.1. Materials

Two types of ratchet microchannel were used in this work, which, as shown in Figure 1a, were composed of 20 consecutive symmetric and asymmetric ratchets, respectively. They were fabricated with polydimethylsiloxane (PDMS) using the standard soft lithography technique. The broadest part of the microchannel was 500 μm wide and the narrowest part between the opposing ratchet tips was 100 μm wide in both channel structures (see the zoomed-in views in Figure 1b). The period at which the ratchet structure repeats itself, i.e., the peak-to-peak distance of two consecutive ratchets, is 250 μm, leading to an overall 5 mm long ratchet region. The total length of each ratchet microchannel is 8 mm, and the depth is uniformly 40 μm. Spherical polystyrene particles of 3, 5, and 10 μm diameter (Sigma-Aldrich Corp., St. Louis, MO, USA) were re-suspended in 1 mM phosphate buffer solution with a measured electric conductivity of 200 μS/cm (Fisher Scientific, Accumet AP85,

Waltham, MA, USA). ATCC9763 yeast cells (Saccharomyces cerevisiae) were cultured at 35 °C in Sabouraud dextrose broth (Becton and Dickinson Co., Franklin Lakes Township, NJ, USA) medium. They were harvested after 24 h and washed three times with phosphate buffered saline (PBS) solution. Prior to use, yeast cells were re-suspended in 1 mM phosphate buffer to a final concentration of around 10^5 cells/mL. They were measured to have an average diameter of around 5 μm. To avoid particle/cell aggregations and adhesions (to microchannel walls), a small amount of Tween 20 (0.5 % v/v, Fisher Scientific, Waltham, MA, USA) was added into each suspension.

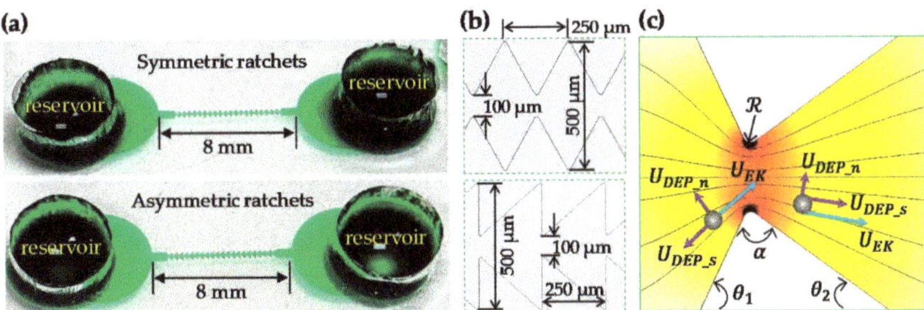

Figure 1. (**a**) Photos of the symmetric (top) and asymmetric (bottom) ratchet microchannels used in the experiment; (**b**) Zoomed-in views of the symmetric (top) and asymmetric (bottom) ratchet structures with their corresponding dimensions highlighted; (**c**) Velocity analysis for a particle traveling towards and away from the ratchet throat, respectively, where the background color shows the electric field contour (the darker, the larger magnitude) and the background lines represent the electric field lines (equivalent to the fluid streamlines).

2.2. Methods

The DC electric field across the ratchet microchannels was generated by a high-voltage DC power supply (Glassman High Voltage Inc., High Bridge, NJ, USA) via platinum electrodes. To avoid Joule heating effects [49], the average field magnitude was kept no more than 500 V/cm (i.e., a 400 V voltage drop over the 0.8 cm long microchannel) in all tests. Prior to every test, the liquid heights in the two reservoirs were carefully balanced to eliminate the flow due to hydrostatic pressure. Moreover, the time of the application of the electric field was limited to no more than 2 min in order to minimize the electroosmosis-induced pressure-driven backflow [50]. Each test was repeated at least three times on different days to ensure the repeatability of the attained results. The motions of particles and cells at different locations of the microchannel were captured using an inverted microscope (Nikon Eclipse TE2000U; Nikon Instruments, Lewisville, TX, USA) with a CCD Camera (Nikon DS-Qi1Mc, Lewisville, TX, USA) at a rate of around 15 frames per second. The obtained digital images were post-processed in the Nikon imaging software (NIS-Elements AR 2.30, Lewisville, TX, USA). The electrokinetic mobility (= electrokinetic velocity/electric field) of the particles was determined by measuring the particle velocity in the region away from the ratchets where the particle DEP was negligible. We found an approximately identical mobility of 1.86×10^{-8} m²/(V·s) for all three sizes of particle used in the experiment.

3. Theory

3.1. Focusing Mechanism

The insulating ratchets create electric field gradients around them (see the contour in Figure 1c) in a microchannel because of: (1) the variation in the cross-sectional area from the channel to the constriction formed by the facing ratchets, which is primarily along the direction of the electric field lines (or equivalently, the fluid streamlines because of their similarity in purely electrokinetic flows under the thin electric double layer assumption [51]); and (2) the variation in the path length for electric current around the ratchet tips, which is primarily normal to the direction of electric field lines. Thus, a dielectrophoretic force is induced by the ratchets, which acts on the suspended particles and cells. As they are less conductive than the suspending medium in our experiment, the polystyrene particles and yeast cells tend to be pushed away from the regions with a higher electric field, i.e., the ratchet tip (see Figure 1c), by negative DEP. Therefore, particles get focused towards the centerline of the microchannel when they travel through the ratchet region electrokinetically. Such a focusing effect via DC DEP can be characterized by the (dimensional) particle deflection that depends on the ratio of the normal component (i.e., perpendicular to the electric field lines in Figure 1c) of the particle velocity to the streamwise component (i.e., tangential to the electric field lines) within one period of the ratchets:

$$deflection = \frac{|U_{DEP\_n}|\mathcal{R}\alpha}{|U_{EK} + U_{DEP\_s}|} \tag{1}$$

where U_{DEP} is the dielectrophoretic particle velocity, with the subscripts n and s denoting, respectively, the normal and stream-wise directions; U_{EK} is the streamwise electrokinetic velocity; and the product $\mathcal{R}\alpha$ measures the working distance for the cross-stream particle DEP, with $\mathcal{R}$ and α being the curvature radius and opening angle (in the unit of radians) of the ratchet tip (see Figure 1c), respectively. Note that velocity magnitudes are used in Equation (1) because both U_{DEP} and U_{EK} can be positive or negative. It is also important to point out that the particle deflection in Equation (1) is not a constant because both U_{DEP} and U_{EK} vary with the particle position.

Following the traditional analysis of electrokinetic phenomena [52], the particle deflection in Equation (1) may be rewritten as

$$deflection = \frac{\left|\mu_{DEP}\nabla_n E^2\right|\mathcal{R}\alpha}{\left|\mu_{EK}E + \mu_{DEP}\nabla_s E^2\right|} = \frac{\mathcal{R}\alpha\left|\mu_{DEP}\frac{2E^2}{\mathcal{R}}\right|}{\left|\mu_{EK}E + \mu_{DEP}\frac{\partial E^2}{\partial s}\right|} = \frac{2\alpha}{\left|\frac{\mu_{EK}}{\mu_{DEP}}\frac{1}{E} + \frac{2}{E}\frac{\partial E}{\partial s}\right|} \tag{2}$$

$$\mu_{DEP} = f_{CM}\frac{d^2\varepsilon}{12\eta} \tag{3}$$

where μ_{DEP} is the dielectrophoretic particle mobility, μ_{EK} is the electrokinetic particle mobility, and E is the electric field magnitude. In the definition of μ_{DEP}, $f_{CM} = (\sigma_p - \sigma)/(\sigma_p + 2\sigma)$ is the Clausius–Mosotti factor, with σ_p and σ being the particle and fluid electric conductivities, respectively; d is the (spherical) particle diameter; ε is the fluid electric permittivity; and η is the fluid viscosity. As illustrated by the particle velocity analysis in Figure 1c, the streamline component of the dielectrophoretic particle velocity, U_{DEP_s}, slows down the electrokinetic particle motion towards the ratchet throat while accelerating it when the particle is traveling away. Its impact on the particle deflection hence becomes a strong function of the ratchet structure as determined by the angles θ_1 and θ_2 (note these two angles are dependent on each other if the height and width of each ratchet are both fixed). Moreover, as $\alpha = \pi - \theta_1 - \theta_2$ (see Figure 1c), the impact of the normal component of the dielectrophoretic particle velocity, U_{DEP_n}, on the particle focusing effect is also a function of the ratchet structure. In addition, Equation (2) predicts an enhanced deflection for larger particles at a higher electric field. All these effects are examined in this work. It is interesting to see that the particle deflection in Equation (2) becomes independent of the curvature radius of the ratchet tip. This is because we assume that particles traveling around the ratchet behave like those traveling through an exactly circular channel [52].

3.2. Numerical Modeling

A two-dimensional numerical model was developed in COMSOL® Multiphysics 5.3a to understand and simulate the observed particle focusing effect in the tested two-dimensional ratchet microchannels. A Lagrangian tracking method was used to trace the motion of particles in the electric field-driven fluid flow under various conditions [53]. Only the electric field was solved using the

"Electric Currents (ec)" module because of the similarity between the electric field lines and fluid streamlines in purely electrokinetic flows [51]. Particle trajectories were plotted using the particle tracing function in COMSOL® via the particle velocity, U_P, which, as shown in Figure 1c, is the vector sum of the electrokinetic and dielectrophoretic velocities:

$$U_P = U_{EK} + \lambda U_{DEP} = \mu_{EK}E + \lambda\mu_{DEP}\nabla E^2 \tag{4}$$

where E is the electric field vector and λ is the correction factor that accounts for the effect of particle size on the dielectrophoretic velocity [54]. It is because the particle's disturbances to the electric field (and as well the flow field) were neglected in the model. Such a treatment has been proved effective in our earlier studies as well as in those from other research groups [55]. To calculate the Clausius–Mosotti factor, f_{CM}, in Equation (3), we assumed that the electric conductivity of polystyrene particles is determined solely by the surface conduction, $\sigma_s = 1$ nS, through $\sigma_p = 4\sigma_s/d$ [56]. The obtained values are hence -0.45, -0.47, and -0.49 for 3, 5, and 10 μm particles, respectively. The fluid permittivity and viscosity were both assumed to be identical to those of water at room temperature, i.e., $\varepsilon = 7.1 \times 10^{-10}$ F/m and $\eta = 9.52 \times 10^{-4}$ Pa·s. The correction factor, λ, was determined by fitting the computed particle trajectories to the experimentally obtained particle images.

4. Results and Discussion

4.1. Effect of Ratchet Structure

Figure 2a shows the experimentally obtained top-view images of 5 μm particles in both the symmetric and asymmetric ratchet microchannels under a fixed DC electric field of 250 V/cm (specifically, a 200 V DC voltage drop averaged over the 0.8 cm long channel). For the asymmetric ratchets, the direction of the DC electric field is also switched to further study the effect of particle movement direction (with respect to the inclined surface of each ratchet) on the dielectrophoretic focusing of particles. Following our earlier study on particle trapping in an asymmetric ratchet microchannel [37], we still define the particle movement direction along which the inclined surface of each ratchet follows its normal surface as the *asymmetric forward* motion and its opposite as the *asymmetric backward* motion. To demonstrate the development of particle focusing in each of these ratchet structures, we present in Figure 2a the particle images at five different locations (specifically, at the 1st, 5th, 10th, 15th, and 20th ratchets) along the length of each ratchet microchannel. As expected, particles are gradually focused towards the channel centerline when they travel through each type of ratchet microchannel. The best particle focusing is achieved in the channel with symmetric ratchets. The worst particle focusing occurs in the *asymmetric backward* motion. These phenomena are reasonably predicted in our numerical model, where the correction factor, λ, for particle DEP in Equation (4) was set to 0.7 for all ratchet structures. This is demonstrated by the visual similarity in Figure 2a between the experimentally and numerically obtained particle trajectories at varying ratchets in every ratchet structure. Note that the numerical results are displayed for only the entrance and exit of the ratchet region in the figure.

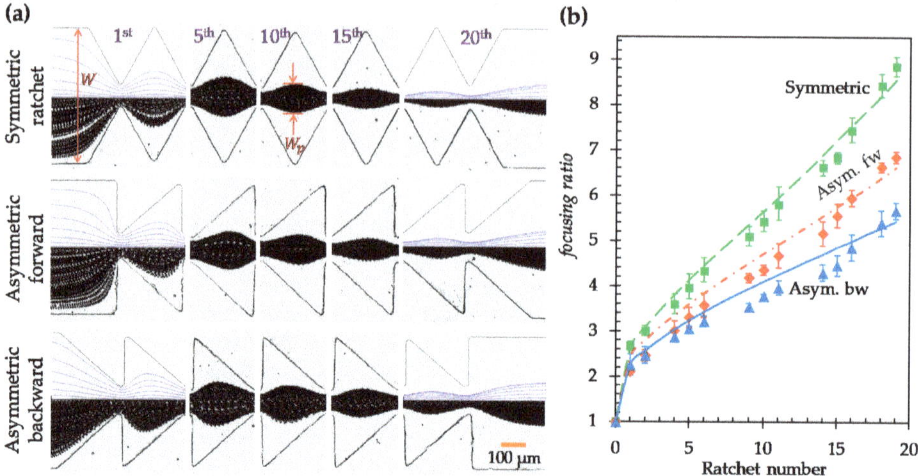

Figure 2. Effect of ratchet structure on the dielectrophoretic focusing of 5 μm diameter particles: (**a**) Comparison of the experimentally obtained and numerically predicted (top half of the left- and right-most images only) particle trajectories (traveling from left to right) at varying locations of the microchannels with symmetric (top row), asymmetric forward (middle row), and asymmetric backward (bottom row) ratchets, respectively; (**b**) Comparison of the experimentally measured (symbols with error bars) and numerically calculated (curves) particle focusing ratios, defined as the channel width, W, over the particle stream width, W_p (see the highlighted dimensions in (a)), among the three ratchet structures.

To quantify the ratchet structure's effect on particle focusing, we define a dimensionless focusing ratio as the microchannel width, W, over the particle stream width, W_p (see the highlighted dimension on the particle image in Figure 2a):

$$focusing\ ratio = \frac{W}{W_p} \tag{5}$$

The comparison of the particle focusing ratios among the three ratchet structures is illustrated in Figure 2b. A good agreement between the experimental and numerical data is obtained in every ratchet structure. The focusing ratio exhibits an approximately linear (with a positive correlation) relationship with respect to the ratchet number (except for the zeroth ratchet, where particle DEP ceases). The slope of the linear trendline for the data points (excluding that at the zeroth ratchet) is approximately 0.34 for the symmetric ratchets. This value is 42% greater than the slope of the linear trendline ($\approx$0.24) for the asymmetric forward motion and 79% greater than that ($\approx$0.19) for the asymmetric backward motion. We attribute the strongest particle focusing effect in the symmetric ratchet microchannel to: (1) the larger opening angle, α (= 64.0°), of the ratchet tip in Equation (1) (see Figure 3a) than that (= 51.3°) in the asymmetric ratchet microchannel (see Figure 3b), and (2) the smaller discrepancy in the upstream and downstream particle dynamics as demonstrated by the symmetry of the electric field (squared) and DEP before and after the ratchet tips in Figure 3. In between the two asymmetric ratchet structures, particle DEP becomes highly asymmetric on the two sides of the ratchet in Figure 3b. Specifically, for the asymmetric forward motion, an increase in the DEP on the side of the ratchet with a normal surface to the microchannel (i.e., the upstream side of the ratchet) significantly enhances the particle deflection because it increases $\left|U_{DEP\_n}\right|$ in the numerator while decreasing the particle velocity, $U_{EK} - \left|U_{DEP\_s}\right|$, in the denominator of Equation (1). By contrast, for the asymmetric backward motion, a stronger DEP on the downstream side of the ratchet does not necessarily enhance the particle deflection because it increases both $\left|U_{DEP\_n}\right|$ in the numerator and the particle velocity, $U_{EK} + \left|U_{DEP\_s}\right|$, in the denominator of Equation (1).

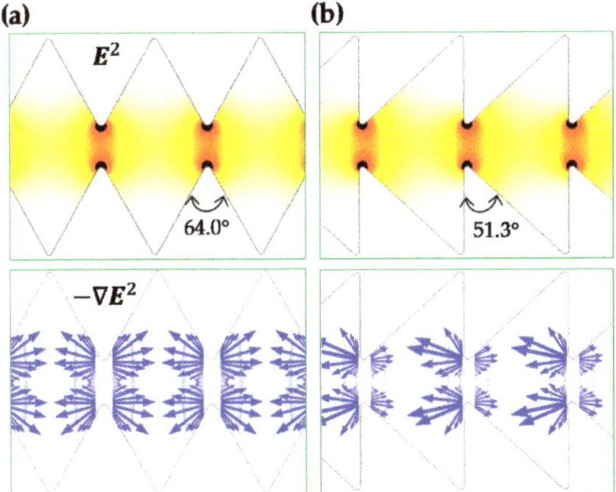

Figure 3. Comparison of the numerically predicted contour of electric field squared (top row), E^2 (the darker color the larger magnitude), and arrows (length proportional to the velocity magnitude) of negative dielectrophoretic particle velocity, U_{DEP}, in terms of $-\nabla E^2$ in between a symmetric (**a**) and an asymmetric (**b**) ratchet microchannel.

4.2. Effect of Electric Field in the Symmetric Ratchet Microchannel

We further study in this section and the next sections the effects of electric field and particle size, respectively, on the DC dielectrophoretic focusing of particles in the symmetric ratchet microchannel. Figure 4a shows the experimental and numerical images of 5 µm particles under 125, 250, and 500 V/cm electric fields, respectively. The correction factor, λ, for the dielectrophoretic particle velocity in the model was set to 0.7 in all three cases. As predicted by Equation (2), the particle deflection increases under a higher electric field, leading to an enhanced focusing towards the channel centerline. Figure 4b compares the experimentally measured and numerical predicted particle focusing ratios that show good agreement in every electric field. Moreover, similar to the observation in Figure 2b, the focusing ratio increases almost linearly with the number of ratchets under all three electric fields (except for the zeroth ratchet). The slopes of the linear trendlines for the particle focusing ratio, i.e., focusing ratio per ratchet, are 0.19, 0.34, and 0.78 under 125, 250, and 500 V/cm electric fields, respectively. Interestingly, the obtained values for the focusing ratio per ratchet also exhibit an approximately linear correlation with the DC electric field, which can be understood as follows. Our numerical simulation indicates that the magnitude of the streamwise dielectrophoretic velocity, U_{DEP_s}, at the throat of the ratchets is no more than 10% of that of the local electrokinetic velocity, U_{EK}, even under the highest electric field of 500 V/cm. Further considering that the direction of U_{DEP_s} alternates before and after any pairs of ratchets, we may safely neglect its contribution to the particle deflection within one period of ratchets in Equation (2) for a symmetric ratchet microchannel, i.e.,

$$deflection = \frac{2\alpha}{\left| \frac{\mu_{EK}}{\mu_{DEP}} \frac{1}{E} + \frac{2}{E} \frac{\partial E}{\partial s} \right|} \sim 2E\alpha \left| \frac{\mu_{DEP}}{\mu_{EK}} \right| \tag{6}$$

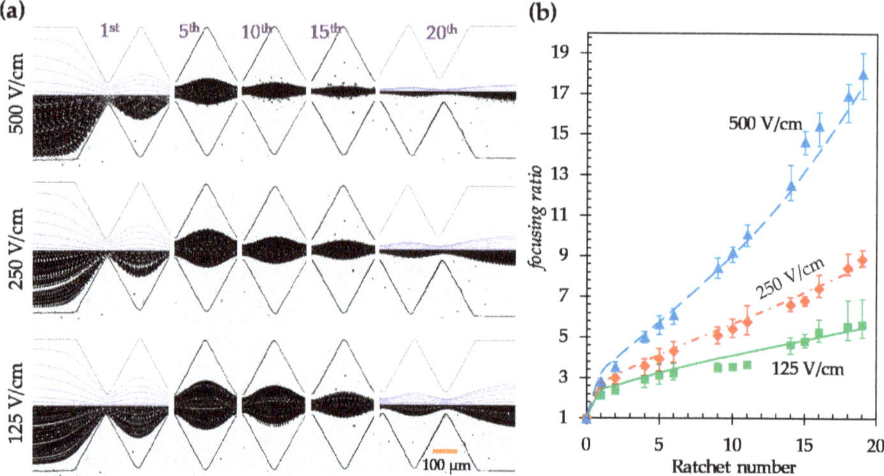

Figure 4. Effect of the electric field on the dielectrophoretic focusing of 5 μm diameter particles in the symmetric ratchet microchannel: (**a**) Comparison of the experimentally obtained and numerically predicted (top half of the left- and right-most images only) particle trajectories (traveling from left to right) at varying locations of the microchannel under 125 (bottom row), 250 (middle row), and 500 V/cm (top row) electric fields, respectively; (**b**) Comparison of the experimentally measured (symbols with error bars) and numerically calculated (curves) particle focusing ratios among the three electric fields.

Thus, neglecting the action of DEP from the ratchets on the other half of the microchannel, which is equivalent to assuming that the channel width $W \to \infty$ or the particle deflection is very small compared to W, we can obtain the half-width of the particle stream as

$$\frac{W_p}{2} \sim \frac{W}{2} - m \times deflection \sim \frac{W}{2} - 2mE\alpha \left| \frac{\mu_{DEP}}{\mu_{EK}} \right| \tag{7}$$

where m is the number of ratchets that particles have traveled through. Then, we can rewrite the particle focusing ratio in Equation (5) as follows:

$$focusing\ ratio \sim \frac{W}{W - 4mE\alpha \left| \frac{\mu_{DEP}}{\mu_{EK}} \right|} \tag{8}$$

The focusing ratio per ratchet is hence determined as

$$focusing\ ratio\ per\ ratchet \sim \frac{W}{W - 4(m+1)E\alpha \left| \frac{\mu_{DEP}}{\mu_{EK}} \right|} - \frac{W}{W - 4mE\alpha \left| \frac{\mu_{DEP}}{\mu_{EK}} \right|}$$

$$= \frac{4WE\alpha \left| \frac{\mu_{DEP}}{\mu_{EK}} \right|}{\left(W - 4(m+1)E\alpha \left| \frac{\mu_{DEP}}{\mu_{EK}} \right| \right) \left(W - 4mE\alpha \left| \frac{\mu_{DEP}}{\mu_{EK}} \right| \right)} \sim \frac{4E\alpha}{W} \left| \frac{\mu_{DEP}}{\mu_{EK}} \right| \tag{9}$$

Note that in this derivation, we have used the assumption of small particle deflection as compared to the channel width. Therefore, the particle focusing ratio per ratchet in Equation (8) becomes a linear function of the applied electric field.

4.3. Effect of Particle Size in the Symmetric Ratchet Microchannel

Figure 5a shows the experimental and numerical images of 3, 5, and 10 μm particles in the symmetric ratchet microchannel under a fixed DC electric field of 250 V/cm. The correction factor, λ, was set to 0.8, 0.7, and 0.6 for 3, 5, and 10 μm particles, respectively, in the simulation. As the dielectrophoretic mobility of particles, μ_{DEP}, (see Equation (3)) is a second order function of particle size, the focusing ratio in Equation (7) should increase for larger particles because of their enhanced deflection. This is supported by the experiment and simulation in Figure 5a, where 10 μm particles attain nearly single-file focusing at the end of the ratchet region, while 3 μm particles experience only slight focusing. Figure 5b compares the experimental and numerical data of the particle focusing ratio, where good agreement is seen for all three types of particle. However, the focusing ratio for 10 μm particles exhibits an apparently nonlinear relationship with the ratchet number, though that for 3 μm particles still follows a linear trend (excluding the data at the zeroth ratchet). It may be because the U_{DEP_s} of 10 μm particles becomes comparable to U_{EK}, which invalidates the scaling analysis in the preceding section. In fact, the focusing ratio for 5 μm particles at 500 V/cm in Figure 4b already displays a visible deviation from the linear trendline because of the same reason. As predicted by Equation (8), the particle focusing ratio per ratchet is proportional to the magnitude of μ_{DEP} and hence a second order function of particle size. This analysis is well supported by the value of 0.16 for 3 μm particles against that of 0.34 for 5 μm particles.

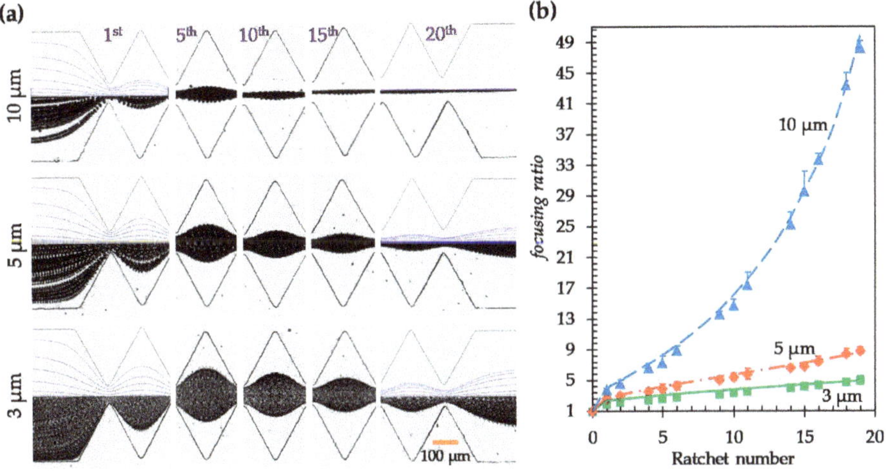

Figure 5. Effect of particle size on the dielectrophoretic focusing of polystyrene particles in the symmetric ratchet microchannel under a fixed DC electric field of 250 V/cm: (**a**) Comparison of the experimentally obtained and numerically predicted (top half of the left- and right-most images only) trajectories (traveling from left to right) of 3 (bottom row), 5 (middle row), and 10 μm (top row) particles, respectively, at varying locations of the microchannel; (**b**) Comparison of the experimentally measured (symbols with error bars) and numerically calculated (curves) particle focusing ratios among the three types of particles.

4.4. Focusing of Yeast Cells in the Symmetric Ratchet Microchannel

To demonstrate the potential biological applications of the passive dielectrophoretic particle focusing method, yeast cells were chosen to replace 5 μm polystyrene particles in a test with the symmetric ratchet microchannel. The superimposed images in Figure 6 show the development of cell focusing along the microchannel under the application of a 250 V DC voltage (i.e., a 312.5 V/cm electric field, on average, over the entire channel length). Since the size of yeast cells is not homogenous, the observed cell focusing is slightly worse than that of 5 μm particles (see Figure 2a). The application of the DC electric field may affect the viability of yeast cells via Joule heating-induced temperature elevation [57] and/or electrical field-induced transmembrane voltage [58]. For the former, we did not notice any significant increase in the electric current through the buffer solution in the microchannel, which indicates an insignificant Joule heating effect during the focusing experiment [49]. To check the impact of the electrical shock, we conducted a viability test using trypan blue, which can stain non-viable cells blue while viable cells remain unstained. Specifically, 100 µL yeast cell suspension was taken from the outlet reservoir of the ratchet microchannel and stained with trypan blue in 1:1 ratio. A hemocytometer slide was then filled with the stained cell suspension and incubated at room temperature for 1–2 min. Live and dead cells were counted under a microscope, and the viability was calculated by dividing the number of live cells by the total number of cells. We confirmed that more than 98% of the yeast cells still remained alive after the dielectrophoretic focusing experiment.

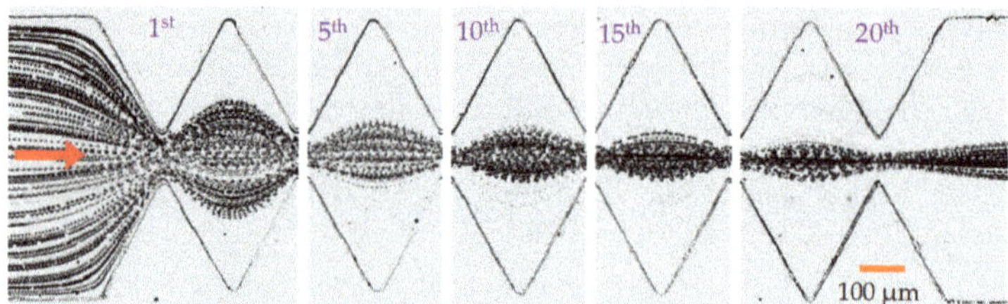

Figure 6. Top-view superimposed images demonstrating the development of yeast cell focusing at varying locations of the symmetric ratchet microchannel under a DC electric field of around 300 V/cm. The block arrow indicates the movement direction of cells.

It is worth mentioning that our group has recently demonstrated a passive focusing of particles [34] and cells [59] in a serpentine microchannel via curvature-induced DEP. Compared to that method, the current dielectrophoretic particle focusing in a ratchet microchannel has the disadvantage of drawing significantly higher electric fields around the ratchet tips, which may cause potential thermal [57] and electrical [58] issues for the sample and/or the microfluidic device as noted above. However, the current method has the capability of focusing much smaller particles because of the much stronger electric field gradients around the ratchet tips than around the corners of a serpentine microchannel. Moreover, the DEP in ratchet-like microchannels offers more diverse applications such as the focusing, concentration [35], patterning [37], electroporation [60], and separation [40] of particles or cells. It therefore has the potential to perform multiple functions in a single microfluidic device.

5. Conclusions

We have performed a combined experimental, numerical, and theoretical study of the DC dielectrophoretic focusing of polystyrene particles in symmetric and asymmetric ratchet microchannels with similar dimensions. The symmetric ratchet microchannel is found to offer better particle focusing than the asymmetric one because of the larger opening angle of the symmetric ratchets. In the asymmetric ratchet microchannel, particles can attain a stronger focusing effect in the forward motion than in the backward motion because of both the asymmetry and the directional switch of particle DEP on the upstream and downstream sides of any pair of ratchets. Moreover, we have investigated the effects of electric field and particle size on the DC dielectrophoretic focusing of polystyrene particles in the symmetric ratchet microchannel. The defined dimensionless particle focusing ratio is found to increase for larger particles under higher electric fields. It also increases almost linearly with the number of ratchets, through which particles have travelled, unless the streamwise dielectrophoretic particle velocity becomes comparable to the electrokinetic velocity at the ratchet region. These phenomena can be reasonably explained by the formulae that are obtained from a theoretical analysis and may serve as a guideline for the design of ratchet microchannels in future particle focusing applications. In addition, we have demonstrated the passive dielectrophoretic focusing of yeast cells in the symmetric ratchet microchannel. The impact of DC electric field exposure on cell viability is found to be minimal under our experimental conditions.

Compared to other *passive* focusing methods, our demonstrated DC dielectrophoretic focusing of particles and cells in ratchet microchannels has the advantages of simplicity, being free of moving parts,

Micromachines **2020**, *11*, 451

and being easy to integrate with other electrically-controlled microfluidic components, etc. It does not require the patterning of microelectrodes that is needed for classical AC DEP-based focusing. While it provides a much smaller throughput than fluid inertia-based hydrodynamic focusing, our electrokinetic method may find a niche application in areas that need to process small amounts of samples. Moreover, if the channel-to-constriction width ratio and/or the number of ratchets becomes sufficient large, our method has the potential to work with submicron particles or even nanoparticles that are usually very hard to control using inertial microfluidics [61]. We are currently working on how to optimize the ratchet structure for particle focusing via DC DEP.

Author Contributions: X.X., T.-R.T., and Y.-N.W. conceived and designed the project; S.-Y.L., A.M., S.B.-B. and W.Z. performed the experiments and analysed the experimental data; A.M. and A.K. performed the theoretical analysis and numerical simulations; S.-Y.L., A.M. and X.X. wrote the paper; all authors commented on the paper. All authors have read and agreed to the published version of the manuscript.

Funding: This work was supported in part by Clemson University through the Creative Inquiry program and by NSF under grant number CBET-1704379 (X.X.).

Conflicts of Interest: The authors declare no conflict of interest.

References

1. Zhang, T.; Hong, Z.-Y.; Tang, S.-Y.; Li, W.; Inglis, D.; Hosokawa, Y.; Yalikun, Y.; Li, M.; Yaxiaer, Y. Focusing of sub-micrometer particles in microfluidic devices. *Lab Chip* **2019**, *20*, 35–53. [CrossRef]
2. Song, Y.; Zhang, J.; Li, D. Microfluidic and Nanofluidic Resistive Pulse Sensing: A Review. *Micromachines* **2017**, *8*, 204. [CrossRef]
3. Nilsson, J.; Evander, M.; Hammarström, B.; Laurell, T. Review of cell and particle trapping in microfluidic systems. *Anal. Chim. Acta* **2009**, *649*, 141–157. [CrossRef]
4. Sajeesh, P.; Sen, A. Particle separation and sorting in microfluidic devices: A review. *Microfluid. Nanofluid.* **2013**, *17*, 1–52. [CrossRef]
5. Yang, R.-J.; Fu, L.-M.; Hou, H.-H. Review and perspectives on microfluidic flow cytometers. *Sensors Actuators B Chem.* **2018**, *266*, 26–45. [CrossRef]
6. Gong, Y.; Fan, N.; Yang, X.; Peng, B.; Jiang, H. New advances in microfluidic flow cytometry. *Electrophoresis* **2018**, *40*, 1212–1229. [CrossRef] [PubMed]
7. Gossett, D.R.; Weaver, W.M.; Mach, A.J.; Hur, S.C.; Tse, H.T.K.; Lee, W.; Amini, H.; Di Carlo, D. Label-free cell separation and sorting in microfluidic systems. *Anal. Bioanal. Chem.* **2010**, *397*, 3249–3267. [CrossRef] [PubMed]
8. Karimi, A.; Yazdi, S.; Ardekani, A.M. Hydrodynamic mechanisms of cell and particle trapping in microfluidics. *Biomicrofluidics* **2013**, *7*, 021501. [CrossRef] [PubMed]
9. Yan, S.; Zhang, J.; Yuan, D.; Li, W. Hybrid microfluidics combined with active and passive approaches for continuous cell separation. *Electrophoresis* **2016**, *38*, 238–249. [CrossRef]
10. Tang, W.; Jiang, D.; Li, Z.; Zhu, L.; Shi, J.; Yang, J.; Xiang, N. Recent advances in microfluidic cell sorting techniques based on both physical and biochemical principles. *Electrophoresis* **2018**, *40*, 930–954. [CrossRef]
11. Shrirao, A.B.; Fritz, Z.; Novik, E.M.; Yarmush, G.M.; Schloss, R.S.; Zahn, J.D.; Yarmush, M.L. Microfluidic flow cytometry: The role of microfabrication methodologies, performance and functional specification. *Technology* **2018**, *6*, 1–23. [CrossRef] [PubMed]
12. Xuan, X.; Zhu, J.; Church, C. Particle focusing in microfluidic devices. *Microfluid. Nanofluid.* **2010**, *9*, 1–16. [CrossRef]
13. Wu, M.; Ozcelik, A.; Rufo, J.; Wang, Z.; Fang, R.; Huang, T.J. Acoustofluidic separation of cells and particles. *Microsyst. Nanoeng.* **2019**, *5*, 32. [CrossRef] [PubMed]
14. Jia, Y.; Ren, Y.; Jiang, H. Continuous dielectrophoretic particle separation using a microfluidic device with 3D electrodes and vaulted obstacles. *Electrophoresis* **2015**, *36*, 1744–1753. [CrossRef]
15. Xuan, X. Recent Advances in Continuous-Flow Particle Manipulations Using Magnetic Fluids. *Micromachines* **2019**, *10*, 744. [CrossRef]
16. Shi, J.; Ahmed, D.; Colletti, A.; Mao, X.; Huang, T.J. Focusing microparticles in a microfluidic channel with standing surface acoustic waves (SSAW). *Lab Chip* **2008**, *8*, 221–223. [CrossRef]

17. Huang, C.-T.; Weng, C.-H.; Jen, C.-P. Three-dimensional cellular focusing utilizing a combination of insulator-based and metallic dielectrophoresis. *Biomicrofluidics* **2011**, *5*, 044101. [CrossRef]

18. Afshar, R.; Moser, Y.; Lehnert, T.; Gijs, M.A.M. Three-Dimensional Magnetic Focusing of Superparamagnetic Beads for On-Chip Agglutination Assays. *Anal. Chem.* **2011**, *83*, 1022–1029. [CrossRef]

19. Di Carlo, D.; Irimia, D.; Tompkins, R.G.; Toner, M. Continuous inertial focusing, ordering, and separation of particles in microchannels. *Proc. Natl. Acad. Sci. USA* **2007**, *104*, 18892–18897. [CrossRef]

20. Martel, J.M.; Toner, M. Inertial focusing in microfluidics. *Annu. Rev. Biomed. Eng.* **2014**, *16*, 371–396. [CrossRef]

21. Zhang, J.; Yan, S.; Yuan, D.; Alici, G.; Nguyen, A.V.; Warkiani, M.E.; Li, W. Fundamentals and applications of inertial microfluidics: A review. *Lab Chip* **2016**, *16*, 10–34. [CrossRef] [PubMed]

22. Liu, C.; Hu, G. High-Throughput Particle Manipulation Based on Hydrodynamic Effects in Microchannels. *Micromachines* **2017**, *8*, 73. [CrossRef]

23. Stoecklein, D.; Di Carlo, D. Correction to Nonlinear Microfluidics. *Anal. Chem.* **2019**, *91*, 12596. [CrossRef] [PubMed]

24. D'Avino, G.; Greco, F.; Maffettone, P.L. Particle Migration due to Viscoelasticity of the Suspending Liquid and Its Relevance in Microfluidic Devices. *Annu. Rev. Fluid Mech.* **2017**, *49*, 341–360. [CrossRef]

25. Lu, X.; Liu, C.; Hu, G.; Xuan, X. Particle manipulations in non-Newtonian microfluidics: A review. *J. Colloid Interface Sci.* **2017**, *500*, 182–201. [CrossRef]

26. Yuan, D.; Zhao, Q.; Yan, S.; Tang, S.-Y.; Alici, G.; Zhang, J.; Li, W. Recent progress of particle migration in viscoelastic fluids. *Lab Chip* **2018**, *18*, 551–567. [CrossRef]

27. Tian, F.; Feng, Q.; Chen, Q.; Liu, C.; Li, T.; Sun, J.; Fei, T.; Qiang, F.; Tiejun, L. Manipulation of bio-micro/nanoparticles in non-Newtonian microflows. *Microfluid. Nanofluid.* **2019**, *23*, 68. [CrossRef]

28. Yang, S.; Kim, J.; Lee, S.J.; Lee, S.S.; Kim, J.M. Sheathless elasto-inertial particle focusing and continuous separation in a straight rectangular microchannel. *Lab Chip* **2011**, *11*, 266–273. [CrossRef]

29. Lim, E.J.; Ober, T.J.; Edd, J.F.; Desai, S.P.; Neal, D.; Bong, K.W.; Doyle, P.S.; McKinley, G.H.; Toner, M. Inertio-elastic focusing of bioparticles in microchannels at high throughput. *Nat. Commun.* **2014**, *5*, 4120. [CrossRef]

30. Choi, S.; Song, S.; Choi, C.; Park, J.-K. Sheathless Focusing of Microbeads and Blood Cells Based on Hydrophoresis. *Small* **2008**, *4*, 634–641. [CrossRef]

31. Aoki, R.; Yamada, M.; Yasuda, M.; Seki, M. In-channel focusing of flowing microparticles utilizing hydrodynamic filtration. *Microfluid. Nanofluid.* **2008**, *6*, 571–576. [CrossRef]

32. Liu, Z.; Li, D.; Song, Y.; Pan, X.; Li, N.; Xuan, X. Surface-conduction enhanced dielectrophoretic-like particle migration in electric-field driven fluid flow through a straight rectangular microchannel. *Phys. Fluids* **2017**, *29*, 102001. [CrossRef]

33. Braff, W.A.; Pignier, A.; Buie, C.R. High sensitivity three-dimensional insulator-based dielectrophoresis. *Lab Chip* **2012**, *12*, 1327. [CrossRef] [PubMed]

34. Zhu, J.; Tzeng, T.-R.J.; Hu, G.; Xuan, X. DC dielectrophoretic focusing of particles in a serpentine microchannel. *Microfluid. Nanofluid.* **2009**, *7*, 751–756. [CrossRef]

35. Lewpiriyawong, N.; Yang, C.; Lam, Y.C. Electrokinetically driven concentration of particles and cells by dielectrophoresis with DC-offset AC electric field. *Microfluid. Nanofluid.* **2011**, *12*, 723–733. [CrossRef]

36. Saucedo-Espinosa, M.A.; Lapizco-Encinas, B.H. Experimental and theoretical study of dielectrophoretic particle trapping in arrays of insulating structures: Effect of particle size and shape. *Electrophoresis* **2015**, *36*, 1086–1097. [CrossRef]

37. Kale, A.; Lu, X.; Patel, S.; Xuan, X. Continuous-flow dielectrophoretic trapping and patterning of colloidal particles in a ratchet microchannel. *J. Micromech. Microeng.* **2014**, *24*, 75007. [CrossRef]

38. Pudasaini, S.; Perera, A.T.K.; Das, D.; Ng, S.H.; Yang, C. Continuous flow microfluidic cell inactivation with the use of insulating micropillars for multiple electroporation zones. *Electrophoresis* **2019**, *40*, 2522–2529. [CrossRef]

39. Lapizco-Encinas, B.H.; Simmons, B.A.; Cummings, E.B.; Fintschenko, Y. Insulator-based dielectrophoresis for the selective concentration and separation of live bacteria in water. *Electrophoresis* **2004**, *25*, 1695–1704. [CrossRef]

40. Pysher, M.D.; Hayes, M.A. Electrophoretic and Dielectrophoretic Field Gradient Technique for Separating Bioparticles. *Anal. Chem.* **2007**, *79*, 4552–4557. [CrossRef]

41. Hawkins, B.G.; Smith, A.E.; Syed, Y.A.; Kirby, B. Continuous-Flow Particle Separation by 3D Insulative Dielectrophoresis Using Coherently Shaped, dc-Biased, ac Electric Fields. *Anal. Chem.* **2007**, *79*, 7291–7300. [CrossRef] [PubMed]

42. Abdallah, B.G.; Roy-Chowdhury, S.; Coe, J.; Fromme, P.; Ros, A. High Throughput Protein Nanocrystal Fractionation in a Microfluidic Sorter. *Anal. Chem.* **2015**, *87*, 4159–4167. [CrossRef] [PubMed]

43. Hill, N.; Lapizco-Encinas, B.H. Continuous flow separation of particles with insulator-based dielectrophoresis chromatography. *Anal. Bioanal. Chem.* **2020**, 1–12. [CrossRef] [PubMed]

44. Srivastava, S.K.; Gencoglu, A.; Minerick, A. DC insulator dielectrophoretic applications in microdevice technology: A review. *Anal. Bioanal. Chem.* **2010**, *399*, 301–321. [CrossRef]

45. Regtmeier, J.; Eichhorn, R.; Viefhues, M.; Bogunovic, L.; Anselmetti, D. Electrodeless dielectrophoresis for bioanalysis: Theory, devices and applications. *Electrophoresis* **2011**, *32*, 2253–2273. [CrossRef]

46. Lapizco-Encinas, B.H. On the recent developments of insulator-based dielectrophoresis: A review. *Electrophoresis* **2018**, *40*, 358–375. [CrossRef]

47. Zhu, J.; Xuan, X. Dielectrophoretic focusing of particles in a microchannel constriction using DC-biased AC electric fields. *Electrophoresis* **2009**, *30*, 2668–2675. [CrossRef]

48. Xuan, X.; Raghibizadeh, S.; Li, N. Wall effects on electrophoretic motion of spherical polystyrene particles in a rectangular poly(dimethylsiloxane) microchannel. *J. Colloid Interface Sci.* **2006**, *296*, 743–748. [CrossRef]

49. Kale, A.; Song, L.; Lu, X.; Yu, L.; Hu, G.; Xuan, X. Electrothermal enrichment of submicron particles in an insulator-based dielectrophoretic microdevice. *Electrophoresis* **2017**, *39*, 887–896. [CrossRef]

50. Yan, D.G.; Yang, C.; Huang, X.Y. Effect of finite reservoir size on electroosmotic flow in microchannels. *Microfluid. Nanofluid.* **2006**, *3*, 333–340. [CrossRef]

51. Cummings, E.B.; Griffiths, S.K.; Nilson, R.H.; Paul, P.H. Conditions for Similitude between the Fluid Velocity and Electric Field in Electroosmotic Flow. *Anal. Chem.* **2000**, *72*, 2526–2532. [CrossRef] [PubMed]

52. Xuan, X. Recent advances in direct current electrokinetic manipulation of particles for microfluidic applications. *Electrophoresis* **2019**, *40*, 2484–2513. [CrossRef] [PubMed]

53. Qian, S.; Ai, Y. *Electrokinetic Particle Transport in Micro-/Nanofluidics: Direct Numerical Simulation Analysis*; CRC Press: Boca Raton, FL, USA, 2012.

54. Kang, K.H.; Xuan, X.; Kang, Y.; Li, D. Effects of dc-dielectrophoretic force on particle trajectories in microchannels. *J. Appl. Phys.* **2006**, *99*, 64702. [CrossRef]

55. Hill, N.; Lapizco-Encinas, B.H. On the use of correction factors for the mathematical modeling of insulator based dielectrophoretic devices. *Electrophoresis* **2019**, *40*, 2541–2552. [CrossRef] [PubMed]

56. Ermolina, I.; Morgan, H. The electrokinetic properties of latex particles: Comparison of electrophoresis and dielectrophoresis. *J. Colloid Interface Sci.* **2005**, *285*, 419–428. [CrossRef] [PubMed]

57. Cetin, B.; Li, N. Effect of Joule heating on electrokinetic transport. *Electrophoresis* **2008**, *29*, 994–1005. [CrossRef]

58. Voldman, J. Electrical Forces for Microscale cell Manipulation. *Annu. Rev. Biomed. Eng.* **2006**, *8*, 425–454. [CrossRef]

59. Church, C.; Zhu, J.; Wang, G.; Tzeng, T.-R.J.; Xuan, X. Electrokinetic focusing and filtration of cells in a serpentine microchannel. *Biomicrofluidics* **2009**, *3*, 44109. [CrossRef]

60. Wang, H.-Y.; Lu, C. Electroporation of Mammalian Cells in a Microfluidic Channel with Geometric Variation. *Anal. Chem.* **2006**, *78*, 5158–5164. [CrossRef]

61. Mutlu, B.R.; Edd, J.F.; Toner, M. Oscillatory inertial focusing in infinite microchannels. *Proc. Natl. Acad. Sci. USA* **2018**, *115*, 7682–7687. [CrossRef]

 *micromachines*

Article

Protein Dielectrophoresis: I. Status of Experiments and an Empirical Theory

Ralph Hölzel [1] and Ronald Pethig [2,*]

[1] Fraunhofer Institute for Cell Therapy and Immunology, Branch Bioanalytics and Bioprocesses (IZI-BB), Am Mühlenberg 13, 14476 Potsdam-Golm, Germany; ralph.hoelzel@izi-bb.fraunhofer.de

[2] School of Engineering, Institute for Integrated Micro and Nanosystems, University of Edinburgh, The King's Buildings, Edinburgh EH9 3JF, UK

* Correspondence: Ron.Pethig@ed.ac.uk

Received: 3 May 2020; Accepted: 20 May 2020; Published: 22 May 2020

Abstract: The dielectrophoresis (DEP) data reported in the literature since 1994 for 22 different globular proteins is examined in detail. Apart from three cases, all of the reported protein DEP experiments employed a gradient field factor ∇E_m^2 that is much smaller (in some instances by many orders of magnitude) than the ~4×10^{21} V^2/m^3 required, according to current DEP theory, to overcome the dispersive forces associated with Brownian motion. This failing results from the macroscopic Clausius–Mossotti (*CM*) factor being restricted to the range $1.0 > CM > -0.5$. Current DEP theory precludes the protein's permanent dipole moment (rather than the induced moment) from contributing to the DEP force. Based on the magnitude of the β-dispersion exhibited by globular proteins in the frequency range 1 kHz–50 MHz, an empirically derived molecular version of *CM* is obtained. This factor varies greatly in magnitude from protein to protein (e.g., ~37,000 for carboxypeptidase; ~190 for phospholipase) and when incorporated into the basic expression for the DEP force brings most of the reported protein DEP above the minimum required to overcome dispersive Brownian thermal effects. We believe this empirically-derived finding validates the theories currently being advanced by Matyushov and co-workers.

Keywords: Clausius–Mossotti function; dielectrophoresis; dielectric spectroscopy; interfacial polarization; proteins

---

1. Introduction

Dielectrophoresis (DEP) studies of biological particles have progressed from the microscopic scale of cells and bacteria, through the much smaller scale of virions to the molecular scale of DNA and proteins [1]. In a pioneering study of 1994, Washizu et al. [2] demonstrated that DEP forces capable of overcoming randomizing Brownian influences could be exerted on protein molecules (avidin, chymotripsinogen, concanavalin and ribonuclease) using micrometer-sized electrodes. The applied fields (0.4–1.0×10^6 V/m) were considered to be substantially lower than standard DEP theory predicts [2]. In fact, the word 'substantially' can be considered as an understatement of the situation. As reviewed elsewhere, at least 22 different globular proteins have now been investigated for their DEP responses [3–7]. In all the analyses by the authors of the cited studies, the so-called Clausius–Mossotti (*CM*) function has been invoked. However, the macroscopic electrostatic concepts and assumptions used in the theoretical derivation of *CM* arguably fail to describe the situation for nanoparticles, such as proteins, that possess a permanent dipole moment, interact with water dipoles of hydration, and possess other physico-chemical attributes at the molecular scale [6–8]. The fact that standard DEP theory does not provide a basis for understanding protein DEP is recognized as "a well-accepted paradigm, repeated in numerous studies" [6]. In another recent review it is correctly stated that protein

DEP remains under development because due to their small size proteins "require greater magnitudes of electric field gradients to achieve manipulation" [7]. Put more bluntly, protein DEP is considered to not have a theoretical leg to stand on!

A new theory is in fact evolving in terms of a description at the molecular level of how a macroscopic dielectric sample responds to an applied electric field [8–10]. This involves a consideration of the actual 'cavity field' experienced by the protein molecule, as well as the time-dependent correlation of the total electric moment of the protein. This moment is a resultant of all the permanent and induced moments of the system comprising the protein molecule's polypeptide chain(s), the protein's hydration sheath, as well as neighboring water molecules under the electrostatic influence of the protein's induced and permanent dipole field.

The purpose of this and an accompanying paper [11] is to critically evaluate the protein DEP literature, to derive an empirical-based theory, and to then describe and summarize the molecular-based theory developed by Matyushov and colleagues [8,10]. In this paper we examine aspects of the reported protein DEP work not covered in previous reviews, and conclude that the reported DEP responses for a range of proteins are largely consistent. Practically all of the DEP data cannot be explained in terms of the induced-dipole moment theory currently employed by the DEP community. The previous proposal [9] that the permanent, intrinsic, dipole moment of a protein, manifested when polarized as a dielectric β-dispersion, should form the underlying basis for a proper theory of protein DEP is repeated here. It is also shown that the reported DEP responses of protein molecules are understandable if the 'cavity' field experienced by the protein is at least 1000-times larger than the local macroscopic field in the surrounding aqueous medium. By linking the β-dispersion (a molecular-scale phenomenon) to the macroscopic phenomenon known as the Maxwell-Wagner interfacial polarization exhibited by colloids, we derive an empirical relationship to describe this amplification of the protein's cavity field. This empirical relationship underscores the fact that the macroscopic *CM* function employed in the present standard DEP theory is an analogue of (but not the same as) the molecular *CM*-relation that formed the bedrock of classical dielectric theory [12] used to describe the electrical polarization of proteins [13]. However, to exploit the potential benefits that protein DEP can offer to basic research needs and clinical applications [6], we require a solid molecular-based theory. In our opinion, the most promising theory currently being developed for protein DEP is that emerging from Matyushov's group [8,10]. An attempt to summarize this is given in the accompanying paper [11], within the frameworks of the development and application of the molecular *CM*-relation in classical dielectric theory; the key dielectric properties of solvated proteins; the published work on protein DEP.

In all of this it is instructive to appreciate how the *CM*-factor is incorporated into present DEP theory. A detailed description is presented elsewhere [9], but in brief it is based on the following sequence of assumptions and derivations:

(i)   The internal electrical field E_i induced in an uncharged (or uniformly charged) spherical particle, of radius R, located in an electric field E_m within a dielectric medium is given by:

$$E_i = \left(\frac{3\varepsilon_m}{\varepsilon_p + 2\varepsilon_m}\right)E_m \tag{1}$$

with ε_p and ε_m the relative permittivity of the particle and surrounding medium, respectively. It is assumed that ε_p and ε_m are well defined. At the molecular scale this requires certain conditions to be met regarding dipole–dipole correlations. Boundary conditions also assume that the electric potential, current density and displacement flux are continuous across an infinitesimally thin surface at the sphere's interface with the surrounding medium. Fine details such as those that occur, for example, at the molecular interface between a protein and its hydration sheath are not considered.

(ii)   The induced polarization P_p per unit volume of the sphere is given by:

$$P_p = \left(\varepsilon_p - \varepsilon_m\right)\varepsilon_0 E_i = 3\varepsilon_0 \varepsilon_m\left(\frac{\varepsilon_p - \varepsilon_m}{\varepsilon_p + 2\varepsilon_m}\right)E_m \tag{2}$$

where ε_0 is the permittivity of vacuum. The macroscopic dielectric concepts involved in this equation and throughout this paper are summarized in Figure 1. It is assumed that the polarization P_m of the surrounding medium remains uniform right up to the particle–medium interface. This assumption requires examination at the molecular scale.

(iii)   The dipole moment m of the sphere is the value of P_p multiplied by the sphere's volume:

$$m = 4\pi R^3 \varepsilon_0 \varepsilon_m\left(\frac{\varepsilon_p - \varepsilon_m}{\varepsilon_p + 2\varepsilon_m}\right)E_m \tag{3}$$

The term in brackets in Equations (2) and (3) is the Clausius–Mossotti (*CM*) function. Depending on the relative values of ε_p and ε_m, *CM* is limited to values between +1.0 ($\varepsilon_p \gg \varepsilon_m$) and −0.5 ($\varepsilon_p \ll \varepsilon_m$). This represents a severe limitation, at the macroscopic scale, to the range of effective dipole moment densities that a particle can assume.

(iv)   For the case where E_m has a gradient, the particle experiences a DEP force given by:

$$F_{DEP} = (m \cdot \nabla)E_m \tag{4}$$

where ∇ is the gradient (del) operator and E_m is assumed irrotational (i.e., $\nabla \times E_m = 0$). This assumption holds if E_m is said to be a conservative field. In our particular case of DEP, this means that moving a polarized particle from location a to b, and then back again to location a, will involve no net expenditure of work by the field. The actual path taken in moving from say a to z is of no relevance. In the language of thermodynamics each infinitesimal change in location is reversible. At the molecular level, the DEP motion of a protein involves the breaking (enthalpy absorbed and entropy increased) and remaking of hydrogen-bonded water networks at the hydrodynamic plane of shear. Some interesting variations of changes in Gibbs free energy ($\Delta G = \Delta H - T\,\Delta S$) might occur. The response of an assembly of dipoles to an external electric field is basically a thermodynamically non-equilibrium process—the thermal energy is never equally distributed among the various degrees of motional freedom of the dipoles. Perhaps, at the molecular level, each infinitesimal change in location is not reversible?

From Equations (3) and (4):

$$F_{DEP} = 4\pi R^3 \varepsilon_0 \varepsilon_m\left(\frac{\varepsilon_p - \varepsilon_m}{\varepsilon_p + 2\varepsilon_m}\right)(E_m \cdot \nabla)E_m = 2\pi R^3 \varepsilon_0 \varepsilon_m\left(\frac{\varepsilon_p - \varepsilon_m}{\varepsilon_p + 2\varepsilon_m}\right)\nabla E_m^2 \tag{5}$$

Equation (5) can be extended to describe oblate and prolate spheroids by introducing a polarization parameter that moderates the internal field, and AC fields are accommodated using a complex *CM* (i.e., contains real and imaginary components) that takes into account the phase difference between charge displacement and ohmic currents in particles exhibiting dielectric losses. The complex conductivity and permittivity are related by $\sigma^* = i\omega\varepsilon_0\varepsilon^*$ where $i = \sqrt{-1}$ and ω is the radian frequency of the applied r.m.s. field E_0. The form of *CM* (the term in brackets) shown in Equation (5) is valid at high frequencies (typically >50 MHz). At DC and below ~100 Hz

$$CM = \left(\frac{\sigma_p - \sigma_m}{\sigma_p + 2\sigma_m}\right) \tag{6}$$

At intermediate frequencies *CM* contains real and imaginary components, with only the real value (Re[*CM*]) employed in Equation (5).

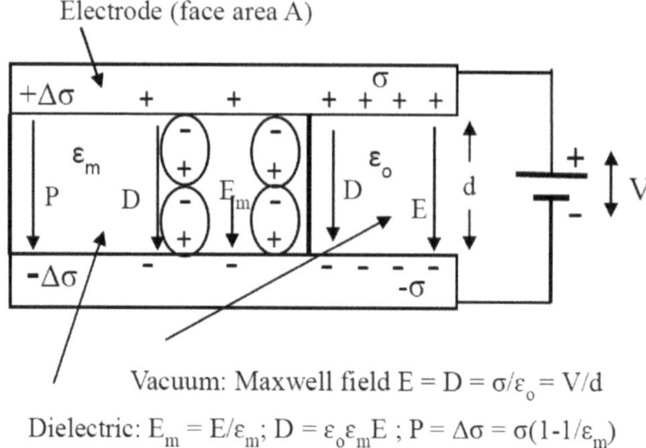

Vacuum: Maxwell field $E = D = \sigma/\varepsilon_o = V/d$

Dielectric: $E_m = E/\varepsilon_m$; $D = \varepsilon_o\varepsilon_m E$; $P = \Delta\sigma = \sigma(1-1/\varepsilon_m)$

Figure 1. A dielectric of relative permittivity ε_m is shown partly inserted between two electrified electrodes. 'Free' charge density σ on the electrodes creates the Maxwell field E and electric displacement D (both $= \sigma/\varepsilon_o$). 'Bound' charge density $\Delta\sigma$ created by polarization (charge displacement) of the dielectric generates the polarization vector P ($\Delta\sigma = P_s \cdot \hat{n} = P$), and equates to the number density of polarized molecules—i.e., the dielectric's dipole moment M per unit volume. These relationships give $D = E + P/\varepsilon_o$, and $P = \varepsilon_o(\varepsilon_m - 1)E_o = \chi_m\varepsilon_o E_o$.

2. The Basic Problem to Be Empirically Resolved

According to the standard induced dipole moment model of DEP, *CM* is limited to the range of values $1.0 > CM > -0.5$, and so the parameters that predominantly determine the magnitude of F_{DEP} are particle size and the magnitude of the field-parameter ∇E_m^2. In a first approximation a cubic root relation is expected regarding the physical dimensions of a globular protein and its molecular weight. However, an empirical relationship, provided by Malvern Panalytical® in their calculator software, gives a good estimate of a protein's hydrodynamic (Stokes) radius. This relationship is plotted in Figure 2, to show that those proteins reported to exhibit DEP responses have radii in the range 2–7 nm. Values of E_m and of ∇E_m^2 in the ranges 10^5–10^8 V/m and 10^{12}–10^{24} V²/m³, respectively, are reported for the DEP translocation and trapping of protein molecules. We can ask to what extent these fields and their gradients are consistent with the expectations of the current theoretical model of DEP when applied to a globular protein molecule. This question can be addressed by considering both the time-averaged free energy ($U_{DEP} = -(mE_m)/2$) of an electrically polarized particle and the work required to overcome the maximum dispersive (diffusional) force acting on it [4] (pp. 352–353). The first approach addresses the extent to which U_{DEP} represents a sufficiently deep 'trap' to compete against thermal energy $(3kT)/2$ associated with Brownian motion. Using the relationship given for the dipole moment m in Equation (3):

$$U_{DEP} = -\frac{1}{2}m \cdot E_m = -2\pi R^3 \varepsilon_o \varepsilon_m [CM]E_m^2 \tag{7}$$

The total free energy U_T of a polarized protein molecule is the sum of the Brownian thermal energy and U_{DEP}:

$$U_T = \frac{3}{2}kT + U_{DEP} = \frac{3}{2}kT - 2\pi R^3 \varepsilon_o \varepsilon_m [CM]E_m^2 \tag{8}$$

For the protein to be trapped by DEP, U_T must have a negative value—it should represent a sufficiently deep potential energy well for the molecule to be trapped for a time equivalent to the inverse of its probability to escape. For proteins such as bovine serum albumin (BSA) and avidin ($R \approx 3.5$ nm, $T = 300$ K, and assigning $CM = 0.5$) suspended in an aqueous medium (i.e., $\varepsilon_m \approx 78$) the required field is $E_m \geq 2.3 \times 10^7$ V/m. The maximum dispersive force per protein molecule is equal to $-kT/2R$. For $R \approx 3.5$ nm and $T = 300$ K, this force is 1.2×10^{-12} N. For the protein molecule to exhibit DEP it must oppose this dispersive force. With the expression for F_{DEP} from Equation (5) or from Equation (7), using the definition $F_{DEP} = -\nabla U_{DEP}$, this implies the following condition must hold:

$$2\pi R^3 \varepsilon_0 \varepsilon_m [CM] \nabla E_m^2 > 1.2 \times 10^{-12} N \tag{9}$$

For $R = 3.5$ nm and $CM = 0.5$ this requires $\nabla E_m^2 > 3.5 \times 10^{21}$ V^2/m^3. As discussed in Section 3.7, only two of the reported values of ∇E_m^2 have exceeded this minimum value. In one reported DEP manipulation of BSA a value of 10^{12} V^2/m^3 is cited! There are also the interesting cases where both positive and negative DEP of BSA have been reported at DC and 1 kHz, and where DEP of opposite polarities have also been reported for the same protein types at DC.

Considering the potential importance that protein DEP can offer, these experimental quirks should be addressed by a critical evaluation of both the validity of Equation (5) for protein DEP and the reported studies themselves. An effort is made here to examine, in broader detail than we consider has been attempted previously by others, the reported DEP literature on proteins and the relevant theory. An assessment is made of possible confounding influences, such as protein aggregation. All of the reported studies of protein DEP appear to be the results of careful work, and so even the more puzzling cases should assist in a better understanding of protein DEP and for the development of a more appropriate theoretical model to describe the DEP of proteins. We show that the protein DEP results reported to date are consistent with a model in which an evaluation of an induced dipole moment through Equations (1)–(3) should not be treated as the sole pertinent consideration. An important step forward regarding Equation (5) should be inclusion of the intrinsic (i.e., permanent) dipole moment possessed and well-studied for globular proteins [13]. Of particular importance is the orientation polarization of this dipole moment [8,10]—a feature overlooked in a previous discourse where the protein was considered to be a rigid dipole [9].

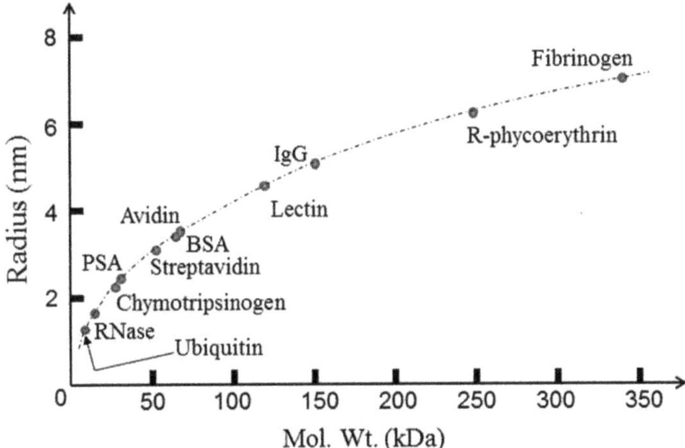

Figure 2. Globular proteins studied for their dielectrophoresis (DEP) response, with their hydrodynamic (Stokes) radii located on the empirical relationship between protein size and molecular weight (dotted curve) provided by Malvern Panalytical® (Zetasizer Nano ZS).

3. The Status of Protein Dielectrophoresis (DEP) Experimentation

3.1. Summary of Protein DEP

The most studied protein for its DEP characteristics is BSA. Figures 3 and 4 provide summaries of the observed DEP polarity, the frequency of the applied electric field and solution conductivity, for the two main situations where the field gradients are generated using either conductive electrode structures (eDEP) or posts/constrictions fabricated from insulator materials (iDEP) [14–42]. Most investigators observed positive DEP, but in two cases negative iDEP is reported [17,25]. Cao et al. [26] observed a transition from positive to negative DEP, with the cross-over frequency located between 1 and 10 MHz. The DEP results obtained [2,16,18,24–42] for proteins other than BSA are summarized in Figure 5. Most research groups report positive DEP for frequencies up to 6 MHz (including direct current), and of particular note is the observation that avidin and prostate specific antigen (PSA) exhibit a DEP cross-over frequency at ~10 MHz [27,37]. As summarized in Figure 6, various concentrations of BSA in aqueous solution have been employed. Included in Figure 6 are the concentrations used for streptavidin which, after BSA, is the most studied protein for its DEP characteristics. The mean separation distances between protein molecules for the various protein concentrations are also shown in Figure 6. This information is of relevance regarding any discussion of possible interaction between molecules. The molecular separation was estimated on the basis that a 1M solution contains Avogadro's number of molecules—i.e., 0.6 molecules/nm^3. The volume occupied per molecule is thus 1.66/C nm^3 for a C molar solution. The separation distances shown in Figure 6 were calculated by taking the cube root of the volume per molecule. A wide range of values for the field gradient factor ∇E^2 has been reported by the various investigators for a range of proteins, or has been estimated by Hayes [5] in his review. These values are shown in Figure 7 for both iDEP and eDEP studies, together with an indication of the minimum value of ∇E_m^2 (3.5×10^{21} V^2/m^3) calculated according to Equation (9), required to attain a DEP force that overcomes the dispersive forces of Brownian motion. The adjusted minimum value (~4×10^{18} V^2/m^3) based on the empirical relationship described in Section 3.4 is also shown in Figure 7.

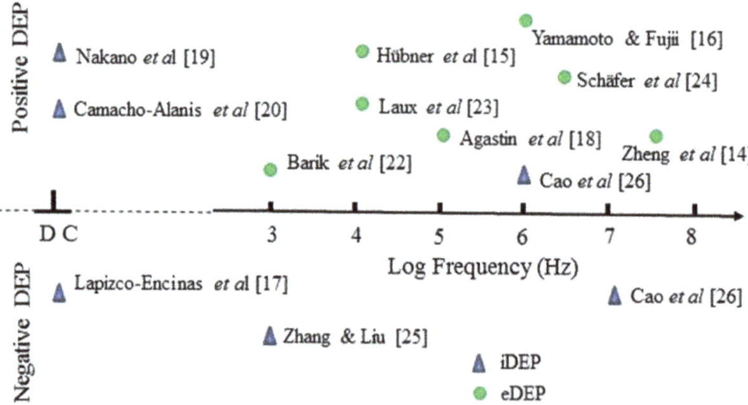

Figure 3. Insulator-based (iDEP) and electrode-based (eDEP) studies reported for bovine serum albumin (BSA). Most groups observed positive DEP, but two cases of negative iDEP have also been reported [17,25]. Cao et al. [26] report a DEP cross-over frequency between 1~10 MHz.

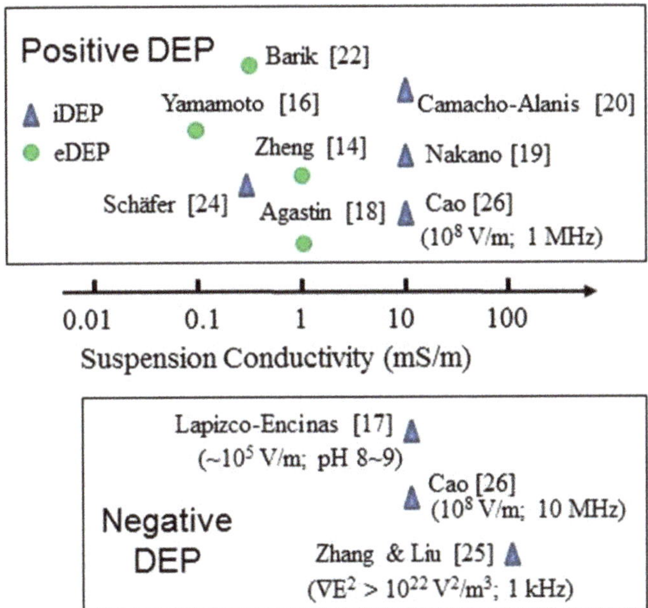

Figure 4. A range of aqueous solvent conductivity has been used in DEP studies of BSA. The experimental factors associated with the two cases [17,25] of negative iDEP and the cross-over of polarity between 1~10 MHz [26] are discussed in Section 3.2.

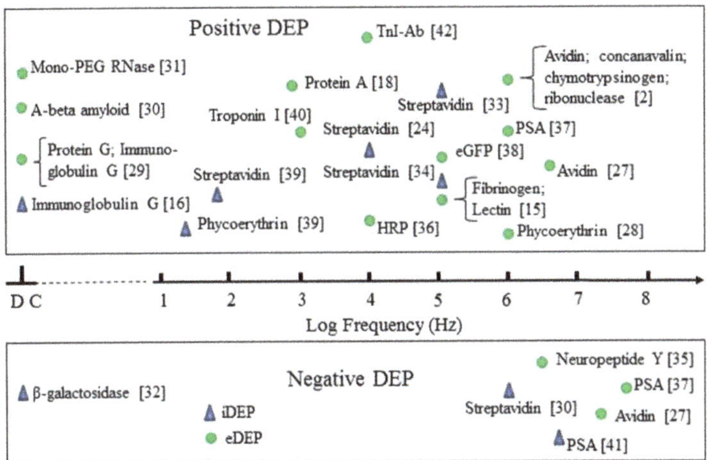

Figure 5. A summary of the DEP responses, at specific frequencies of the applied field, for proteins other than BSA. (HRP: horse radish peroxidase; PSA: prostate specific antigen; eGFP: enhanced green fluorescent protein; TnI-Ab: troponin I antibody).

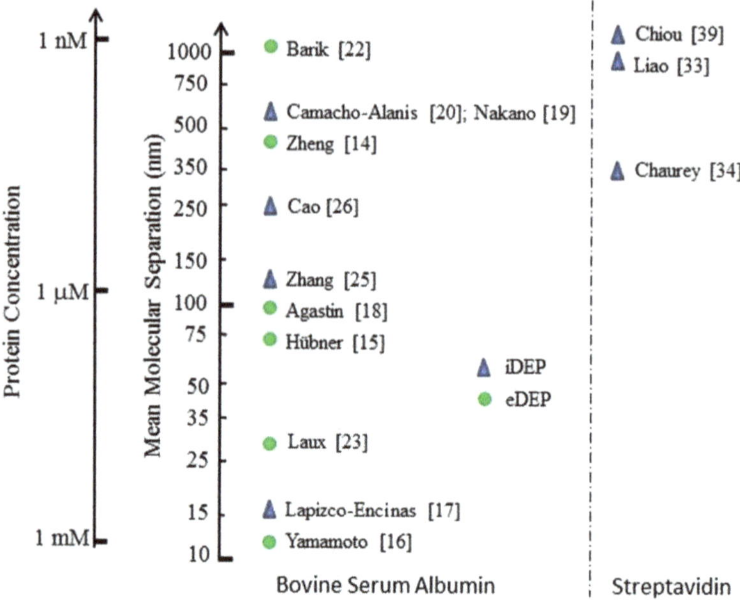

Figure 6. Mean distance between BSA and streptavidin molecules for the reported sample concentrations, estimated as the cube root of the volume occupied per protein molecule.

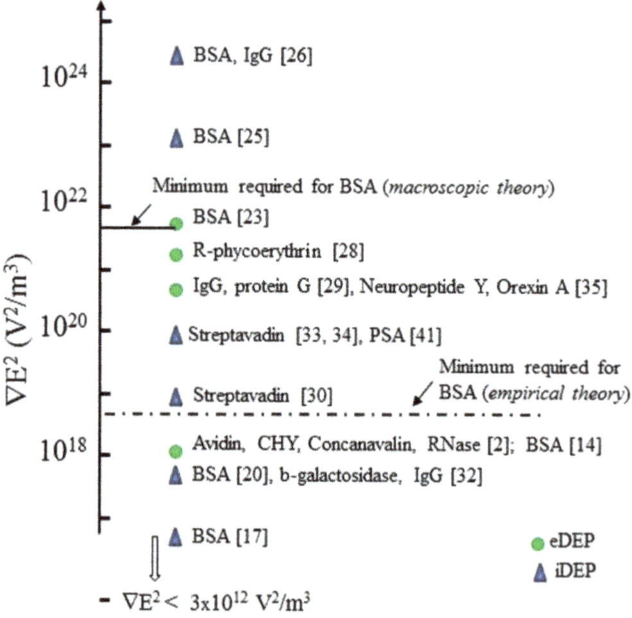

Figure 7. Values of the field gradient factor ∇E_m^2 as reported by the investigators or estimated by Hayes [5]. The minimum value of ∇E_m^2 required to compete against Brownian diffusive effects, calculated according to Equation (9), is shown for the case of BSA. The adjusted value shown for this is based on the empirical relationship described in Section 3.4.

3.2. Bovine Serum Albumin (BSA)

BSA is a well-studied, water soluble, protein. It has a molecular weight of 66.5 kDa; is composed of 583 amino acid residues (54% of which form six α-helices); takes the form of a prolate ellipsoid of dimensions 14 nm $\times$ 4 nm $\times$ 4 nm; has an isoelectric point in water of 4.7 at 25 °C [43,44]. Most DEP experiments on BSA have employed pH buffers to maintain a pH of 7.4 or higher. For such studies the protein molecules were thus negatively charged (the number of ionized acidic side-groups exceeded that of basic ones). Unbuffered solutions of 0.1 mM concentration and lower typically have a pH of 5.0–6.0 and are less negatively charged, close to having an equal number of, but not uniformly distributed, ionized acidic and basic groups. The BSA monomer contains 17 disulphide bridges between adjacent cysteine groups of its polypeptide chain, whilst bonding of the one free cysteine (Cys34) between interacting monomers leads to the formation of a dimer. BSA adopts its normal globular form between pH 4.5 and 7.0, but partially unfolds as the pH approaches the range 8.0–9.0 [45,46]. This unfolding involves the breaking and rearrangement of disulphide bonds, which is temperature sensitive and can lead to a loss of α-helix content [47,48] and irreversible self-aggregation [49]. It should also be noted that monomer, dimer and other aggregates typically exist in commercial samples of BSA [50]. Proteins in general follow first and second order aggregation kinetics [51]. Conformational change is the rate limiting step in the first order kinetics, making the rate of reaction independent of initial protein concentration. The second order reaction rate does depend on concentration, because molecular collision frequency limits formation of dimers, trimers, etc., and heat-induced aggregation. The suggestion by Nakano et al. [19] that 'most iDEP manipulations of proteins may require the control of protein aggregation' is well-founded, and as discussed in Section 3.4 may be of relevance to understanding the two conspicuous cases [17,26] of negative iDEP observed for BSA (Figure 3).

3.3. The Dielectric β-Dispersion

Of particular relevance to protein DEP is the fact that globular proteins possess an intrinsic dipole moment. The magnitude of this moment is given by the resultant of the moments of the amino acids in the polypeptide chain (especially the additive effect of those forming α-helices), the moments of the charged acidic and basic groups about the molecule's hydrodynamic center, and polarizations of the surrounding water molecules [52]. If the protein molecule is free to rotate about its prolate major and minor axes, this dipole moment manifests itself as a large dielectric dispersion (known as the β-dispersion)—the form of which for BSA is shown in Figure 8. By analyzing this dispersion, Moser et al. [53] computed dipole moment values for the BSA monomer and dimer as 384 D (1.28×10^{-27} Cm) and 636 D (2.12×10^{-27} Cm), respectively. The angle between the dipole moment and the long axis of the monomer was determined to be 50°. Moser et al. [53] performed dielectric and transient birefringence measurements on BSA solutions of concentrations in the range 0.2–1.4 mM and observed the effect of strong intermolecular interactions. In their measurement of the β-dispersion, Grant et al. [54] considered that the BSA concentrations (0.6–5.5 mM) were "high enough to permit molecular interaction".

3.4. Empirical Relationship Connecting Clausius–Mossotti (CM) and the β-Dispersion

For dielectric and impedance spectroscopy measurements on cell suspensions of sufficiently low volume concentrations c_v, the dielectric increment $\Delta\varepsilon$ depicted in Figure 8, as well as the conductivity increment $\Delta\sigma$ characterizing this dispersion in terms of the increase of conductivity of the suspension with increasing frequency, are given by:

$$\Delta\varepsilon = 3c_v\varepsilon_m CM; \quad \Delta\sigma = \frac{1}{\tau}\Delta\varepsilon \tag{10}$$

The relationship between $\Delta\varepsilon$ and $\Delta\sigma$ results from application of the Kramers–Kronig transforms, where τ is the characteristic relaxation time of the Maxwell-Wagner interfacial polarization giving rise to the β-dispersion [4] (Chapter 9). Equation (10) can be extended to accommodate larger values of c_v and to derive multi-shell models for analyzing impedance and DEP measurements on cell suspensions [4]. However, as stated elsewhere [9] (without the following explanation), Equation (10) is not applicable to protein suspensions. According to the Maxwell–Wagner mixture theory for particle suspensions, the measured effective permittivity ε_{eff} of a dilute particle suspension is given by [4] (pp. 222–223):

$$\frac{\varepsilon_{eff} - \varepsilon_m}{\varepsilon_{eff} + 2\varepsilon_m} = c_v \frac{\varepsilon_p - \varepsilon_m}{\varepsilon_p + 2\varepsilon_m} \tag{11}$$

with ε_p and ε_m the particle and medium relative permittivity, respectively. The term effective permittivity is used to signify that a defined volume of a particle suspension may be replaced conceptually with an equal volume of a homogeneous medium of smeared-out bulk properties.

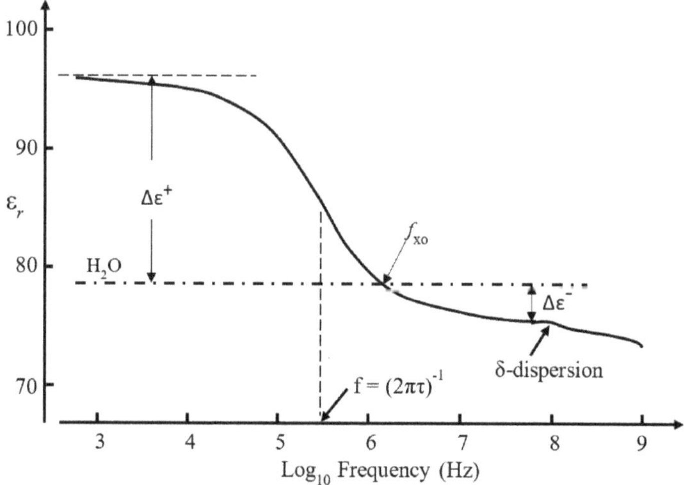

Figure 8. The β-dispersion and δ-dispersion, arising from orientation polarization of the protein and protein-bound water, respectively, exhibited by 0.18 mM BSA (based on Moser et al. [50] and Grant et al. [51]). The radian frequency of orientation relaxation for BSA is given by the reciprocal of its relaxation time τ. For frequencies below $f$xo (~1 MHz) the relative permittivity ε_r of the BSA solution exceeds that of pure water, and is less than this above $f$xo. According to Equation (11) the dielectric increment $\Delta\varepsilon^+$ and decrement $\Delta\varepsilon^-$, respectively, specify the frequency ranges where positive and negative DEP, respectively, should be observed for monomer BSA in aqueous solution.

Substitution of one volume with the other is assumed to not alter the electric field in the surrounding medium. The assumption is thus made that $\varepsilon_{eff} \approx \varepsilon_m$, implying that for a sufficiently large observation scale a heterogeneous compound material can be considered as a homogeneous one. Inserting this approximation into the denominator of the left-hand side of Equation (11) leads to the expression for $\Delta\varepsilon$ in Equations (10). However, an instructive result is obtained if this is applied to form a relationship between the Clausius–Mossotti factor *CM* and the dielectric increments depicted in Figure 8. For a dilute protein suspension, this relationship should thus be of the form:

$$CM = \frac{\Delta\varepsilon}{3c_v\varepsilon_m} = \frac{\Delta\varepsilon}{3\varepsilon_m}\left(\frac{C_w\rho_p}{C_p\rho_w}\right) \tag{12}$$

where C_w and C_p, ρ_w and ρ_p, respectively, are the molar concentration and mass density of pure water and the protein, respectively. The concentration C_w of pure water is taken as 55.5 M (1000 g/L divided by its molecular weight of 18 g/mol), and protein density values can be derived using the molecular-weight-depending function derived by Fischer et al. [55]. Equation (12) qualitatively predicts that in a frequency range where there is a dielectric increment $\Delta\varepsilon^+$ a positive value for CM and a positive DEP response will result. As indicated in Figure 8 the opposite case should also hold for the high-frequency range where a dielectric decrement is exhibited. But can the CM-factor of Equation (12) simply be inserted into Equation (5) to describe protein DEP? Moser et al. [53] obtained $\Delta\varepsilon/C_p = 1.11$ per mM for a BSA monomer concentration, so that with $\varepsilon_m = 78.4$ and $\rho_p = 1.41$ gm/cm^3, Equation (12) yields the result $CM = 369$. This is not possible according to the definition and limited range of values $(1.0 > CM > -0.5)$ of the macroscopic CM factor derived from Equation (2) for the induced polarization P_p per unit volume of a particle. Furthermore, based on the work of Takashima and Asami [56] who obtained values for $\Delta\varepsilon$ (per mM protein concentration) of 5.06 and 37.24 for cytochrome-c and carboxypeptidase, respectively, the corresponding values obtained for CM are 1745 and 12,480, respectively! This is the basis for stating [9] that the macroscopic theory leading to Equation (10) cannot be employed at the molecular level.

If, instead of the assumption $\varepsilon_{eff} \approx \varepsilon_m$, the identity $\varepsilon_{eff} = \kappa\varepsilon_m$ is inserted into the denominator of the left-hand side of Equation (11) we obtain the relationship:

$$(\kappa + 2)CM = \frac{\Delta\varepsilon}{\varepsilon_m}\left(\frac{C_w\rho_p}{C_p\rho_w}\right) \tag{13}$$

Values for the parameter $(\kappa + 2)CM$ are given in Table 1, based on values of $\Delta\varepsilon/C_p$ obtained experimentally [56–62] for a range of globular proteins. No obvious relationship can be seen to link the value of a protein's effective polarization factor $(\kappa + 2)CM$ (per unit volume) with its molecular weight. Based on Equations (10) and (13) and the data given in Table 1, the following empirical relationship is proposed that links the molecular- (micro-) and macro-scales:

$$CM_{\text{micro}} = (\kappa + 2)CM_{\text{macro}} \tag{14}$$

For the DEP of macro-particles, such as mammalian cells and bacteria, the plane of hydrodynamic shear of the particle, as it undergoes DEP through its suspending medium, can be considered to coincide with its 'mathematical' boundary at the particle–medium interface. At the molecular scale applicable to protein DEP, however, the situation is far more complicated. The plane of shear is most likely to lie within the outer boundary of the protein's hydration shell, whose total extent is defined when the protein is stationary. We have, as shown schematically in Figure 9, the equivalent of a molecular 'Russian doll'. The protein with its permanent dipole moment and most strongly 'attached' water that can rotate with it, occupies the inner cavity. The protein's dipole field extends beyond an outer 'macroscopic' boundary at which the macroscopic boundary conditions of classical electrostatics can be applied. The medium polarization P_m must be uniform right up to this boundary. Located within this mathematical boundary is the hydrodynamic plane of shear (defining the zeta-potential determined by electrophoresis) and the protein's outer hydration sheath. It is tempting to propose a conceptual equivalence of Equation (14) in terms of the ratio of two polarizations and interfacial dipole moment free energies:

$$\frac{P_i}{P_m} \equiv \frac{\chi_i E_i}{\chi_m E_m} \equiv \frac{\langle M_i \rangle \cdot E_i}{\langle M_m \rangle \cdot E_m} \propto (\kappa + 2)CM_{\text{macro}} \tag{15}$$

where suffices i, m identify the polarization, susceptibility, induced moment and local field in the protein cavity and bulk medium, respectively. These ratios will be sensitive to the physico-chemical attributes of a particular protein (e.g., peptide chain folding, net charge and the distribution of polar and hydrophobic groups on the protein surface) and could explain the very wide range of values of the parameter $(\kappa + 2)CM$ given in Table 1. At this stage it is of interest to note that the large values given for ribonuclease (7000–11,000) and concanavalin (~15,000) would place these proteins above the minimum required level indicated in Figure 7 for BSA. DEP measurements for the other proteins cited in Table 1 would be of considerable value in this speculative argument.

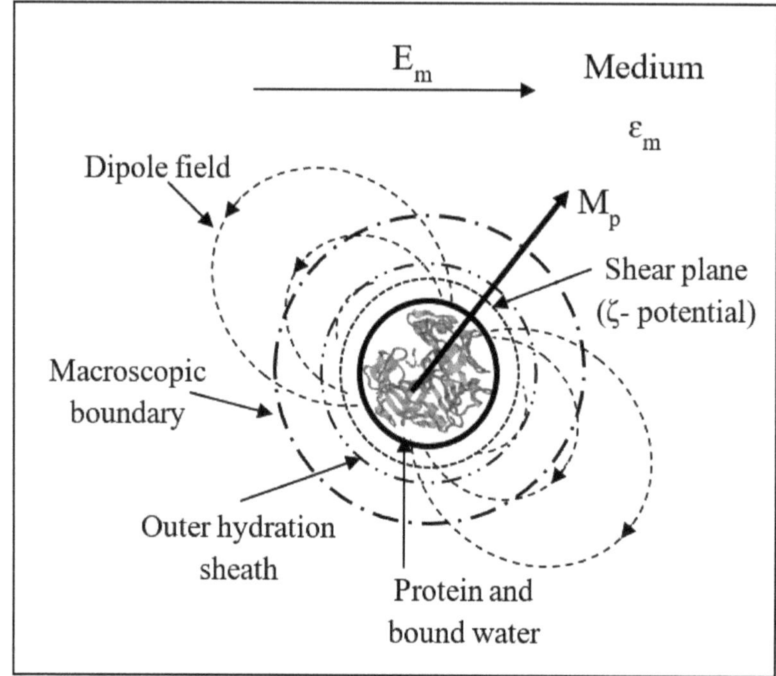

Figure 9. Schematic of a 'Russian doll' model for a protein with a permanent dipole moment M_p, which occupies the innermost cavity together with its strongly bound water molecules. The protein's dipole field extends beyond an outer macroscopic, 'mathematical', surface where the classical boundary conditions of electrostatics can be applied. Within this mathematical surface is a boundary that contains the protein's outer hydration sheath, and the hydrodynamic plane of shear that defines the zeta-potential within the protein's diffuse electrical double-layer.

Table 1. Values of the factor $(\kappa + 2)\,CM$ given by Equation (13) for various globular proteins, derived from reported $\Delta\varepsilon$ and corresponding protein concentration values. The protein density values were derived from the weight-depending function given by Fischer et al. [55].

Protein	Mol. Wt.	Density (g/cm³)	$\Delta\varepsilon/c_p$ (c_p: mM)	$(\kappa + 2)CM$ Equation (13)	Reference
Ubiquitin	8600	1.49	3.82	4020	[58]
RNAse SA	10,500	1.48	15.00	15,720	[57]
Phospholipase	13,000	1.46	1.82	189	[56]
Cytochrome-c	13,000	1.46	5.06	5240	[56]
Ribonuclease	13,700	1.46	11.0	11,400	[59]
			7.12	7350	[56]
Lysozyme	14,300	1.46	1.34	1390	[56]
Myoglobin	17,000	1.45	0.07	2090	[60]
			1.79	1440	[61]
Trypsin	23,000	1.43	6.74	6810	[56]
Carboxypeptidase	34,000	1.42	37.24	37,440	[56]
Hemoglobin	64,000	1.41	1.29	1290	[62]
BSA	66,000	1.41	1.11	1110	[53]
Concanavaline	102,000	1.41	15.31	15,270	[56]

3.5. The β-Dispersion and Dipole Moment Density

The β-dispersion can also conceptually be linked to the DEP frequency response of the BSA monomer in terms of the polarization (dipole moment density) of the medium and protein molecule. Two approaches can be adopted. The first involves the 'book-keeping' exercise of calculating the change ΔU of free energy stored in the field as a result of the following three actions: (i) Increase the field E_m from zero (where $D_m = 0$) to its final value ($D_m = \varepsilon_0\varepsilon_rE_m$) in the medium, that has a total volume V_m; (ii) reduce D_m by removing from the medium a cavity of volume v_p large enough to contain the hydrated protein molecule; (iii) account for the incremental change (either positive or negative) of the medium polarization resulting from its interaction with the field of the protein's induced and permanent dipole moment. These three actions can be expressed in the form [4] (pp. 87–89):

$$\Delta U = \frac{1}{2}\int_{V_m}\int_0^D E_m \cdot \delta D dv - \frac{1}{2}\int_{v_p} E_m \cdot D_m dv - \delta U \tag{16}$$

Volume V_m is very much larger than v_p and so the first integral in Equation (16) represents a significantly larger contribution to ΔU than the second integral. The δU term thus plays a significant role. In the macroscopic derivation of the Maxwell–Wagner mixture theory that leads to Equation (10) the assumption is made that $\varepsilon_{eff} \approx \varepsilon_m$. This effectively removes the requirement for calculating the δU term in Equation (16), which at a molecular scale is a significant weakness. Evaluation of δU can conceptually, for our present purpose, be accomplished by assuming the applicability of the boundary condition regarding continuity of the normal component of displacement flux ($D = \varepsilon_r\varepsilon_0E_m$) across the interface between the solvated protein and the bulk medium. The free energy change δU is then given by an integral of the following form [4] (p. 89), taken over the protein's effective cavity volume v_p:

$$\delta U = \frac{1}{2}\int_{v_p}\left(\varepsilon_m - \varepsilon_p\right)E_i \cdot E_m dv \tag{17}$$

For the frequency range where the dielectric increment $\Delta\varepsilon^+$ has a finite value in Figure 8, the protein's effective permittivity ε_p can be regarded as being greater than ε_m. The integral in Equation (14) thus yields a negative value for δU. According to the work-energy theorem, for frequencies lower than f_{xo}, work will be required on the particle by the field to withdraw it from the medium. Furthermore, this free energy is further reduced if the field E_m increases. The protein monomer or dimer will attempt to minimize its electrostatic free energy by moving up a field gradient to a maximum value of this gradient. This describes the action of positive DEP. For frequencies lower than f_{xo} (i.e., where the dielectric decrement $\Delta\varepsilon^-$ has a finite value), the protein's effective permittivity is less than that of the medium. The protein will move down a field gradient to search for a field minimum. Work is required by the field to insert the protein into the medium. This describes negative DEP. It is tempting to consider the cross-over of DEP polarity at 1–10 MHz for BSA, observed by Cao et al. [26], as experimental evidence for this scenario, because such cross-over is expected from inspection of the β-dispersion shown in Figure 8.

A second approach to linking the β-dispersion to protein DEP is to consider the time-averaged potential energy of the polarized protein particle in terms of its polarizability α per unit volume in unit field [4] (p. 89): $\langle U \rangle = -\frac{1}{2}\alpha E_m^2$ (per unit volume). From the fundamental relationships between the fields E, D, P and the dipole moment M per unit volume (see Figure 1) we have the following expression for δU in terms of the surface polarization P and induced dipole moment M_p of the solvated protein:

$$\delta U = \frac{1}{2}\langle M_p \rangle \cdot E_m, \text{ where } M_p = \int_{v_p} P_s \cdot \hat{n} dv \tag{18}$$

The magnitude of M_p will give the strength of the DEP force, whilst its polarity will also define the F_{DEP} polarity. A negative value for M_p will indicate it is directed against the direction of E_m. Work will be required to insert the polarized particle into the field E_m within the aqueous medium. This describes negative DEP. A polarized protein possessing a positive value for M_p will be aligned with E_m and exhibit positive DEP.

Defining the protein's effective cavity volume v_p to be used in the integrals of Equations (17) and (15) is not straightforward. Different protein molecules have from 0.20 to 0.70 g strongly associated (bound) water per g protein, contributing to its effective radius of rotation by up to one to two water molecule diameters [63]. From their studies, Moser et al. [53] determined a hydration of 0.64 g of H_2O per g of BSA. Grant et al. [54] confirmed the existence of a subsidiary dispersion (δ-dispersion) in the frequency range 200–2000 MHz, and concluded that this dispersion is probably due to the rotational relaxation of water 'bound' to the protein. The term 'bound water' is taken to mean water bound to the protein by bonding of greater strength than the water–water bonding that exists in pure bulk water. This characteristic water structure that is formed near the surfaces of solvated proteins arises not only through hydrogen bonding of the water molecules to available proton donor and proton acceptor sites on the protein surface, but also through electrostatic forces associated with the water molecule's electric dipole moment. The protein molecule and the water around it thus form a strongly coupled system, involving mechanical damping of the protein motion by adsorbed water, together with a dynamic electrical coupling between the tumbling electric dipole of the protein and the fluctuating dipoles of the adsorbed and bulk water. With such heterogeneity of the dielectric medium and also possibly of E_i within the effective volume v_p, computation of the integrals in Equations (17) and (18) thus involves some 'interesting' challenges. Not least of which is defining the effective volume v_p of the protein, and how the normal components of displacement flux D and polarization P vary within the heterogeneous boundary between the protein's surface and the bulk aqueous medium.

3.6. Interfacial Polarizations

The formation of defect dipoles in both amorphous and crystalline polymers is known to influence their dielectric properties [64]. Examples of possible relevance to protein DEP are depicted in Figure 10. These are suggested examples where the standard boundary conditions of Maxwell-based electrostatics may not apply—the implications of which have been described by Martin et al. for the specific case of a 'Rossky cavity' [65]. The example shown in Figure 10a could, for example, depict the disruption of the network of hydrogen bonds at a protein–water interface—possibly resulting in the creation of nanodomains that have the capability of dynamically freezing into a ferroelectric glass [66]. Ferroelectric materials are known to develop structures with curls on their faces where the field is no longer conservative [67]. This of relevance to Equation (4) in which E_m is assumed to be irrotational. Boundaries of the form depicted in Figure 10b between dielectrics of different permittivity have been shown, through theory and classical molecular dynamics simulations of hydrated cytochrome c, to exist in the hydration shells of proteins [68]. The large dispersion strength ($\Delta\varepsilon \sim 2400$) shown in Figure 10c for a suspension of polystyrene microspheres was analyzed and determined not to arise from classical Maxwell–Wagner interfacial polarization, electrophoretic particle acceleration, or the presence of a frequency-independent surface conductance [69]. The most likely origin was considered to be a frequency-dependent surface conductance that varies with the ionic strength of the suspending aqueous electrolyte. Interfacial polarizations of these types should be included in the exercise to find a molecular-based DEP theory. It is also pertinent to mention that excised samples of biological tissue can exhibit large $\Delta\varepsilon$ values [52], a good example being skeletal muscle with measured relative permittivity $\varepsilon_r \approx 10^7$ at 10 Hz [70]. This is known as the α-dispersion and, according to the convention used in assigning Greek letters to dielectric dispersions, occurs in a frequency range below that of the β-dispersion.

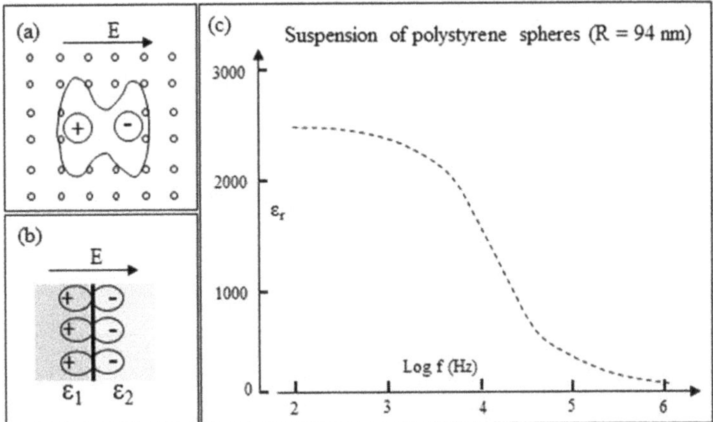

Figure 10. (**a**) Schematic of a dipole formed at the site of a structural defect in a molecular lattice. An example could be the disruption of the hydrogen bond network in bulk water at a protein–water interface—with the possible creation of ferroelectric nanodomains [66]. (**b**) Dipole polarization at a boundary of dielectric inhomogeneity. A solvated protein, with its bound water and surrounding bulk water, represents an inhomogeneous dielectric [68]. (**c**) Dielectric dispersion exhibited by an aqueous suspension of polystyrene nanospheres (R = 94 nm) (based on Schwan et al. [69]).

3.7. Protein Dipole Polarization

Other paths to formulation of the DEP force acting on a protein permanent dipole are through either Equation (4) or, as follows, the relationship between U_{DEP} and F_{DEP} given by Equation (7). In the absence of an electric field, the orientations of the dipole moments of proteins in solution will on average be distributed with the same probability over all directions. On average their net dipole moment in any specific direction is zero. On application of a field each dipole will experience a field alignment torque $m \times E_i$, so that net polarization results. The electrical potential energy U of each dipole is given by $U = -(mE_i\cos\theta)$, where θ is the angle between the dipole moment and the local field vector E_i. From Boltzmann–Maxwell statistics the probability of finding a dipole oriented in an element of solid angle $d\Omega$ is proportional to $\exp(-U/kT)$, with k the Boltzmann constant and T in kelvin. A moment pointing in the same direction as $d\Omega$ has a component $(m\cos\theta)$ in the direction of E_i. As detailed elsewhere [4,9] the thermal average of $\cos\theta$ is given by the derivation of the so-called Langevin function:

$$\langle\cos\theta\rangle = \frac{\int \exp(-U/kT)\cos\theta d\Omega}{\int \exp(-U/kT)d\Omega} = \frac{m}{3kT}\left(1 - \frac{1}{15}\left(\frac{mE_i}{kT}\right)^2\right) \tag{19}$$

For a monomer BSA dipole moment of m = 384 D and $E_i \approx 3 \times 10^5$ V/m (e.g., Lapizco-Encinas et al. [17], assuming $E_i \approx E_o$) the factor $(mE_i/kT) \approx 0.01$. So, to a good approximation $m\langle\cos\theta\rangle = m^2E_i/3kT$. For $E_i > 3 \times 10^7$ V/m (e.g., Cao et al. [26]) the full expression for the thermal average of $\cos\theta$ should be used. With an applied field less than 10^6 V/m, then through Equation (7) the average orientational DEP force (F_{oDEP}) acting on a protein's dipole is given by:

$$F_{oDEP} = -\nabla U_{DEP} = (m\langle\cos\theta\rangle \cdot \nabla)E_i = \frac{m^2}{3kT}(E_i \cdot \nabla)E_i = \frac{m^2}{6kT}\nabla E_i^2 \tag{20}$$

This expression for F_{oDEP}, which also follows from Equation (4), has two important features. The first is that the DEP force exerted on a polarized protein molecule possessing a permanent dipole moment is directly proportional to ∇E^2. Previously, one of the authors [9] has concluded that for frequencies below $f$xo (see Figure 8) proteins with a permanent dipole should exhibit positive DEP directly proportional to ∇E, and not ∇E^2, whereas negative DEP should be expected solely above $f$xo and have a ∇E^2 dependence. This conclusion is only valid for a 'rigid' protein molecule whose dipole is constrained from responding to the alignment torque $m \times E_i$, or where the relaxation time of the protein's permanent dipole is too slow to respond to a high-frequency oscillating field. Based on Equation (20) the ratio of the DEP force exerted on an orientationally polarized dipole moment to that on an induced dipole moment (Equation (5)) is:

$$\frac{F_{oDEP}(orientation)}{F_{DEP}(induced)} = \frac{m^2}{12\pi R^3 kT\varepsilon_0\varepsilon_m[CM]} = 1.85 \times 10^{28} \frac{m^2}{R^3}\frac{E_i}{E_m} \qquad (21)$$
$$(k = 1.38 \times 10^{-23}\ \text{J·K}^{-1};\ T = 300\ \text{K};\ \varepsilon_m = 80;\ CM = 0.5)$$

For monomer BSA ($m = 1.28 \times 10^{-27}$ Cm; $R = 3.5$ nm), and assuming $E_i = E_m$, this gives near equality of F_{oDEP} and F_{DEP} ($F_{oDEP} = 0.71\ F_{DEP}$). This result indicates that unless $E_i >> E_m$, Equation (20) does not offer a theoretical basis to explain why the majority of experimental ∇E^2 values shown in Figure 7 fall well below the minimum requirement of $\nabla E^2 > 3.5 \times 10^{21}$ V^2/m^3. It also implies that we require a better understanding of the relationship between the DEP force and the β-dispersion shown in Figure 8. Qualitatively, a protein molecule will exhibit positive DEP if the polarization per unit volume (i.e., total dipole moment) of the bulk water it displaces is less than that of the protein and its associated water molecules of solvation. A quantitative understanding should include a molecular-level description of short- and long-range interactions of the dipoles (protein–water and water–water) and the nature of the interfacial and/or dipole charges that can create the situation $E_i >> E_m$. A route to this might be offered through the suggested empirical relationship given in Equation (15), that relates the protein's local cavity field and its polarization to the large values of the effective polarization factor ($\kappa + 2$)CM given in Table 1. Of the proteins listed in Table 1, only three (BSA, concanavalin, ribonuclease) appear to have been investigated for their DEP characteristics. It is of interest to compare the locations of these proteins in the ∇E^2 'ranking' of Figure 7, with their relative values of ($\kappa + 2$) CM given in Table 1 (~1000: BSA; ~11,000: ribonuclease; ~ 15,000: concanavalin). If the macroscopic CM factor is replaced by the proposed microscopic version ($\kappa + 1$)CM in Equation (9), then the DEP results cited for ribonuclease and concanavalin lie well above the 'minimum required' level in Figure 7.

3.8. Protein Stability

Concerning the interesting cases [17,25] of negative iDEP indicated for BSA in Figure 3, both studies were carefully performed and analyzed, so there is no intent here to label their experiments as 'wrong'. It is often the case in biological work that the 'odd' finding is the very one to pursue further. Lapizco-Encinas et al. [17]—the first to report protein iDEP—employed a BSA concentration of 0.46 mM, buffered at high pH (8 and 9) and ionic conductivities (10 mS/m). This brings their situation to within the bounds of protein conformational change and unfolding, as well as loss of α-helix content and self-aggregation [45–49]. A concentration of 0.46 mM is also within the range (0.2–0.6 mM) where dielectric studies [53,54] provided evidence of strong intermolecular interactions (see also Figure 6). As a general rule, in an aqueous environment with a high ionic strength (i.e., high conductivity) the solvated ions compete with the protein molecules in binding with water, to such an extent that the protein molecules tend to associate with each other. This is because protein–protein interactions become energetically more favorable than protein–solvent interaction [71]. The result is the precipitation of the least soluble solute—namely the protein. This could easily have been interpreted by Lapizco-Encinas et al. as collection of the protein by negative DEP. There is also the possibility that true iDEP of aggregates, rather than precipitation, was observed. This would explain why a very small field ($\sim 10^5$ V/m) could be employed, and might also provide insights into the DEP behavior of a test sample as it makes the transition from the molecular- to the macro-scale. In their studies,

Zhang et al. [25] employed low sample concentrations (0.78 μM) but high conductivities (0.1 S/m). The likelihood of molecular interactions and self-aggregation was thus low (Figure 6) but with such a high ionic strength the precipitation of the BSA was likely.

3.9. Other Experimental Details

Comprehensive details of electrode and chamber designs for both iDEP and eDEP devices have been reviewed elsewhere [5,7,72–74] and are not considered here. Also, for some studies thorough consideration may not have been given to the possible confounding influence of electrothermal effects. We consider these to be relatively minor considerations for the bigger picture. The following experimental aspects are, however, suggested for further consideration.

For a quantitative interpretation of the published results one has to be aware that the reported experimental parameters are often not given or might be somewhat uncertain. One reason for this is the high surface-to-volume ratio of the microfluidic system. This is required because microscope-aided observation of protein DEP calls for flat observation chambers with typical heights between 20 and 200 μm, ranging down to 2 μm [24,39] and even 200 nm [33,34]. This relatively large surface area can result in uncertainties concerning conductivity, pH value and solute concentration. At initially low ionic strength tiny amounts of contamination can lead to a substantial increase in conductivity. This holds, to a lesser extent, also for the pH value. Depending on the experimental arrangement, diffusion of CO_2 from the environment can lead to an increased conductivity and lowered pH value. In DC-DEP, artificial pH gradients might also be generated in a way similar to the preparation of pH gradients for isoelectric focusing. Due to adsorption at the surface of the measuring chamber, as well as within fluidic tubing, solute concentrations can decrease even in the course of the actual experiment. Often, counter-measures are taken using buffers or surface modifications [19,32]. Published results should thus be compared and interpreted carefully.

Another cause of uncertainty is the determination of electrical parameters. Sometimes it is not clear whether voltages are given as peak-to-peak or as root-mean-square (rms) values. In about half the work on protein DEP, values of either $|E|$ or $\nabla |E|^2$ are calculated. Both values are given for only a few of the studies cited here [16,20,38]. The spatial distribution of just $|E|$ is given by Agastin et al. [18], that of $\nabla |E|^2$ in rather more cases [20,22,26,28,32,37] and sometimes the distribution of both values is given [22,26,38]. Owing to experimental limitations actual measurements of $|E|$ or $\nabla |E|^2$ have not been carried out in any of these works. All calculations have been performed numerically by commercial software based on finite-element-methods (FEM). Although this is not specified by any of the authors, it is very probable that the spatial models of these simulations were based on simple geometrical bodies like cuboids and cylinders. This means that in essence the edges are modelled with infinitesimal radius of curvature. This should lead to infinite values of both $|E|$ and $\nabla |E|^2$ since both are calculated as spatial derivatives of the potential distribution. In practice, this is not the case because the calculations are performed on a mesh or grid with finite resolution. This means that the field distributions are qualitatively correct. However, the maximal values of $|E|$ and $\nabla |E|^2$ are now dependent on the spatial resolution of the mesh. It might well be that in several cases the resolution is not known because the software automatically adapts the mesh locally. In only two reports have the resolutions been given—namely, values of 50 nm^3 [38] and 100 nm^3 [23]. In order to determine the impact of the chosen resolution we have calculated the field distribution for two basic electrode arrangements, i.e., for co-planar interdigitated electrodes and for arrays of cylindrical pins. Using the FEM software Maze (Field Precision, Albuquerque, USA) the resolution of the Cartesian grid was varied from 120 nm down to 12 nm. This produced a roughly linear increase of both $|E|$ and $\nabla |E|^2$ (data not shown) with resolution (i.e., with the inverse of the linear voxel dimensions). For interdigitated electrodes $|E|$ and $\nabla |E|^2$ increase by a factor of 4 and 10, respectively, whilst for cylindrical arrays these factors amount to 2 and 60, respectively. As a consequence, the currently available data on $|E|$ and $\nabla |E|^2$ should only serve as a more or less rough estimate when comparing them with physical theory.

4. Concluding Comments

Commencing with the first reported studies in 1994 of the DEP responses of avidin, chymotripsinogen, concanavalin and ribonuclease [2] at least 22 different globular proteins have now been investigated for their DEP responses [2,14–42]. Aspects of this work are examined here, covering details not encompassed in previous reviews [1,3–7] of protein DEP. Apart from a few cases, whether through insulator-based (iDEP), electrode-based (eDEP) investigations, at DC or with applied field frequencies ranging from 20 Hz to 30 MHz, the reported results are largely consistent. In their DEP analyses all the authors employ the standard induced-dipole moment theory that employs the Clausius-Mossotti (*CM*) factor derived from macroscopic electrostatics. However, apart from the three studies of Laux et al. [23], Zhang et al. [25] and Cao et al. [26], none of the reported DEP responses can be explained in terms of the limitations set by this classical theory. As shown in Figure 7, only these three studies employed a gradient field factor $\nabla E_m^2 > 3.5 \times 10^{21}$ V^2/m^3 required, according to Equation (9), to overcome the dispersive forces associated with the Brownian motion of the protein molecules. All of the other studies fell far short of this requirement. In one reported DEP manipulation of BSA, a value of 10^{12} V^2/m^3 is cited [17].

Of particular relevance to protein DEP is the fact that globular proteins possess an intrinsic dipole moment. If the protein molecule is not rigid, but free to rotate about a major or minor axis when subjected to an applied AC field, this dipole moment manifests itself as a large dielectric dispersion known as the β-dispersion. The form of this dispersion for BSA is shown in Figure 8. For the frequency range where the β-dispersion exhibits a dielectric increment $\Delta\varepsilon^+$, the protein's effective permittivity ε_p can be regarded as being greater than the value ε_m for the surrounding medium. This should result in a positive DEP response. A negative DEP response should then be exhibited on increasing the field frequency to the part of the β-dispersion where a dielectric decrement occurs, as shown in Figure 8. There are three examples where a DEP cross-over (transition from positive to negative DEP with increasing frequency) has been observed at 1–10 MHz, namely: that reported for BSA by Cao et al. [26] as shown in Figure 3; for avidin (Bakewell et al. [27]) and PSA (Kim et al. [37]) as shown in Figure 5. This is consistent with the DEP responses of these proteins resulting from polarization of their permanent dipole moment, and not only as the result of an induced dipole moment.

The DEP force arising from a permanent dipole moment is given by Equation (20), and is shown to be directly proportional to ∇E^2. This corrects a previous conclusion [9], based on the presumption of a rigid rather than rotationally free permanent dipole, that the DEP force arising from a permanent dipole would be proportional to ∇E. However, as shown by Equation (21), the contribution of the DEP force expected for a BSA from its permanent dipole moment is predicted (according to current accepted theory) to be slightly less than the contribution of its induced moment. This indicates that, unless the 'cavity' field experienced by the protein molecule is very much larger than the field existing within the surrounding bulk medium, we have is no explanation in terms of the standard DEP theory (even if modified to encompass both an induced plus a permanent dipole moment) why the majority of experimental ∇E^2 values shown in Figure 7 fall well below the minimum requirement of $\nabla E^2 > 3.5 \times 10^{21}$ V^2/m^3 to overcome thermal dispersion effects. As shown in Figure 7, the minimum required ∇E^2 value is lowered by a factor of ~1000-fold for BSA, if the macroscopic *CM*-factor is replaced in Equations (5) and (9) by the empirically based molecular version $CM_{\text{micro}} = (\kappa + 2)CM_{\text{macro}}$ formulated in Section 3.4, and tabulated for various proteins in Table 1. Of the proteins listed in Table 1, only three (BSA, concanavalin, ribonuclease) are cited in Figure 7. The location of these proteins in the ∇E^2 'ranking' of Figure 7 is significant. Their relative values of $(\kappa + 2)CM$ given in Table 1, namely: ~11,000 for ribonuclease and ~15,000 for concanavalin, would place them above the minimum requirement level indicated in Figure 7 for BSA. It would clearly be of value to populate Table 1 with as yet unavailable dielectric spectroscopy data for the other proteins cited in Figure 5, and vice versa. With this information protocols could be developed to spatially manipulate or selectively sort targeted protein molecules, so bringing protein DEP in line with the achievements and promise enjoyed by the more established DEP of cells and bacteria, for example [75].

Finally, Equation (15) is offered for the following relationships between the ratios of the polarization of a protein in its cavity field and of the surrounding medium:

$$\frac{P_i}{P_m} \equiv \frac{\chi_i E_i}{\chi_m E_m} \equiv \frac{\langle M_i \rangle \cdot E_i}{\langle M_m \rangle \cdot E_m} \propto (\kappa + 2)CM_{\text{macro}}$$

These ratios will be sensitive to the physico-chemical attributes of a particular protein (e.g., peptide chain folding, net charge, and the distribution of polar and hydrophobic groups on the protein surface) and could explain the very wide range of values for the parameter $(\kappa + 2)CM$ given in Table 1. This empirical-based suggestion mirrors various theoretical findings of Matyushov and co-workers [8,10]. The possible significance of this for further development of a robust theory for protein DEP is discussed in an accompanying paper [11].

Author Contributions: Conceptualization, R.P.; methodology, R.P.; software, R.H.; writing—original draft preparation, R.P. and R.H.; writing—review and editing, R.P. and R.H.; funding acquisition, R.H. All authors have read and agreed to the published version of the manuscript.

Funding: Parts of this work were financially supported by the European Regional Development Fund (ERDF) and by The Brandenburg Ministry of Sciences, Research and Cultural Affairs (MWFK) within the framework StaF.

Acknowledgments: We acknowledge valuable discussions with Antonio Ramos (regarding the possible influence of electrohydrodynamic effects on protein DEP) and with Dmitry Matyushov on various aspects of his theories regarding cavity fields and interfacial polarizations we consider to be of relevance to protein DEP.

Conflicts of Interest: The authors declare no conflict of interest.

References

1. Kim, D.; Sonker, M.; Ros, A. Dielectrophoresis: From molecular to micrometer-scale analytes. *Anal. Chem.* **2019**, *91*, 277–295. [CrossRef] [PubMed]
2. Washizu, M.; Suzuki, S.; Kurosawa, O.; Nishizaka, T.; Shinohara, T. Molecular dielectrophoresis of biopolymers. *IEEE Trans. Ind. Appl.* **1994**, *30*, 835–843. [CrossRef]
3. Nakano, A.; Ros, A. Protein dielectrophoresis: Advances, challenges, and applications. *Electrophoresis* **2013**, *34*, 1085–1096. [CrossRef] [PubMed]
4. Pethig, R. *Dielectropohoresis: Theory, Methodology and Biological Applications*; John Wiley & Sons: Hoboken, NJ, USA, 2017.
5. Laux, E.M.; Bier, F.F.; Hölzel, R. Electrode-based AC electrokinetics of proteins: A mini review. *Bioelectrochemistry* **2018**, *120*, 76–82. [CrossRef]
6. Hayes, M.A. Dielectrophoresis of proteins: Experimental data and evolving theory. *Anal. Bioanal. Chem.* **2020**, *115*, 7144.
7. Lapizco-Encinas, B.H. Microscale electrokinetic assessments of proteins employing insulating structures. *Curr. Opin. Chem. Eng.* **2020**, *29*, 9–16. [CrossRef]
8. Seyedi, S.S.; Matyushov, D.V. Protein Dielectrophoresis in Solution. *J. Phys. Chem. B* **2018**, *122*, 9119–9127. [CrossRef]
9. Pethig, R. Limitations of the Clausius-Mossotti function used in dielectrophoresis and electrical impedance studies of biomacromolecules. *Electrophoresis* **2019**, *40*, 2575–2583. [CrossRef]
10. Matyushov, D.V. Electrostatic solvation and mobility in uniform and non-uniform electric fields: From simple ions to proteins. *Biomicrofluidics* **2019**, *13*, 064106:1–064106:15. [CrossRef]
11. Hölzel, R.; Pethig, R. Protein Dielectrophoresis: II. Key dielectric parameters and evolving theory. *Micromachines* **2020**. to be submitted.
12. Debye, P. *Polar Molecules*; The Chemical Catalog Co.: New York, NY, USA, 1929.
13. Oncley, J.L. The investigation of proteins by dielectric measurements. *Chem. Rev.* **1942**, *30*, 433–450. [CrossRef]
14. Zheng, L.; Brody, J.P.; Burke, P.J. Electronic manipulation of DNA, proteins, and nanoparticles for potential circuit assembly. *Biosens. Bioelectron.* **2004**, *20*, 606–619. [CrossRef] [PubMed]

15. Hübner, Y.; Hoettges, K.F.; McDonnell, M.B.; Carter, M.J.; Hughes, M.P. Applications of dielectrophoretic/electrohydrodynamic "zipper" electrodes for detection of biological nanoparticles. *Int. J. Nanomed.* **2007**, *2*, 427–431.

16. Yamamoto, T.; Fujii, T. Active immobilization of biomolecules on a hybrid three-dimensional nanoelectrode by dielectrophoresis for single-biomolecule study. *Nanotechnology* **2007**, *18*, 495503:1–495503:7. [CrossRef] [PubMed]

17. Lapizco-Encinas, B.H.; Ozuna-Chacón, S.; Rito-Palomares, M. Protein manipulation with insulator-based dielectrophoresis and direct current electric fields. *J. Chromat. A* **2008**, *1206*, 45–51. [CrossRef] [PubMed]

18. Agastin, S.; King, M.R.; Jones, T.B. Rapid enrichment of biomolecules using simultaneous liquid and particulate dielectrophoresis. *Lab Chip* **2009**, *9*, 2319–2325. [CrossRef] [PubMed]

19. Nakano, A.; Chao, T.-C.; Camacho-Alanis, F.; Ros, A. Immunoglobulin G and bovine serum albumin streaming dielectrophoresis in a microfluidic device. *Electrophoresis* **2011**, *32*, 2314–2322. [CrossRef]

20. Camacho-Alanis, F.; Gan, L.; Ros, A. Transitioning streaming to trapping in DC insulator based dielectrophoresis for biomolecules. *Sens. Actuators B Chem.* **2012**, *173*, 668–675. [CrossRef]

21. Liao, K.-T.; Chou, C.-F. Nanoscale Molecular Traps and Dams for Ultrafast Protein Enrichment in High-Conductivity Buffers. *J. Am. Chem. Soc.* **2012**, *134*, 8742–8745. [CrossRef]

22. Barik, A.; Otto, L.M.; Yoo, D.; Jose, J.; Johnson, T.W.; Oh, S.-H. Dielectrophoresis-enhanced plasmonic sensing with gold nanohole arrays. *Nano Lett.* **2014**, *14*, 2006–2012. [CrossRef]

23. Laux, E.M.; Knigge, X.; Bier, F.F.; Wenger, C.; Hölzel, R. Dielectrophoretic immobilization of proteins: Quantification by atomic force microscopy. *Electrophoresis* **2015**, *36*, 2094–2101. [CrossRef] [PubMed]

24. Schäfer, C.; Kern, D.P.; Fleischer, M. Capturing molecules with plasmonic nanotips in microfluidic channels by dielectrophoresis. *Lab Chip* **2015**, *15*, 1066–1071. [CrossRef] [PubMed]

25. Zhang, P.; Liu, Y. DC biased low-frequency insulating constriction dielectrophoresis for protein biomolecules concentration. *Biofabrication* **2017**, *9*, 045003:1–045003:11. [CrossRef] [PubMed]

26. Cao, Z.; Zhu, Y.; Liu, Y.; Dong, S.; Chen, X.; Bai, F.; Song, S.; Fu, J. Dielectrophoresis-based protein enrichment for a highly sensitive immunoassay using Ag/SiO$_2$ nanorod arrays. *Small* **2018**, *14*, 17032265. [CrossRef]

27. Bakewell, J.G.; Hughes, M.P.; Milner, J.J.; Morgan, H. Dielectrophoretic manipulation of avidin and DNA. In Proceedings of the 20th Annual International Conference of the IEEE Engineering in Medicine and Biology Society, Hong Kong, China, 1 November 1998; Volume 20, pp. 1079–1082.

28. Hölzel, R.; Calander, N.; Chiragwandi, Z.; Willander, M.; Bier, F.F. Trapping single molecules by dielectrophoresis. *Phys. Rev. Lett.* **2005**, *95*, 128102:1–128102:4. [CrossRef]

29. Clarke, R.W.; White, S.S.; Zhou, D.J.; Ying, L.M.; Klenerman, D. Trapping of proteins under physiological conditions in a nanopipette. *Angew. Chem. Int. Ed.* **2005**, *44*, 3747–3750. [CrossRef]

30. Staton, S.J.R.; Jones, P.V.; Ku, G.; Gilman, S.D.; Kheterpal, I.; Hayes, M.A. Manipulation and capture of A beta amyloid fibrils and monomers by DC insulator gradient dielectrophoresis (DC-iGDEP). *Analyst* **2012**, *137*, 3227–3229. [CrossRef]

31. Mata-Gomez, M.A.; Gallo-Villanueva, R.C.; Gonzalez-Valdez, J.; Martinez-Chapa, S.O.; Rito-Palomares, M. Dielectrophoretic behavior of PEGylated RNase A inside a microchannel with diamond-shaped insulating posts. *Electrophoresis* **2016**, *37*, 519–528. [CrossRef]

32. Nakano, A.; Camacho-Alanis, F.; Ros, A. Insulator-based dielectrophoresis with β-galactosidase in nanostructured devices. *Analyst* **2015**, *140*, 860–868. [CrossRef]

33. Liao, K.T.; Tsegaye, M.; Chaurey, V.; Chou, C.F.; Swami, N.S. Nano-constriction device for rapid protein preconcentration in physiological media through a balance of electrokinetic forces. *Electrophoresis* **2012**, *33*, 1958–1966. [CrossRef]

34. Chaurey, V.; Rohani, A.; Su, Y.H.; Liao, K.T.; Chou, C.F.; Swami, N.S. Scaling down constriction-based (electrodeless) dielectrophoresis devices for trapping nanoscale bioparticles in physiological media of high-conductivity. *Electrophoresis* **2013**, *34*, 1097–1104. [CrossRef] [PubMed]

35. Sanghavi, B.J.; Varhue, W.; Chavez, J.L.; Chou, C.F.; Swami, N.S. Electrokinetic Preconcentration and Detection of Neuropeptides at Patterned Graphene-Modified Electrodes in a Nanochannel. *Anal. Chem.* **2014**, *86*, 4120–4125. [CrossRef] [PubMed]

36. Laux, E.-M.; Kaletta, U.C.; Bier, F.F.; Wenger, C.; Hölzel, R. Functionality of dielectrophoretically immobilized enzyme molecules. *Electrophoresis* **2014**, *35*, 459–466. [CrossRef] [PubMed]

37. Kim, H.J.; Kim, J.; Yoo, Y.K.; Lee, J.H.; Park, J.H.; Wang, K.S. Sensitivity improvement of an electrical sensor achieved by control of biomolecules based on the negative dielectrophoretic force. *Biosens. Bioelectron.* **2016**, *85*, 977–985. [CrossRef]

38. Laux, E.-M.; Knigge, X.; Bier, F.F.; Wenger, C.; Hölzel, R. Aligned Immobilization of Proteins Using AC Electric Fields. *Small* **2015**, *12*, 03052:1–03052:7. [CrossRef]

39. Chiou, C.-H.; Chien, L.-J.; Kuo, J.-N. Nanoconstriction-based electrodeless dielectrophoresis chip for nanoparticle and protein preconcentration. *Appl. Phys. Express* **2015**, *8*, 085201:1–085201:3. [CrossRef]

40. Sharma, A.; Han, C.-H.; Jang, J. Rapid electrical immune assay of the cardiac biomarker troponin I through dielectrophoretic concentration using imbedded electrodes. *Biosens. Bioelectron.* **2016**, *82*, 78–84. [CrossRef]

41. Rohani, A.; Sanghavi, B.J.; Salahi, A.; Liao, K.T.; Chou, C.F.; Swami, NS. Frequency-selective electrokinetic enrichment of biomolecules in physiological media based on electrical double-layer polarization. *Nanoscale* **2017**, *9*, 12124–12131. [CrossRef]

42. Han, C.-H.; Woo, S.Y.; Bhardwaj, J.; Sharma, A.; Jang, J. Rapid and selective concentration of bacteria, viruses, and proteins using alternating current signal superimposition on two coplanar electrodes. *Sci. Rep.* **2018**, *8*, 14942:1–14942:10. [CrossRef]

43. Axelsson, I. Characterization of proteins and other macromolecules by agarose gel chromatography. *J. Chromatogr. A* **1978**, *152*, 21–32. [CrossRef]

44. Majorek, K.A.; Porebski, P.J.; Dayal, A.; Zimmerman, M.D.; Jablonska, K.; Stewart, A.J.; Chruszcz, M.; Minor, W. Structural and immunologic characterization of bovine, horse, and rabbit serum albumins. *Mol. Immunol.* **2012**, *52*, 174–182. [CrossRef] [PubMed]

45. Li, Y.; Lee, J.; Lal, J.; An, L.; Huang, Q. Effects of pH on the interactions and conformation of bovine serum albumin: Comparison between chemical force microscopy and small-angle neutron scattering. *J. Phys. Chem. B* **2008**, *112*, 3797–3806. [CrossRef]

46. Barbosa, L.R.S.; Ortore, M.G.; Spinozzi, F.; Mariani, P.; Bernstorff, S.; Itri, R. The importance of protein-protein interactions on the pH-induced conformational changes of bovine serum albumin: A small-angle X-Ray scattering study. *Biophys. J.* **2010**, *98*, 147–157. [CrossRef] [PubMed]

47. Takeda, K.; Wada, A.; Yamamoto, K.; Moriyama, Y.; Aoki, K. Conformational change of bovine serum albumin by heat treatment. *J. Protein Chem.* **1989**, *8*, 653–659. [CrossRef] [PubMed]

48. Murayama, K.; Tomida, M. Heat-induced secondary structure and conformation change of bovine serum albumin investigated by Fourier transform infrared spectroscopy. *Biochemistry* **2004**, *43*, 11526–11532. [CrossRef] [PubMed]

49. Pindrus, M.A.; Cole, J.L.; Kaur, J.; Shire, S.; Yadav, S.; Kalonia, D.S. Effect of aggregation on the hydrodynamic properties of bovine serum albumin. *Pharm. Res.* **2017**, *34*, 2250–2259. [CrossRef] [PubMed]

50. de Frutos, M.; Cifuentes, A.; Díez-Masa, J.C. Multiple peaks in HPLC of proteins: Bovine serum albumin eluted in a reversed-phase system. *J. High Resol. Chromatogr.* **1998**, *21*, 18–25. [CrossRef]

51. Chi, E.Y.; Krishnan, S.; Randolph, T.W.; Carpenter, J.F. Physical stability of proteins in aqueous solution: Mechanism and driving forces in non-native protein aggregation. *Pharm. Res.* **2003**, *20*, 1325–1336. [CrossRef]

52. Pethig, R. Dielectric properties of biological materials: Biophysical and medical applications. *IEEE Trans. Electr. Insul.* **1984**, *EI-19*, 453–474. [CrossRef]

53. Moser, P.; Squire, P.G.; O'Konski, C.T.O. Electric polarization in proteins—Dielectric dispersion and Kerr effect studies of isionic bovine serum albumin. *J. Phys. Chem.* **1966**, *70*, 744–756. [CrossRef]

54. Grant, E.H.; Keefe, S.E.; Takashima, S. The dielectric behavior of aqueous solutions of bovine serum albumin from radiowave to microwave frequencies. *J. Phys. Chem.* **1968**, *72*, 4373–4380. [CrossRef] [PubMed]

55. Fischer, H.; Polikarpov, I.; Craievich, A.F. Average protein density is a molecular-weight-dependent function. *Protein Sci.* **2004**, *13*, 2825–2827. [CrossRef] [PubMed]

56. Takashima, S.; Asami, K. Calculation and measurement of the dipole moment of small proteins: Use of protein data base. *Biopolymers* **1993**, *33*, 59–68. [CrossRef] [PubMed]

57. Chari, R.; Singh, S.N.; Yadav, S.; Brems, D.N.; Kalonia, D.S. Determination of the dipole moments of RNAse SA wild type and a basic mutant. *Proteins* **2012**, *80*, 1041–1052. [CrossRef]

58. Knocks, A.; Weingärtner, H. The dielectric spectrum of ubiquitin in aqueous solution. *J. Phys. Chem. B* **2001**, *105*, 3635–3638. [CrossRef]

59. Keefe, S.E.; Grant, E.H. Dipole moment and relaxation time of ribonuclease. *Phys. Med. Biol.* **1974**, *9*, 701–707. [CrossRef]

60. Schlecht, P. Dielectric properties of hemoglobin and myoglobin. II. Dipole moment of sperm whale myoglobin. *Biopolymers* **1969**, *8*, 757–765. [CrossRef]

61. South, G.P.; Grant, E.H. Dielectric dispersion and dipole moment of myoglobin in water. *Proc. R. Soc. Lond. A* **1972**, *328*, 371–387.

62. Takashima, S. Use of protein database for the computation of the dipole moments of normal and abnormal hemoglobins. *Biophys. J.* **1993**, *64*, 1550–1558. [CrossRef]

63. Pethig, R. Protein-water interactions determined by dielectric methods. *Annu. Rev. Phys. Chem.* **1992**, *43*, 177–205. [CrossRef]

64. Hedvig, P. *Dielectric Spectroscopy of Polymers*; Akadémiai Kiadó: Budapest, Hungary, 1977.

65. Martin, D.R.; Friesen, A.D.; Matyushov, D.V. Electric field inside a "Rossky cavity" in uniformly polarized water. *J. Chem. Phys.* **2011**, *135*, 084514. [CrossRef] [PubMed]

66. Martin, D.R.; Matyushov, D.V. Dipolar nanodomains in protein hydration shells. *J. Phys. Chem. Lett.* **2015**, *6*, 407–412. [CrossRef] [PubMed]

67. Naumov, I.I.; Bellaiche, L.; Fu, H. Unusual phase transitions in ferroelectric nanodisks and nanorods. *Nature* **2004**, *432*, 737–740. [CrossRef] [PubMed]

68. Seyedi, S.; Matyushov, D.V. Dipolar susceptibility of protein hydration shells. *Chem. Phys. Lett.* **2018**, *713*, 210–214. [CrossRef]

69. Schwan, H.P.; Schwarz, G.; Maczuk, J.; Pauly, H. On the low-frequency dielectric dispersion of colloidal particles in electrolyte solution. *J. Phys. Chem.* **1962**, *66*, 2626–2635. [CrossRef]

70. Foster, K.R.; Schwan, H.P. Dielectric properties of tissues. In *Handbook of Biological Effects of Electromagnetic Fields*; CRC Press Inc.: Boca Raton, FL, USA, 1996; pp. 118–122.

71. Medda, L.; Monduzzi, M.; Salis, A. The molecular motion of bovine serum albumin under physiological conditions is ion specific. *Chem. Commun.* **2015**, *51*, 6663–6666. [CrossRef]

72. Saucedo-Espinosa, M.A.; Lapizco-Encinas, B.H. Design of insulator-based dielectrophoretic devices: Effect of insulator posts characteristics. *J. Chromatogr. A* **2015**, *1422*, 325–333. [CrossRef]

73. Xuan, X. Recent advances in direct current electrokinetic manipulation of particles for microfluidic applications. *Electrophoresis* **2019**, *40*, 2484–2513. [CrossRef]

74. Lapizco-Encinas, B.H. On the recent development of insulator based dielectrophoresis (iDEP) by comparing the streaming and trapping regimes. *Electrophoresis* **2019**, *40*, 358–375. [CrossRef]

75. Pethig, R. Review—Where is dielectrophoresis going? *J. Electrochem. Soc.* **2017**, *164*, B3049–B3055. [CrossRef]

 *micromachines*

Article

Highly Localized Enrichment of *Trypanosoma brucei* Parasites Using Dielectrophoresis

Devin Keck, Callie Stuart, Josie Duncan, Emily Gullette and Rodrigo Martinez-Duarte *

Multiscale Manufacturing Laboratory, Department of Mechanical Engineering, Clemson University, Clemson, SC 29634, USA; dkeck@g.clemson.edu (D.K.); cdstuar@g.clemson.edu (C.S.); josied@g.clemson.edu (J.D.); egullet@g.clemson.edu (E.G.)
* Correspondence: rodrigm@clemson.edu

Received: 10 June 2020; Accepted: 25 June 2020; Published: 26 June 2020

Abstract: Human African trypanosomiasis (HAT), also known as sleeping sickness, is a vector-borne neglected tropical disease endemic to rural sub-Saharan Africa. Current methods of early detection in the affected rural communities generally begin with general screening using the card agglutination test for trypanosomiasis (CATT), a serological test. However, the gold standard for confirmation of trypanosomiasis remains the direct observation of the causative parasite, *Trypanosoma brucei*. Here, we present the use of dielectrophoresis (DEP) to enrich *T. brucei* parasites in specific locations to facilitate their identification in a future diagnostic assay. DEP refers to physical movement that can be selectively induced on the parasites when exposing them to electric field gradients of specific magnitude, phase and frequency. The long-term goal of our work is to use DEP to selectively trap and enrich *T. brucei* in specific locations while eluting all other cells in a sample. This would allow for a diagnostic test that enables the user to characterize the presence of parasites in specific locations determined *a priori* instead of relying on scanning a sample. In the work presented here, we report the characterization of the conditions that lead to high enrichment, 780% in 50 s, of the parasite in specific locations using an array of titanium microelectrodes.

Keywords: sleeping sickness; Human African trypanosomiasis; trypanosoma; titanium; dielectrophoresis

---

1. Introduction

Human African trypanosomiasis (HAT), also known as sleeping sickness, is a vector-borne neglected tropical disease endemic to rural sub-Saharan Africa. The disease is caused by infection of the protozoan *Trypanosoma brucei*, which is transmitted by the tsetse fly. Early diagnosis of the presence of *T. brucei* at the first stage of infection can have a significant impact on patient outcome by enabling timely and adequate treatment before the disease moves into a second stage. This is important because at this later stage the parasite penetrates the central nervous system, which leads to neuropsychiatric manifestations. such as sleep disorders, derangement or deep sensory disturbances that severely compromise the quality of life of the patient [1,2]. This second stage is fatal if untreated and drugs used to treat it are expensive and/or highly toxic. In contrast, drug therapy for early-stage HAT is effective and only mildly toxic [1,2].

Current methods of early detection in the affected rural communities generally begin with general screening using the card agglutination test for trypanosomiasis (CATT), a serological test. However, the gold standard for confirmation of trypanosomiasis remains the direct observation of the parasite [3]. Therefore, positive CATT readings are subsequently followed up through the direct observation of trypanosomes in blood, lymph node aspirates or cerebrospinal fluid (of note, examination of the cerebrospinal fluid after lumbar puncture is required to differentiate between

HAT stages). In all cases, enrichment of the parasite in specific locations is crucial to facilitating their identification and different methods have been used to this end.

The mini hematocrit centrifugation technique (mHCT), the quantitative buffy coat (QBC) and the miniature anion-exchange centrifugation technique (mAECT) are all techniques that use centrifugation, but for two different purposes. Centrifugal fractionation is used in mHCT and QBC to enrich the parasite by exploiting their difference in density with respect to blood cells. A hematocrit centrifuge is first used to fraction the blood sample (~50 µL of finger-prick blood) into plasma, buffy coat and red blood cell (RBC) layers in a capillary tube. The capillary tubes are then placed in a special holder and examined under a microscope by a trained eye to scan for the presence of parasites [4,5]. If present, they are expected to be concentrated near the interface between the buffy coat and the plasma layers. The difference between mHCT and QBC is that acridine orange, a fluorescent stain, and UV light, are used in QBC to facilitate parasite identification. However, both techniques can suffer from low specificity since enrichment is done only based on particle density. Contamination of the enriched sample with white blood cells (WBC) of similar density significantly complicates the identification of the parasite. In contrast, mAECT utilizes surface charge to enrich the parasite from the sample. At pH 6–9 *T. brucei* have been shown to have a neutral charge or be less negatively charged than blood cells. While centrifugation is also used in mAECT, this is done for the sole purpose of flowing the sample through a positively-charged column. Hence, the vast majority of the blood cells are retained in the column while the parasites are eluted and retrieved after the column. The use of mAECT has been shown to increase sensitivity over mHCT by 30% to 40% and significantly less contamination of the sample with other cells facilitates direct observation and identification of the parasite [6–8]. However, scanning the eluted sample for parasites is still required, and there remains the possibility that parasites are physically trapped in the column.

Here, we present results on the use of dielectrophoresis (DEP) to enrich *T. brucei* parasites in specific locations to facilitate their identification in a diagnostic assay. DEP refers to physical movement that can be selectively induced on the parasites when exposing them to electric field gradients of specific magnitude, phase and frequency. This occurs thanks to the interaction between the electric field gradient and the electrical dipole induced on the parasite when this is exposed to such a field gradient. DEP can be generally classified as conventional or traveling-wave DEP [9–14], where conventional refers to short-distance movement and traveling wave to long-distance; and as positive DEP, occurring when targeted species move towards the electric field gradient, or negative DEP, when the targeted species move away from the gradient. The long-term goal of our work is to use conventional positive DEP to selectively trap and enrich *T. brucei* in specific locations while eluting all other cells in the sample. This would allow for a user to characterize the presence of parasites in specific locations determined *a priori* instead of relying on scanning a sample. In the work presented here, we report the characterization of the conditions that lead to high enrichment of the parasite in specific locations.

DEP methods are advantageous over density-based techniques, such as centrifugation, due to their increase in specificity. In fact, different DEP signatures have been reported for blood cells and parasites [12,15,16]. Of most relevance to the work presented here is the separation of *T. brucei* from RBCs demonstrated by Kremer et al. using a light-induced DEP (LiDEP) setup [17], and the localization of *T. brucei* reported by Menachery et al. using spiral gold electrodes and traveling-wave DEP (twDEP) [18]. As previously detailed by one of us, there are different techniques to implement the field gradient required for DEP [19]. For instance, LiDEP relies on a light modulator; usually, a digital micromirror device and digital light processing technology coupled to optimized optics; and a photoconductive substrate to generate said gradient. Although Kremer et al. developed a portable LiDEP setup that was demonstrated for the manipulation of *T. brucei*, the complexity and cost of the instrumentation might not yield a practical application in the affected zones. TwDEP relies on a mobile or traveling electric field gradient, implemented through polarizing an electrode array with an AC field of alternating phases, and offers the ability for long-distance cell transportation through the interaction between the moving field and the polarized cell [20]. For example, twDEP could be desirable to move the *T. brucei* away from the original sample in an attempt to eliminate any background noise from remaining cells, or could eliminate the need for external forces for cell transport, i.e., centrifugation or micropumps. However, twDEP requires more sophisticated electronics than conventional DEP or LiDEP to apply AC fields of alternating phases. Furthermore, the low transport velocity achievable in twDEP may compromise the practicality of using such a technique in a diagnostic assay [13,18].

Building upon existent work on the use of DEP for *T. brucei* enrichment, we demonstrate the use of titanium microelectrodes to induce conventional DEP on *T. brucei* towards its enrichment in specific locations. Ti is a low-cost alternative to gold and other more expensive metals, which also offers biocompatibility and desirable mechanical properties for microfluidic devices [21–24]. The device presented here uses titanium planar microelectrodes that are patterned on a silicon wafer using batch processes standard in microfabrication. We expect that the relatively low cost of Ti and the straightforward fabrication approach will lead to inexpensive devices and a practical diagnostic assay in the future.

2. Materials and Experimental Methods

2.1. Fabrication of the Experimental Device

Different electrode designs were considered for this work. The traditional interdigitated fingers, where targets are trapped along the entire length of the electrode, were first considered, but were quickly discarded since such design would not afford for punctual locations to enrich the parasite and facilitate their observation at specific and pre-determined spots. We then considered triangular electrodes because they are known to create punctual and strong electric field gradients [25], but abandoned this geometry for fear that the field strength at the sharp vertices of the electrodes would damage the integrity of the parasites (see Figure S1 in supplementary information). We finally settled on semi-circular electrodes with the rationale that semi-circles would offer weaker field gradients than the triangular electrodes but maintain the ability to safely enrich the parasites at specified locations. To this end, the semi-circles were arbitrarily positioned 3000 μm apart center-to-center to allow for separate, well-defined locations for potential enrichment (Figure 1H).

The titanium electrodes were fabricated on a silicon oxide surface through a lift-off process, as shown in Figure 1A–F and detailed next. Silicon wafers (100 mm) featuring a 500 nm-thick thermal oxide (Noel Technologies, Inc. Campbell, CA, USA) were first cleaned in oxygen plasma (20 μTorr) for 15 s. A layer of LOR resist (Microchem, Newton, MA, USA) was spin coated for 45 s at 2000 rpm on the silicon substrate and baked at 150 °C on a hotplate for 150 s (Figure 1B) (Brewer Sciences Cee Spin Coater and integrated hotplate). A layer of AZ701 photoresist was then spin coated for 45 s at 3000 rpm on top of the LOR, baked at 110 °C for 75 s and exposed to a light with $\lambda = 365$ nm and an intensity of 6 mW/cm^2 for 20 s to generate a pattern (Quintel Ultra i-line Series). Post exposure bake was done on the hotplate for 60 s at 110 °C. The exposed AZ and LOR layers were then immersed in a 2.3% tetramethylammonium hydroxide/97.7% water bath to develop the AZ layer and underetch the LOR (Figure 1D). The immersion time was manually adjusted to around 2 min following visual inspection until obtaining an underetch of the AZ layer of about 2 μm. After rinsing and drying, the patterned silicon substrate was transferred to a metal evaporator to deposit 350 nm of Ti (CCS CA-40 E-beam Evaporator). After the deposition process, the arrangement was immersed in remover NMP (1-methyl-2-pyrrolidone) to dissolve the AZ and LOR layers and effectively lift-off Ti from undesired regions of the substrate (Figure 1F).

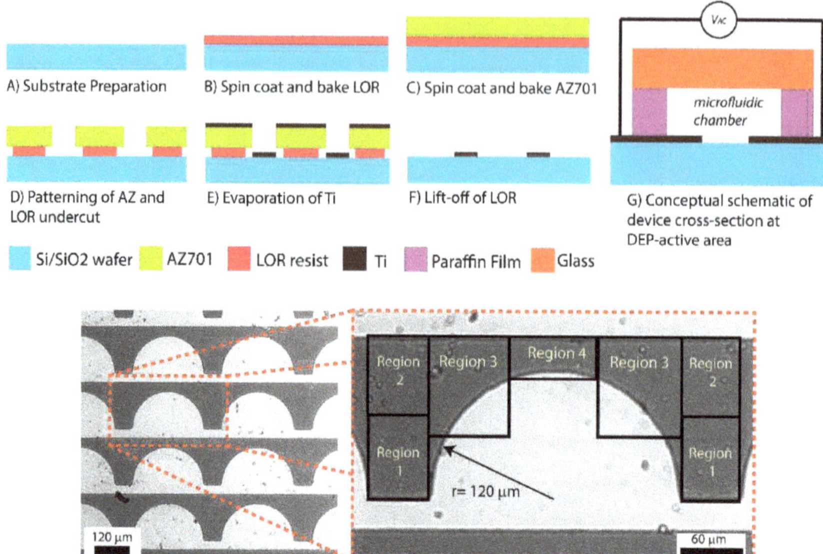

H) Experimental device and detail of electrode dimensions and regions of interest

Figure 1. Fabrication of Ti electrodes: (**A**) A six-inch silicon substrate was descummed with oxygen plasma treatment at 20 uTorr. (**B**) LOR resist was spin coated at 2000 rpm for 45 s on to the silicon substrate and a soft bake was performed at 150 °C for 150 s. (**C**) AZ701 resist was spin coated at 3000 rpm for 45 s on top of the LOR resist layer and a soft bake was performed at 110 °C for 75 s. (**D**) A Quintel Ultra i-line Series machine was used to pattern the resist layers using UV light with λ = 365 nm at an intensity of 6 mW/cm^2 for 20 s. Pattern development was performed via immersion in a 2.3% tetramethylammonium hydroxide/97.7% water bath (**E**). The patterned silicon substrate was transferred to a CCS CA-40 E-beam Evaporator to deposit 350 nm of Ti. (**F**) Lastly, the wafer was immersed in NMP (1-methyl-2-pyrrolidone) to dissolve the AZ and LOR layers and effectively lift-off Ti from undesired regions of the substrate (Figure 1F). (**G**) Conceptual schematic of the cross section of an experimental device. (**H**) Experimental device and details of electrode dimensions and predefined regions of interest surrounding a single semicircular electrode.

To ready the device for experimentation, a microfluidic chamber was created manually by cutting a rectangular shape from a paraffin film (PARAFILM® M) and positioning the film around the Ti electrode array. For each experiment, 10 µL of cell sample was introduced into the chamber via micropipetting and a glass slide was used to cover the sample. Pressure was manually applied to the glass slide to compress the paraffin film and ensure the chamber was sealed. The cross section of an experimental device at the DEP region is shown in Figure 1G.

2.2. Sample Preparation and Viability Study

Samples of procyclic form (PCF) *T. brucei brucei* were obtained from the Morris Laboratory at the Eukaryotic Pathogens Innovation Center (EPIC) in Clemson University. *T. brucei* were cultured at 29 °C in 5% CO$_2$ in SDM-79 media with a target density between 5×10^5 and 1×10^7 cell/mL [26]. *T. brucei* are known to display a worm like morphology reaching 20–40 µm in length by 1–3 µm in width [27]. Their normal behavior in culture media is that of a well dispersed population of individual parasites that are highly motile.

The procedure to prepare experimental samples was optimized as follows. In order to induce the trapping of parasites in specific locations using positive DEP, the parasite must feature higher electrical polarization than its surrounding medium at a given frequency. Practically speaking and at the frequencies used in this work, the electrical conductivity of the medium must thus be as low as possible while also supporting the viability of the parasite during experiments. The requirement for low media conductivity is because complex permittivity ε^* of the parasite ε_p^* and media ε_m^* are described by the same general equation

$$\varepsilon_{p,m}^* = \varepsilon_{p,m} - j\frac{\sigma_{p,m}}{\omega}$$

(1)

where j is the imaginary unit vector; ω is the angular frequency of the applied electric field; and $\varepsilon_{p,m}$ and $\sigma_{p,m}$ are the permittivity and conductivity of the parasite or suspending media respectively. Analysis of Equation (1) illustrates how the complex permittivity is directly related to the electrical conductivity, and that to induce a positive DEP force on the parasite the conductivity of the media, σ_m, must be lower than that of the parasite, σ_p.

To this end, viability studies of the parasite in media with decreasing values of electrical conductivity were then conducted as follows. A sugar solution (9% sucrose, 0.5% dextrose and 0.3% bovine serum albumin by weight) widely used as buffer for DEP experiments was used as a base and its electrical conductivity adjusted to specific values using phosphate buffered saline (PBS). Conductivity values tested were 120 µS/cm (conductivity of the DEP experimental media), 408 µS/cm and 504 µS/cm (OAKTON PC700 conductivity meter). The control experiment was in growth media, SDM-79. In all cases, including the control, the parasites were washed and re-suspended three times into the designated media using centrifugation (Hermle Z200A). The different samples containing the parasites were then individually placed in 35 mm sterile petri dishes. A randomized area of each sample was observed and recorded for 30 min. This time was chosen as the maximum time for our viability study because our experiments were designed to be time efficient such that the results could compete with current clinical practices. Total clump area was measured to assess the health of the culture, since clumping is a common result of cell lysis due to factors such as environmental stress and overgrowth. An optimized media for DEP experiments was assumed to be that with the smallest electrical conductivity and lowest density of parasite clumps.

2.3. Computational Modeling

ANSYS Electronics Desktop running on a DELL XPS 15 with an intel Core i7-6700HQ CPU and 16 GB of RAM was utilized to model the distribution of the electric field (E) and the square of the electric field gradient (∇E^2) using the built-in Maxwell 2D electrostatic solvers. The magnitude and spatial distribution of the electric field were modeled for four different values of polarization voltage in the range 5–20 V_{pp} towards selecting a voltage that would enable DEP forces but prevent electrical lysis of the parasite. Upon selecting such voltage, the corresponding ∇E^2 was modeled to estimate the strength of the DEP force throughout the device and predict the regions we expected to lead to parasite enrichment.

2.4. Protocol for the Dielectrophoretic and Regional Enrichment Characterization of the Parasites

The experimental protocol did not feature any flow and all experiments were done in stationary flow conditions. Each experiment featured the following stages: (1) 10 µL of the experimental sample was injected into the microfluidic chamber of the device using a micropipette; (2) the chamber was sealed, and the sample allowed time to stabilize; (3) a video recording (Andor Zyla Camera coupled to a LV100 Nikon Eclipse Microscope) of the experiment was started; (4) after 10 s the titanium electrode array was polarized using an AC signal of specific frequency (100 kHz–20 MHz) and magnitude (5 V_{pp}) using a BK Precision 4040B voltage generator. The response of the parasites to the polarized electrodes was recorded for 110 s.

All 120 s-long videos were analyzed using ImageJ software [28]. The field of view for all videos recorded included 8 electrodes (Figure 1H). Only individual *T. brucei* that were on the same plane as the electrode, i.e., those that were in focus, were included in the analyses.

3. Analytical Methods

3.1. Dielectrophoretic Characterization of the Parasites

The DEP response of the parasites was characterized based on the percentage of the parasites that were attracted to any part of the electrodes at the time mark of 50 s. The percentage of attracted parasites was calculated using Equation (2) by comparing the total number of parasites visible in the field of view at 50 s, or $C_{T,t=50}$, to the number of parasites that were attached (assessed visually by their characteristic perpendicular alignment to the electrode edge) to any of the 8 monitored electrodes, or $C_{A,t=50}$.

$$[C_{A,t=50}/C_{T,t=50}] \times 100, \tag{2}$$

This analysis was done for three videos per each of the 6 frequencies investigated in the range of 0.1–20 MHz. A higher percentage of attached cells was assumed to indicate a stronger positive DEP response induced on the parasites.

3.2. Assessment of the Regional Enrichment of the Parasites

The ability to enrich the *T. brucei* parasites in specific locations was measured by monitoring the number of them over time in 4 unique and pre-defined regions of interest for each electrode, as illustrated in Figure 1H. Parasite enrichment was measured for each region and reported as a percentage increase or decrease in parasites from time $t = 0$ to $t = 50$ s. The average regional enrichment was plotted for each of the 4 defined regions to determine their enrichment potential. The percent enrichment for each region was calculated as a percentage change in the number of parasites from time $t = 0$ s to time $t = 50$ s using Equation (3).

$$[(C_{RT,t=50} - C_{RT,t=0})/C_{RT,t=0}] \times 100, \tag{3}$$

where $C_{RT,t=50}$ is the regional total count of parasites in at 50 s and $C_{RT,t=0}$ is the regional total count of parasites at 0 s for the same region. Only single, living parasites that were on the same focus plane of the electrode were considered. This consideration was practical and towards an eventual tool to facilitate direct observation of the parasites in specific locations. In this study positive values in enrichment percentage indicate a tendency for parasites to migrate towards the region of interest, while negative values indicate the opposite.

4. Results and Discussion

4.1. Optimizing the Experimental DEP Media for T. brucei

As previously noted, the parasites are highly motile and maintain a single parasite dispersion when immersed in culture media. Since single parasite dispersion is necessary for adequate characterization of their DEP response, the ideal experimental DEP media would feature low electrical conductivity and lead to the least clump formation. Results of total clump area in the culture depending on the electrical conductivity of the media are shown in Figure 2 after an immersion time of 30 min. A clump of parasites was defined as three or more parasite sharing a single junction, and the 2D surface area of each clump was measured in ImageJ software [28]. The reported clump area is the summation of all clumps in the measured area. It is clearly observed that the total clump area decreased as the conductivity of the buffer media increased. The 504 µS/cm buffer was selected for our DEP experiments due to offering the best compromise between maintaining a low clump area and a conductivity that is low enough to potentially induce a positive DEP response within the parasites.

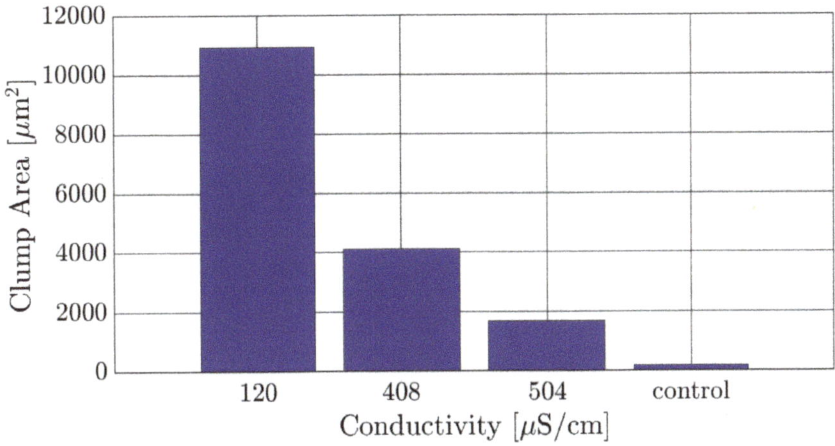

Figure 2. The total clump area of *Trypanosoma brucei* measured in different samples featuring a standard dielectrophoresis (DEP) experimental medium (120 µS/cm) and in experimental media with increasing electrical conductivity. The control was a sample of *T. brucei* in their standard culture media. An experimental medium with electrical conductivity of 504 µS/cm was chosen as a compromise between maintaining a suspension of individual parasites and a conductivity value that can lead to a strong positive DEP response.

4.2. Optimizing Polarization Voltage for DEP Experiments

Figure 3 depicts the distribution of E in the microfluidic device at different polarization voltages in the range 5–20 V_{pp}. As expected, the magnitude of E is directly proportional to the polarization voltage (Figure 3A–D). In this work we targeted an electric field magnitude less than 10^5 V/m throughout the entire device in order to maintain cell viability, by following the work by Glasser et al., who observed that electric fields of this magnitude showed minimal effects on cell viability during short term exposure to strong ac fields in the frequency range of our experiments [29]. Hence, we performed experiments using a polarization voltage of 5 V_{pp}. In this case, the magnitude of E throughout the device would be expected to be $<7 \times 10^4$ V/m.

Figure 3E illustrates the modeled distribution of ∇E^2 in the microfluidic device when electrodes are polarized using 5 V_{pp}. When assuming induction of a positive DEP force on the parasites, the parasites would migrate towards regions with the highest ∇E^2, or the orange–red regions in the figure. Based on the results by Menachery et al., a magnitude of ∇E^2 above 10^{13} V^2/m^3 would be enough to induce movement on the parasites [18]. Hence, from this computational model we would expect that parasites under the effect of a DEP force would migrate to the leading edges of the semicircular electrodes, which are included in regions 3 and 4 (see Figure 1H).

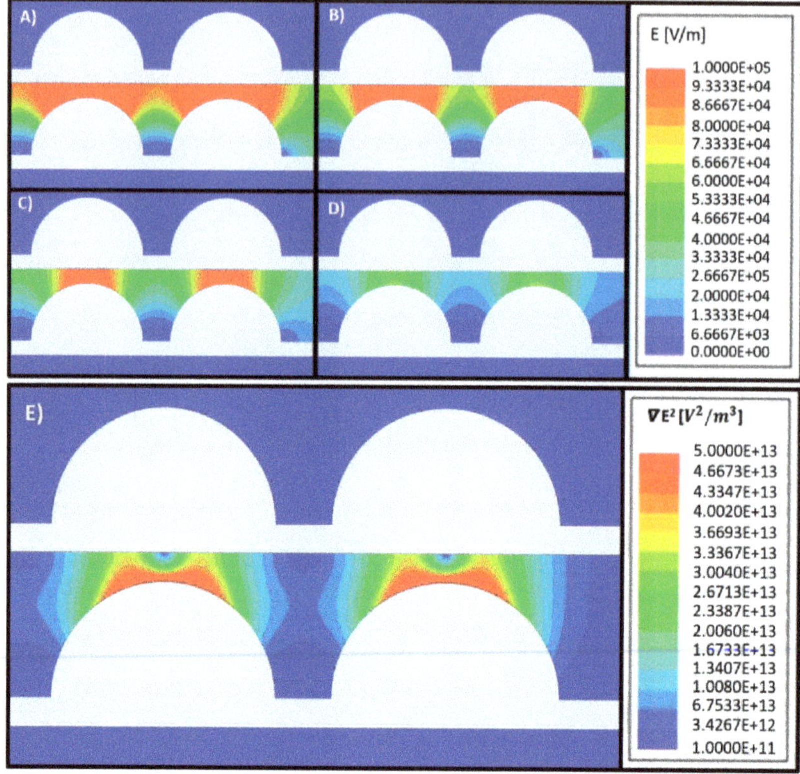

Figure 3. Modeling of the electric field E for an array of titanium electrodes (white geometries) polarized using different voltages: (**A**) 20 V_{pp} (**B**) 15 V_{pp} (**C**) 10 V_{pp} and (**D**) 5 V_{pp}. The modeled media around electrodes was water with an electrical conductivity of 504 μS/cm. As expected, the magnitude of the electric field increases proportionally to the magnitude of the polarizing voltage. A magnitude of $E < 10^5$ V/m is desired to minimize the risk of electrically lysing the parasites. (**E**) Modeling of ∇E^2 in an array of titanium electrodes (white geometries) polarized using 5 V_{pp}. The modeled media around electrodes was water with an electrical conductivity of 504 μS/cm. If the parasites experience a positive DEP force, they are expected to migrate to the regions of highest ∇E^2, shown as orange–red in the figure.

4.3. Characterizing the Dielectrophoretic Response of T. brucei

Electrical charges naturally exist within the cell structure and these can become redistributed and aligned upon exposure of the cell to an electric field. This leads to cell polarization, and the inductance of an electric dipole and motion of the cell due to DEP [30]. The electrical double-layer that develops at the interface between the cell outer envelope and the suspending electrolyte will yield a membrane capacitance that depends on the cell size, shape and composition of said outer envelope. Such capacitance will dominate the DEP response of the cell at low frequency values of the applied electric field, i.e., the leftmost region of the DEP curve in Figure 4. In addition to this interface, the organelles and entities within a cell will yield their own dipole depending on their unique structure and function. At higher frequencies of the applied electric field, the cell's DEP response becomes a function of the electrical properties of the interfaces inside the cell, such as those originating from different organelles in the cytoplasm and their volume relative to that of the cytoplasm [30,31].

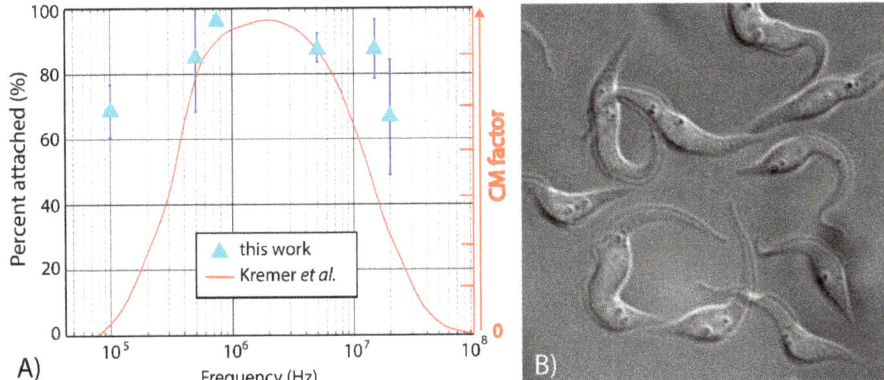

Figure 4. (**A**) Characterization of the DEP response of *T. brucei* across a broad frequency range, 100 kHz to 20 MHz. Dark blue bars represent the standard deviation between experiments ($n = 3$). (**B**) *T. brucei* cultured at 29 °C in 5% CO_2 in SDM-79 media with a target density between 5×10^5 and 1×10^7 cell/mL. Courtesy of Christina Wilkinson and James Morris.

Figure 4 depicts the DEP response of *T. brucei* to an applied voltage excitation of 5 V_{pp} at six different frequencies in the range 100 kHz–20 MHz. The light-blue triangles in Figure 4A represent the average ($n = 3$) attachment percentage of *T. brucei* to the electrode at a given frequency. A theoretical DEP response of *T. brucei*, shown as the smooth red curve, was previously reported by Kremer et al. [17] and overlaid over our results for comparison and discussed next. It can be clearly observed that the percentage of attachment in the experiments performed here was at least 50% and that a polarizing frequency of 750 kHz leads to the strongest positive DEP force and highest percent attachment, thereby providing the most potential to rapidly enrich the parasites in a specific location. These results also indicate that *T. brucei* shows a strong positive DEP response across the entire frequency range tested—a fact that is partially confirmed by Kremer et al. [17]. While experimental and theoretical results highly overlap at frequencies above 500 kHz, the theoretical DEP trapping seems to sharply decrease at frequencies below 400 kHz. Disparities between our experiments and Kremer et al. at these lower frequencies can be accredited to the shape simplifications of the parasite. Their assumptions simplify the parasite into a prolate elliptical shape, composed of two concentric shells, a membrane and cytoplasm. As illustrated in Figure 4B and reported by other authors, the parasites feature a worm-like morphology vastly more complex in shape and internal structure [32] than what was assumed by Kremer and co-authors [17]. Large organelles present inside the parasite, such as the nucleus and the kinetoplast, each would have their own dielectric properties, and together with their spatial distribution, would contribute to the overall dielectric response of the parasite. Other organelles within the parasite are also likely to have an effect on its DEP response, especially at high polarizing frequencies. Hence, the broadening of our experimental curve when compared to Kremer et al. is likely due to the slender shape and size of the parasite at frequencies in the kHz range and the contributions of the different organelles at frequencies beyond 10^7 Hz. A detailed study on the impact of organelles on the DEP response of *T. brucei* is out of the scope of this paper, which focuses on determining the conditions that will lead to rapid and strong enrichment of the parasite in specific locations. Envisioned future work includes such a detailed study and the effects of parasite motility and age on the DEP behavior of *T. brucei*.

4.4. Determining the Region that Yields the Highest Enrichment of T. brucei

Figure 5 depicts the results of the enrichment study, where the enrichment of parasites is expressed as a percentage of increase or decrease in parasites from time $t = 0$ to $t = 50$ s within each of the four pre-defined regions of interest. Region 4 yields close to 800% enrichment while region 1 is actually depleted of parasites (negative enrichment). Importantly, the percentage of enrichment reported is a combination of both enrichment due to the sideways migration of parasites from one region to another, and enrichment from parasites migrating from the bulk of the sample into the observation plane.

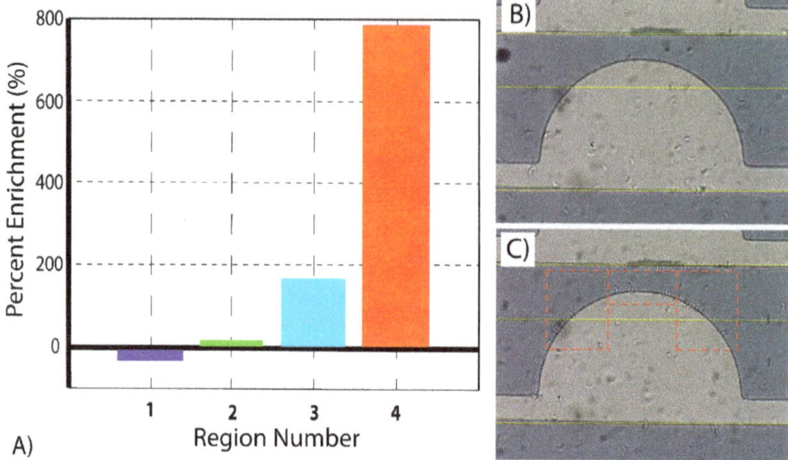

Figure 5. (**A**) Regional enrichment study of parasites from time t = 0 to t = 50 s for four predefined regions of interest shown in Figure 1H. All enrichment experiments were performed at a frequency of 750 kHz, as such frequency yields the strongest positive DEP response of *T. brucei* under the conditions studied in this work. (**B**) Single electrode at t = 0 illustrating low attachment of parasites to electrode edges. (**C**) Single electrode at t = 50 illustrating high attachment of parasites to electrode edges, particularly in regions 3 and 4 (dashed rectangles).

During analysis it was confirmed that the parasites tended to concentrate in regions of higher electric field gradient. Experiments showed that parasites tended to concentrate in region 4, a square area of dimensions 40 μm $\times$ 85 μm, at a rate of 780%, higher than any other region. In fact, the second highest rate was region 3, which increased the concentration of parasites by 163% over the same time span. Region 1 resulted in an average decrease in the concentration of parasites by $-$29%, while region 2 saw the parasite concentration increase by only 12%. The computational model for ∇E^2 presented here further validates these results due to the fact that the regions with the highest enrichment correlated to the regions of the electrode with the highest ∇E^2. These results indicate a significant potential for the use of DEP to position and enrich the concentration of *T. brucei* in specific locations. Particularly, our microfluidic chamber facilitated the ability to increase parasite concentration in region 4, a square area of dimensions of 40 μm $\times$ 85 μm, by 780% in 50 s.

5. Concluding Remarks

In this work we contribute a study of the conditions that led to the enrichment of *T. brucei* in specific locations using DEP. A frequency of 750 kHz at a polarizing voltage of 5 V_{pp} induced the strongest positive DEP response from the *T. brucei* parasites. This frequency was subsequently utilized to position and enrich *T. brucei* within a square planar region 40 $\times$ 85 μm. The positioning proved to be highly efficient, resulting in a 780% enrichment of parasites in less than a minute.

Early diagnosis of the presence of *T. brucei* at the first stage of infection can have a significant impact on patient outcome by enabling timely and adequate treatment before the disease moves into the second stage, which causes neuropsychiatric manifestations, such as sleep disorders, derangement and eventually death. Upon positive results from a CATT, assessing the presence of *T. brucei* in locations determined *a priori* can facilitate their detection and thus lead to an easy-to-use and robust assay.

The use of arrays of semicircular titanium electrodes to enrich parasites in desired locations using DEP shows high potential to achieve this end. However, further work is needed to characterize the specificity of DEP in regard to enriching *T. brucei* in a practical scenario. More specifically, the ability to use DEP to isolate parasites from WBCs and other species, i.e., microorganisms and parasites, might be present in a buffy coat and/or plasma portions of a centrifuged blood sample.

Supplementary Materials: The following are available online at http://www.mdpi.com/2072-666X/11/6/625/s1, Figure S1: Modeling of the electric field *E* for an array of triangular titanium electrodes (white geometries) with the same footprint than the semi-circular electrodes presented in the main text.

Author Contributions: Conceptualization, C.S., J.D., E.G. and R.M.-D.; data curation, D.K.; formal analysis, D.K., C.S., J.D., E.G. and R.M.-D.; funding acquisition, R.M.-D.; investigation, C.S., J.D. and E.G.; methodology, D.K., C.S., J.D., E.G. and R.M.-D.; project administration, R.M.-D.; software, D.K.; supervision, R.M.-D.; writing—original draft, D.K., C.S., J.D., E.G. and R.M.-D.; writing—review and editing, D.K. and R.M.-D. All authors have read and agreed to the published version of the manuscript.

Funding: This research received no external funding.

Acknowledgments: The authors thank Christina Wilkinson for the detailed images of *T. brucei* and the Morris Laboratory in EPIC at Clemson University for facilitating *T. brucei* samples. The authors acknowledge the financial support from Clemson University Creative Inquiry Undergraduate Research Program, project #791. We also thank Mary Grace Heustess, Kayla Wallace, Natalie Hanson, Meredith Hammer, Allison Mills and Emily Kluttz for aiding in the preliminary experiments and discussions leading to this work.

Conflicts of Interest: The authors declare no conflict of interest.

References

1. Babokhov, P.; Sanyaolu, A.; Oyibo, W.; Fagbenro-Beyioku, A.F.; Iriemenam, N.C. A current analysis of chemotherapy strategies for the treatment of human African trypanosomiasis. *Pathog. Glob. Health* **2013**, *107*, 242–252. [CrossRef] [PubMed]

2. Chappuis, F.; Loutan, L.; Simarro, P.; Lejon, V.; Büscher, P. Options for Field Diagnosis of Human African Trypanosomiasis. *Clin. Microbiol. Rev.* **2005**, *18*, 133–146. [CrossRef] [PubMed]

3. Holm, S.H.; Beech, J.P.; Barrett, M.P.; Tegenfeldt, J.O. Separation of parasites from human blood using deterministic lateral displacement. *Lab Chip* **2011**, *11*, 1326. [CrossRef] [PubMed]

4. Bonnet, J.; Boudot, C.; Courtioux, B. Overview of the Diagnostic Methods Used in the Field for Human African Trypanosomiasis: What Could Change in the Next Years? *BioMed. Res. Int.* **2015**, *2015*, 1–10. [CrossRef]

5. Schwarz, N.G.; Loderstaedt, U.; Hahn, A.; Hinz, R.; Zautner, A.E.; Eibach, D.; Fischer, M.; Hagen, R.M.; Frickmann, H. Microbiological laboratory diagnostics of neglected zoonotic diseases (NZDs). *Acta Trop.* **2017**, *165*, 40–65. [CrossRef]

6. Lumbala, C.; Biéler, S.; Kayembe, S.; Makabuza, J.; Ongarello, S.; Ndung'U, J.M. Prospective evaluation of a rapid diagnostic test for Trypanosoma brucei gambiense infection developed using recombinant antigens. *PLoS Negl. Trop. Dis.* **2018**, *12*, e0006386. [CrossRef]

7. Lutumba, P.; Robays, J.; Miaka, C.; Kande, V.; Mumba, D.; Büscher, P.; Dujardin, B.; Boelaert, M. Validité, coût et faisabilité de la mAECT et CTC comme tests de confirmation dans la détection de la Trypanosomiase Humaine Africaine. *Trop. Med. Int. Health* **2006**, *11*, 470–478. [CrossRef]

8. Ngoyi, D.M.; Ekangu, R.A.; Kodi, M.F.M.; Pyana, P.P.; Balharbi, F.; Decq, M.; Betu, V.K.; Van Der Veken, W.; Sese, C.; Menten, J.; et al. Performance of Parasitological and Molecular Techniques for the Diagnosis and Surveillance of Gambiense Sleeping Sickness. *PLOS Negl. Trop. Dis.* **2014**, *8*, e2954. [CrossRef]

9. Gimsa, J. A comprehensive approach to electro-orientation, electrodeformation, dielectrophoresis, and electrorotation of ellipsoidal particles and biological cells. *Bioelectrochemistry* **2001**, *54*, 23–31. [CrossRef]

10. Parkash, O.; Kumar, D.; Prasad, C.D. A unified theory of dielectrophoresis and travelling wave dielectrophoresis. *J. Phys. D Appl. Phys.* **1994**, *27*, 1509–1512. [CrossRef]

11. Jones, T.B. Basic theory of dielectrophoresis and electrorotation. *IEEE Eng. Med. Biol. Mag.* **2003**, *22*, 33–42. [CrossRef] [PubMed]

12. Gascoyne, P.R.C.; Vykoukal, J. Dielectrophoresis-Based Sample Handling in General-Purpose Programmable Diagnostic Instruments. *Proc. IEEE Inst. Electr. Electron. Eng.* **2004**, *92*, 22–42. [CrossRef] [PubMed]

13. Wang, X.-B.; Hughes, M.P.; Huang, Y.; Becker, F.; Gascoyne, P. Non-uniform spatial distributions of both the magnitude and phase of AC electric fields determine dielectrophoretic forces. *Biochim. et Biophys. Acta (BBA) Gen. Subj.* **1995**, *1243*, 185–194. [CrossRef]

14. Green, N.G.; Ramos, A.; Morgan, H. Numerical solution of the dielectrophoretic and travelling wave forces for interdigitated electrode arrays using the finite element method. *J. Electrost.* **2002**, *56*, 235–254. [CrossRef]

15. Gagnon, Z.R. Cellular dielectrophoresis: Applications to the characterization, manipulation, separation and patterning of cells. *Electrophoresis* **2011**, *32*, 2466–2487. [CrossRef]

16. Fernandez, R.E.; Rohani, A.; Farmehini, V.; Swami, N.S. Review: Microbial analysis in dielectrophoretic microfluidic systems. *Anal. Chim. Acta* **2017**, *966*, 11–33. [CrossRef]

17. Kremer, C.; Neale, S.L.; Menachery, A.; Barrett, M.; Cooper, J.M. Optoelectronic tweezers for medical diagnostics. *SPIE BiOS* **2012**, *8212*, 82120. [CrossRef]

18. Menachery, A.; Kremer, C.; Wong, P.E.; Carlsson, A.; Neale, S.L.; Barrett, M.P.; Cooper, J.M. Counterflow Dielectrophoresis for Trypanosome Enrichment and Detection in Blood. *Sci. Rep.* **2012**, *2*, 1–5. [CrossRef]

19. Martinez-Duarte, R. Microfabrication technologies in dielectrophoresis applications—A review. *Electrophoresis* **2012**, *33*, 3110–3132. [CrossRef]

20. Sun, T.; Morgan, H.; Green, N.G. Analytical solutions of ac electrokinetics in interdigitated electrode arrays: Electric field, dielectrophoretic and traveling-wave dielectrophoretic forces. *Phys. Rev. E* **2007**, *76*, 1–18. [CrossRef]

21. Zhang, Y.T.; Bottausci, F.; Rao, M.P.; Parker, E.R.; Mezić, I.; Macdonald, N.C. Titanium-based dielectrophoresis devices for microfluidic applications. *Biomed. Microdevices* **2008**, *10*, 509–517. [CrossRef] [PubMed]

22. Zhang, Y.; Parker, E.R.; Rao, M.P.; Aimi, M.F.; Mezic, I.; MacDonald, N.C. Titanium bulk micromachining for biomems applications: A DEP device as a demonstration. In Proceedings of the IMECE: 2004 ASME International Mechanical Engineering Congress and Rd & D Expo, Anaheim, CA, USA, 13–19 November 2004.

23. Tanaka, Y.; Endo, T.; Yanagida, Y.; Hatsuzawa, T. Design and fabrication of a dielectrophoresis-based cell-positioning and cell-culture device for construction of cell networks. *Microchem. J.* **2009**, *91*, 232–238. [CrossRef]

24. Puttaswamy, S.V.; Sivashankar, S.; Chen, R.-J.; Chin, C.-K.; Chang, H.-Y.; Liu, C.-H. Enhanced cell viability and cell adhesion using low conductivity medium for negative dielectrophoretic cell patterning. *Biotechnol. J.* **2010**, *5*, 1005–1015. [CrossRef] [PubMed]

25. Natu, R.; Islam, M.; Martinez-duarte, R. Carbon cone electrodes for selection, manipulation and lysis of single cells. In Proceedings of the ASME 2015 International Mechanical Engineering Congress and Exposition IMECE 2015, Houston, TX, USA, 13–19 November 2015; pp. 1–5.

26. Brun, R.; Jenni, L.; Tanner, M.; Schönenberger, M.; Schell, K.F. Cultivation of vertebrate infective forms derived from metacyclic forms of pleomorphic Trypanosoma brucei stocks. *Acta Trop.* **1979**, *36*, 387–390. [PubMed]

27. Sharma, R.; Peacock, L.; Gluenz, E.; Gull, K.; Gibson, W.; Carrington, M. Asymmetric Cell Division as a Route to Reduction in Cell Length and Change in Cell Morphology in Trypanosomes. *Protist* **2008**, *159*, 137–151. [CrossRef]

28. Schneider, C.A.; Rasband, W.S.; Eliceiri, K.W. NIH Image to ImageJ: 25 years of image analysis. *Nat. Methods* **2012**, *9*, 671–675. [CrossRef]

29. Glasser, H.; Fuhr, G. Cultivation of cells under strong ac-electric field—Differentiation between heating and trans-membrane potential effects. *BioElectrochem. Bioenerg.* **1998**, *47*, 301–310. [CrossRef]

30. Pethig, R.; Menachery, A.; Pells, S.; De Sousa, P. Dielectrophoresis: A Review of Applications for Stem Cell Research. *J. Biomed. Biotechnol.* **2010**, *2010*, 1–7. [CrossRef]

31. Pethig, R. *Encyclopedia of Surface and Colloidal Science*; Taylor & Francis: New York, NY, USA, 2006.

32. Field, M.C.; Allen, C.L.; Dhir, V.; Goulding, D.; Hall, B.; Morgan, G.W.; Veazey, P.; Engstler, M. New Approaches to the Microscopic Imaging of Trypanosoma brucei. *Microsc. Microanal.* **2004**, *10*, 621–636. [CrossRef]

Article

Frequency Response of Induced-Charge Electrophoretic Metallic Janus Particles

Chong Shen [1,2], **Zhiyu Jiang** [1,2], **Lanfang Li** [1,2], **James F. Gilchrist** [3] **and H. Daniel Ou-Yang** [1,2,4,*]

[1] Department of Physics, Lehigh University, Bethlehem, PA 18015, USA; chs514@lehigh.edu (C.S.); zhj216@lehigh.edu (Z.J.); lal419@lehigh.edu (L.L.)
[2] Emulsion Polymers Institute, Lehigh University, Bethlehem, PA 18015, USA
[3] Department of Chemical & Biomolecular Engineering, Lehigh University, Bethlehem, PA 18015, USA; gilchrist@lehigh.edu
[4] Department of Bioengineering, Lehigh University, Bethlehem, PA 18015, USA
* Correspondence: hdo0@lehigh.edu

Received: 15 February 2020; Accepted: 21 March 2020; Published: 24 March 2020

Abstract: The ability to manipulate and control active microparticles is essential for designing microrobots for applications. This paper describes the use of electric and magnetic fields to control the direction and speed of induced-charge electrophoresis (ICEP) driven metallic Janus microrobots. A direct current (DC) magnetic field applied in the direction perpendicular to the electric field maintains the linear movement of particles in a 2D plane. Phoretic force spectroscopy (PFS), a phase-sensitive detection method to detect the motions of phoretic particles, is used to characterize the frequency-dependent phoretic mobility and drag coefficient of the phoretic force. When the electric field is scanned over a frequency range of 1 kHz–1 MHz, the Janus particles exhibit an ICEP direction reversal at a crossover frequency at ~30 kH., Below this crossover frequency, the particle moves in a direction towards the dielectric side of the particle, and above this frequency, the particle moves towards the metallic side. The ICEP phoretic drag coefficient measured by PFS is found to be similar to that of the Stokes drag. Further investigation is required to study microscopic interpretations of the frequency at which ICEP mobility switched signs and the reason why the magnitudes of the forward and reversed modes of ICEP are so different.

Keywords: induced charge electrophoresis (ICEP); Janus particles; optical trapping; phase-sensitive detection; phoretic force spectroscopy; ICEP motility reversal; micro-robotics

---

1. Introduction

Microrobots are considered as a potential future workforce. Some examples of such applications include the use of active Janus particles to enhance optical resolution for measurements of molecular interactions in biological samples [1], enabling the optimization of transportation or navigation [2], self-assembly and formation of microscopic smart materials [3,4], serving as cargo movers for medicine delivery [5], and functioning as micromanipulators or micromixers [6]. Today, robotic devices are made at increasing smaller scales, reaching that of colloidal particles [5]. Among such endeavors are creative efforts to make active colloidal particles that convert energy provided by an external source to kinetic energy in order to move persistently [7,8]. Such active colloids can be powered by a variety of different mechanisms, such as self-thermophoresis [9], chemical decomposition [10], electric fields [11], and magnetic fields [12]. In this work, we apply an AC electric field and use induced-charge-electrophoresis (ICEP) to drive particle motion and address the possibilities of controlling such colloid-based microrobots by their electric frequency response.

Control of active colloids by external fields has been demonstrated in terms of precise alignment [13], directed control of micro motors [6], and cargo transport [3] for potential microrobot applications.

Gangwal et al. [14] used a uniform distributed alternating current (AC) electric field in the kHz frequency range to produce a propulsive metal-dielectric Janus particle via ICEP. A theory for ICEP was first introduced by Squires and Bazant [15]. Based on the theory, an AC electric field could induce different electroosmotic flows around the dielectric surface and around the metallic surface, causing the particle to move. According to the model, the flow around the metallic surface was stronger than the flow around the dielectric surface; the pushing force on the metal surface was higher than the force on the dielectric surface. The net force pushes the particle in the direction from the metal side toward the dielectric side (normal direction), shown in Figure 1.

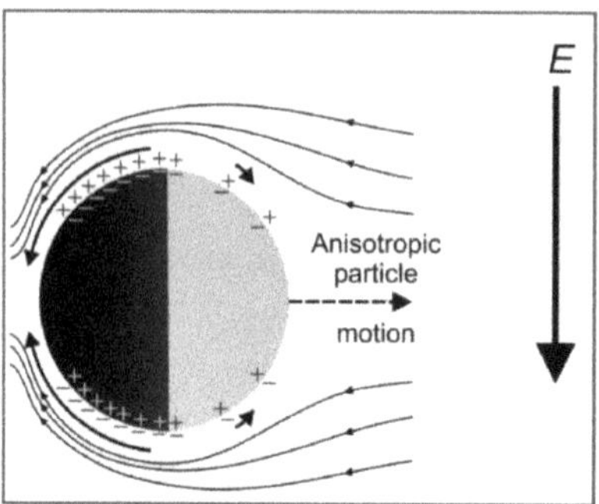

Figure 1. Reproduction of a schematic by Gangwal et al. of a particle in one-half cycle of an AC electric field in the stable configuration. The electric double layer on the gold side (black hemisphere) is more strongly polarized and thus drives a stronger induced-charge-electroosmosis slip (arrows) than the polystyrene side, resulting in induced-charge electrophoresis (ICEP) motion in the direction of the dielectric side. Reprinted with permission [14].

According to the model by Squires and Bazant [15], the ICEP mobility should decay to zero as the frequency of the AC electric field exceeds the characteristic times for the ions in the electric double layers to respond to the applied electric field. In Mano et al.'s experiments [16], however, a phoretic motion reversal was observed during a frequency scan, i.e., the Janus particle was found to be moving in a direction led by the metallic side. One possible explanation for such a mobility reversal was the reversal of induced-charge electroosmosis (ICEO) flow [15], i.e., the external electric field-induced flows of liquid around the two sides of the particles changed direction [17]. In an alternating current electrophoresis (ACEO) setting [18], the additional nonlinear mechanisms including Faradic reactions [19,20] and steric effects [17] due to ion crowding within the electrical double layers (EDLs), were considered to explain the phenomena. In traveling wave electroosmosis (TWEOF), the motility reversal could be expressed as the electrical body force [21,22]. All these explanations neither can be applied to the Janus particle system [21,22] nor need unreasonable ionic density [17,19,20].

A better understanding of the ICEP motility reversal requires better experiments that can determine the crossover frequency and the phoretic drag coefficient accurately. Using traditional DC measurements to determine the crossover frequency is difficult because the Brownian motion often swamps measurements of the null phoretic motion near the crossover frequency [23]. It is impossible to determine the drag coefficient of any phoretic motion by DC measurements, because the phoretic motion, absent of acceleration, is subject to zero net force, as the drag force perfectly opposes the phoretic force [24].

It is, however, possible to determine the drag coefficient by an AC measurement from the phase delay of the particle's phoretic motion relative to that of the harmonically varying electrophoretic force, as demonstrated in earlier work [24]. We have used such a detection method to measure the crossover frequency and phoretic force of dielectrophoresis with high precision [23,25]. Briefly, our detection method, phoretic force spectroscopy (PFS), measures the harmonic response of an optically trapped phoretic particle in an amplitude-modulated AC electric field. A lock-in amplifier analyzes the motion of the particle to yield the particle motion's amplitudes, and phase lags relative to the phase of the harmonic field. The crossover frequency is the frequency at which the phase lag changes by 180 degrees. In this paper, we describe how we use PFS to study the frequency dependence and phoretic drag coefficient of ICEP.

2. Materials and Methods

2.1. Fabrication of Metal-Dielectric Janus Particles

We made metallic Janus particles by depositing a thin film of metal on a monolayer of dry silica particles (SS05N, Bangs Laboratories, Inc., Fishers, IN, USA) [26] with a process shown in Figure 2a. A droplet of 3 micron diameter silica particle suspension was deposited on a glass substrate to create a monolayer by a vibration-assisted convective deposition method. The slide covered with a monolayer of particles was loaded into an e-beam evaporator (Indel E-beam Evaporator, International Delta Systems, LLC, Tucson, AZ, USA) to receive first a coating of 50 nm Ni and then a coating of 10 nm Au on the top half of the particle surface. The slide was then placed in a water-filled centrifuge tube and sonicated in a sonication bath (Branson 1510, Branson Ultrasonics Co., Danbury, CT, USA) for 6 h to release the particles into the water for further treatment. Schematics of the Janus particles and a scanning electron microscope (SEM) (JEOL JXA-8900 EPMA Microprobe, JEOL USA, Inc., Pleasanton, CA, USA) micrograph of the Janus particles are shown in Figure 2b.

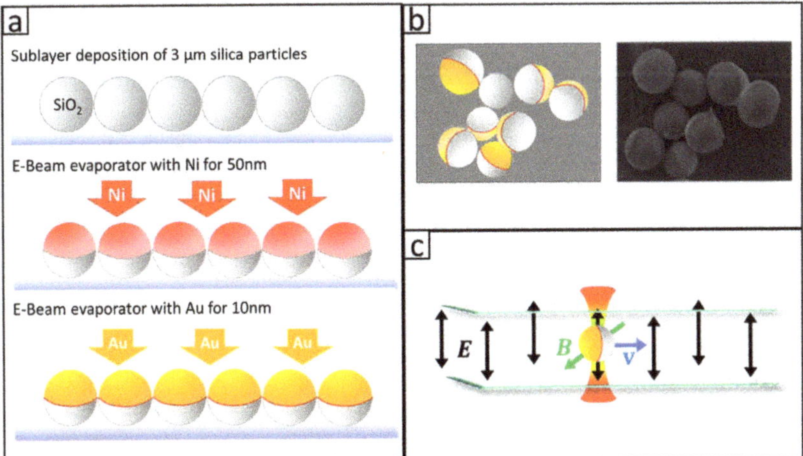

Figure 2. (**a**) Fabrication process of Janus particles. Silica particles were deposited from an aqueous suspension to form a sub-monolayer on a glass substrate, followed by E-beam metal evaporation deposition of Ni and Au. (**b**) On the left is a cartoon showing the shape and geometry of resultant Janus particles, and on the right is a scanning electron microscope (SEM) micrograph of actual particles. (**c**) A depiction of a trapped Janus particle in the sample chamber. The directions of the particle's motion, the E and B fields, are shown relative to the sample chamber.

2.2. Application of AC Electric Field to Drive Janus Particles Based on ICEP

Figure 2c shows the schematic of the sample chamber in which the particle's motions were examined. A pair of indium-tin-oxide (ITO) coated glass slides, separated by a 50 um spacer of polycarbonate tape (3M 980, 3M, St. Paul, MN, USA), were fixed together by wrapping with the same tape. Copper tapes were attached to the ITO coating on each glass slide to form an electrical contact. An AC electric signal was applied across the pair of ITO slides to create a uniform electric field in the vertical direction. In the presence of the AC electric field, the Janus particles were driven into drift motions in the horizontal direction by ICEP. Since Janus particles settled very close to the bottom of the sample chamber, the particles moved in a two dimensional (2D) plane.

2.3. Application of a Magnetic Field to Fix the Direction of the ICEP Driven Phoretic Motion

While the vertically applied electric field exerted a torque on the Janus particles and aligned the metal cap edge parallel to the E field, the polar axis of the particles still had the freedom to orient freely in the plane perpendicular to the E field. Since the metal cap of the Janus particles exhibited magnetic dipole moment in the polar direction, an external magnetic field was applied to align the Janus particle orientation in the 2D plane perpendicular to the E field. The ICEP force was aligned with the polar direction of the Janus particles, and a magnetic field was used to control the direction of particle movement, as shown by Lin et al. [27] and Han et al. [28] to control and manipulate the metallic Janus particles, respectively. In this work, the magnetic field was generated by four 15 mm × 6.5 mm × 3 mm rectangular commercial Neodymium magnets (rectangular magnets, Theodora LLC, Seattle, WA, USA). The four magnets were divided into two groups, with two magnets in each group. Two groups of magnets were placed 10 cm apart and with the S and N poles facing each other in the horizontal plane. The sample chamber was placed at the center of the two groups of magnets. The intensity of the magnetic field at the location of the sample chamber was measured by Gauss meter (SJ200, Guilin Senjie Technology Co., Ltd., Guilin, China) to be 1.0 ± 0.1 mT.

2.4. Position Detection of Phoretic Particles in 2D Using Image Analysis

The particle's motion was measured by video imaging using a 20× objective lens on an inverted microscope (Olympus IX-81, Olympus Corporation, Tokyo, Japan). The videos of the particle movement were recorded by a charge-coupled device (CCD) camera (Sony Sscm256, Sony Corp., Tokyo, Japan) at a frame rate of 1 frame per second for a typical duration of a few minutes. The video recordings were analyzed using a particle tracking program (MOSAIC Suit on ImageJ, Max Planck Institute of Molecular Cell Biology and Genetics (MPI-CBG), Dresden, Germany.) [29] to extract the trajectories of particles in the experimental chamber. Mean-square-displacement (MSD), defined as $< \left| x(t)^2 - x(t + t_0)^2 \right| >$, was calculated by taking the average of the MSD for all particles in the field of view. The experimental MSD vs. time for 3 μm metallic Janus particles at varying applied voltage over a fixed gap distance (50 μm) between the electrodes is shown in Figure 3.

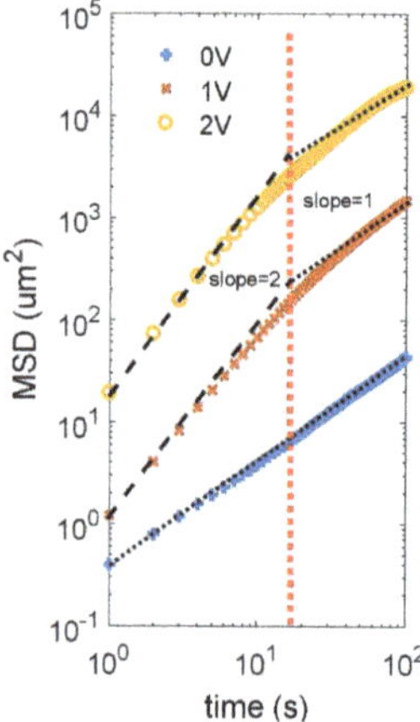

Figure 3. Mean-squared-displacement (MSD) of ICEP-driven metallic Janus particle under 5 kHz AC electric field. The extrapolated dash lines are from ballistic motion (the slope of 2) to the active diffusive motion (the slope of 1). The red vertical dotted line represents the particle's rotational relaxation time.

2.5. Trapping and Manipulation of a Janus Particle

A 1064 nm wavelength laser (Nd:YVO$_4$ 1064nm diode-pumped solid-state CW laser, Spectra-Physics, Santa Clara, CA, USA) was used to create an optical trap, as shown in Figure 4a. The trap confines the Janus particle in a quadratic potential well [30]. The displacement of the particle in a trap, too small for video analysis to pick up, was tracked by a tracking beam (980 nm CL-2000 diode-pumped crystal laser, CrystaLaser, Reno, NV, USA) aligned collinearly with the 1064 nm laser. The tracking beam was received by a quadrant photodiode (QPD) that measured the position of the particle to give 5 nm spatial resolution at 1 kHz sampling rate. We used this setup to measure coefficients of the Stokes' drag [31] An AC voltage was applied using a function generator (Stanford Research DS345, Stanford Research Systems, Sunnyvale, CS, USA) across the electrodes in the experimental chamber. The electric field produced by the electric potential across the electrodes drives ICEP and makes the particles an active Brownian particle. Active driving, trapping, and tracking of a metal-dielectric Janus particle was achieved in the same setup. A magnetic field was added to control the ICEP motion, so the particle moved linearly.

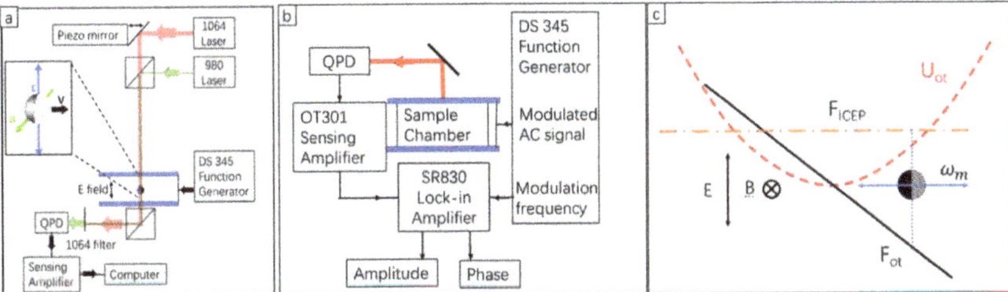

Figure 4. Phoretic force spectroscopy (PFS) set-up and force diagram. (**a**) Lightpath diagram of PFS. A 1064 nm laser was used to trap a Janus particle. A 980 nm tracking laser was co-aligned with a 1064 nm trapping laser. The 980 nm tracking laser was detected by a quadrant photodiode (QPD). A magnet provided an in-plane magnetic field to further align the Janus Particles (insert). (**b**) Signal chain diagram. A lock-in amplifier analyzed the signal of particle position with the modulation signal as a reference. (**c**) Force diagram of an optically trapped Janus particle under a modulated AC electric field. U$_{ot}$ is the potential of a particle in the trap. F$_{ICEP}$ (active force) and F$_{ot}$ (trap force) are in balance at a non-center spot. Electric field (E) is in the z-direction, and the magnetic field is in the y-direction. The orientation of the Janus particle is in the x-direction. The particle oscillates, and the oscillation frequency is the same as the frequency of the modulation signal.

2.6. Phoretic Force Spectroscopy

Phoretic force spectroscopy (PFS) was used to measure the phoretic force of the Janus particles in the electric field [23]. Over a range of frequencies of the ICEP driving field, the motion of a particle in an optical trap was analyzed with an amplitude-modulated AC electric field. By using a lock-in amplifier, we measured the amplitudes and phase lags of an oscillated particle with the modulation signal as a reference. Not only the amplitude of the motion but also the phase relative to the driving force were obtained. The directional change of ICEP was detected when a 180-degree phase change was observed.

The lock-in amplifier (Stanford Research SR830 DSP Dual Phase Lock-In Amplifier, Stanford ResearchSystems, Sunnyvale, CA, USA) required a signal input and a reference input, as shown in Figure 4b. The reference was from the modulation frequency output of the function generator (Stanford Research, DS345, Stanford Research Systems, Sunnyvale, CA, USA). The signal was from the quadrant photodiode (homemade QPD) which measured the position of the particle. Here, the signal was amplified by a sensing amplifier (On-Trak OT301, On-Trak Photonics, Irvine, CA, USA) before it went into the lock-in amplifier. The amplitude (A) and the phase lag (δ) of the particle motion were measured by the lock-in amplifier.

The motion of the particle was analyzed as follows. Janus particles are assumed to have a constant speed, v, in an AC electric field with amplitude E. The speed is proportional to the square of the AC electric field strength, $v = \beta E^2$, where β is ICEP mobility. In the water, the drag force equals the ICEP forces, $F_e = \zeta_{ICEP} v = \zeta_{ICEP}\beta E^2$, where ζ_{ICEP} is the drag coefficient. At any instance, tweezer force, drag force, and resistance are in balance, so the Langevin equation of the position, x, of the particle is:

$$\frac{dx}{dt} + \frac{k_{ot}x}{\zeta_{ICEP}} = \beta E^2 \tag{1}$$

where k_{ot} is the spring constant of the optical trap. In our experiment, the electric field with 30% depth modulation was $E = E_0[0.85 + 0.15\cos(\Omega_M t)]\cos(\Omega_B t)$, where the Ω_M is amplitude modulation frequency. Equation (1) can be rewritten to be

$$\frac{dx}{dt} + \frac{k_{ot}x}{\zeta_{ICEP}} = \beta E_0^2[0.85 + 0.15\cos(\Omega_M t)]^2 \cos(\Omega_B t)^2$$
$$= \beta E_0^2\{0.832 + 0.255\cos(\Omega_M t) + 0.01125\cos(2\Omega_M t)\}\left(\tfrac{1}{2} + \tfrac{1}{2}\cos(2\Omega_B t)\right)$$
$$= \beta E_0^2\{0.416 + 0.1775\cos(\Omega_M t) + 0.005625\cos(2\Omega_M t) + 0.416\cos(2\Omega_B t) + \tag{2}$$
$$\tfrac{1}{2}[0.1775\cos((2\Omega_B - \Omega_M)t) + 0.1775\cos((2\Omega_B + \Omega_M)t)] + \tfrac{1}{2}[0.005625\cos(2(\Omega_B - \Omega_M)t) +$$
$$0.005625\cos(2(\Omega_B + \Omega_M)t)]\}$$

The amplitudes of the high-frequency terms (higher than Ω_B) are small due to water damping [32]. Thus, we can ignore the higher frequency term and only consider the first three terms (in boldface in Equation (2)). The steady-state solution for Equation (2), $x(t)$ has a term of a DC offset, a first harmonic term and a second harmonic term, each with amplitude $A(\Omega_M)$, and $A\prime(2\Omega_M)$ and phase delay $\delta(\Omega_M)$ and $\delta\prime(2\Omega_M)$, respectively, i.e.,

$$x(t) = x_{DC\,offset} + A(\Omega_M)e^{i(\Omega_M t - \delta)} + A'(2\Omega_M)e^{i(2\Omega_M t - \delta')} \tag{3}$$

The frequency-dependent amplitudes A, A' and the phase delays δ, $\delta\prime$ are

$$A(\Omega) = \frac{0.255\beta E_0^2\zeta_{ICEP}}{\sqrt{k_{ot}^2 + (\zeta_{ICEP}\Omega_M)^2}} \tag{4}$$

$$A'(2\Omega) = \frac{0.01125\beta E_0^2\zeta_{ICEP}}{\sqrt{k_{ot}^2 + (2\zeta_{ICEP}\Omega_M)^2}} \tag{5}$$

$$\delta(\Omega) = \tan^{-1}\frac{\zeta_{ICEP}\Omega_M}{k_{ot}} \tag{6}$$

$$\delta'(2\Omega) = \tan^{-1}\frac{2\zeta_{ICEP}\Omega_M}{k_{ot}} \tag{7}$$

Since the second harmonic ($2\Omega_M$) terms have a much smaller amplitude than that of the fundamental frequency terms we use the Ω_M term for further calculation. The two parameters of interest of this study, the mobility β and the phoretic coefficient ζ_{ICEP}, can be determined in terms of the experimental measurables, $A(\Omega_M)$, $\delta(\Omega_M)$ and k_{ot}. According to Equations (4) and (6) above, we have the following relationships:

$$\beta = \frac{A(\Omega_M)\Omega_M}{0.255 E_0^2 \sin\delta(\Omega_M)} \tag{8}$$

$$\zeta_{ICEP} = \frac{k_{ot}}{\Omega_M} tan\delta(\Omega_M) \tag{9}$$

Although the $A(\Omega_M)$ and $\delta(\Omega_M)$ are a function of modulation frequency, according to our model, the β *and* ζ_{ICEP} should be independent of the modulation frequency. The β *and* ζ_{ICEP} were tested to be the same with modulation frequency from 1 Hz to 3 Hz. However, the $A(\Omega_M)$ and $\delta(\Omega_M)$ decreased to about zero when Ω_M was larger than 3 Hz. Thus, we set the modulation frequency Ω_M to be 1 Hz in our experiment for the best signal and use a lock-in amplifier to measure $A(\Omega_M)$ and $\delta(\Omega_M)$ from which we calculated the mobility β and the phoretic coefficient ζ_{ICEP} of ICEP-driven metallic Janus particles

The force diagram of a metallic Janus particle under the modulated AC electric field was shown in Figure 4b. The optical force, F_{ot}, was generated from the optical trap. The ICEP force was generated from the AC electric field. Since the electric field and magnetic field were applied to fix the orientation of the Janus particle, the direction of the ICEP force was fixed. The average ICEP force was non-zero due to the 30% amplitude modulation. The particle oscillated around an offset that was predicted by Equation (3). The 30% amplitude modulation generated an offset which became the balance point between the force from the trap and the force from ICEP.

We conducted an independent experiment (passive microrheology [33]) to measure the trap stiffness k_{ot} using the same optically trapped particle without the presence of the AC electric field. According to the equal partition relationship, we have

$$\frac{1}{2}k_{ot}x^2 = \frac{1}{2}k_BT \tag{10}$$

where $\langle x^2 \rangle$ is the mean-squared displacement of the particle in the trap, k_B the Boltzmann constant, and T the absolute temperature. We determined k_{ot} by measuring the mean-squared displacement $\langle x^2 \rangle$.

2.7. Stokes' Drag Coefficient of the Particle near the Bottom of the ITO Glass Chamber

We used the active microrheology to determined the Stokes' drag coefficient, ζ_{Stokes} [31]. We applied an oscillatory optical force to a trapped bead and analyzed the oscillatory motion of the particle to determine the Stokes' drag coefficient of the particle in 2D near the bottom of the glass chamber. The trapped bead was forced to oscillate along the x-direction by the oscillating tweezers driven by the piezoelectric lead zirconate titanate (PZT)-controlled mirror. The equation of motion of the trapped bead can be written as

$$\zeta_{Stokes}\frac{dx}{dt} = k_{ot}[A_0 \cos(\omega t) - x] \tag{11}$$

where A_0 is the maximum displacement of the trap and ω is the angular velocity of the oscillation. The steady-state solution of the motion is:

$$x = D(\omega) \cos(\omega t + \delta'') \tag{12}$$

where $D(\omega)$ is the amplitude of displacement and δ'' is the phase lag. Both values were from the lock-in amplifier. The frequency-dependent amplitudes D and the phase delays δ'' are

$$D(\omega) = \frac{A_0 k_{ot}}{\sqrt{k_{ot}^2 + (\zeta_{Stokes}\Omega)^2}} \tag{13}$$

$$\delta''(\omega) = tan^{-1}\frac{\zeta_{Stokes}\Omega}{k_{ot}} \tag{14}$$

Thus,

$$\frac{\zeta_{Stokes}\omega}{k_{ot}} = tan\delta\prime \tag{15}$$

Since we already know k_{ot} from above, we can determine the Stokes drag coefficient ζ_{Stokes} of the Janus particle near the bottom of the glass plate.

We estimate the distance between the particle bottom and the ITO surface with the Faxen equation. Faxen law describes the drag coefficient change near a flat surface due to the non-slipping boundary condition, which can be written as [34]

$$\frac{\zeta_{Stokes}}{6\pi\eta r} = \frac{1}{1 - \frac{9}{16}\frac{r}{h} + \frac{1}{8}\left(\frac{r}{h}\right)^3 - \frac{45}{256}\left(\frac{r}{h}\right)^4 - \frac{1}{16}\left(\frac{r}{h}\right)^5} \tag{16}$$

where η is the viscosity of water, r is the radius of Janus particle and h is the distance between the center of particle and the surface. Since we already know ζ_{Stokes}, we can determine the gap $(h - r)$ between the particle bottom and the ITO surface.

3. Results and Discussions

3.1. ICEP Movement of an Unconfined Janus Particle in 2D

Due to the high-mass density and micron sizes, these Janus particles tended to sediment to the bottom of the sample chamber. These particles did not stick to the substrate presumably due to Coulomb repulsion between the same negatively charged surface of the indium–tin–oxide (ITO) coated glass slide and that of the Janus particle surfaces. The electric repulsion was sufficiently strong to elevate the particles at a small distance above the bottom surface. We estimated the electrical repulsive force using Derjaguin–Landau–Verwey–Overbeek (DLVO) theory [35] and Derjaguin and superposition approximations [36],

$$F = 4\pi\varepsilon\kappa r\phi_{Janus}^{eff}\phi_{ITO}^{eff}exp(\kappa z) \tag{17}$$

where ε is the permittivity of water, κ is one over the Debye screening length, r is the radius of Janus particle, ϕ_{Janus}^{eff} and ϕ_{ITO}^{eff} are the effective potential of the Janus particle and ITO surface, respectively; z is the distance between the bottom surface of the Janus particle and the ITO surface, and

$$\phi_{Janus,\ ITO}^{eff} = \left(\frac{4k_B T}{e}\right)\tan h\left(\frac{e\phi_{Janus,ITO}}{4k_B T}\right) \tag{18}$$

We measured the zeta potential of the Janus particle (ϕ_{Janus}) to be -17.01 ± 1.13 mV (ZetaPlus Brookhaven instrument). We used a published value for the zeta potential of ITO surface (ϕ_{ITO}) in DI water to be -32.7 ± 0.2 mW [37] to calculate effective potential [36]. We used the conductance of DI water 167 ms (ZetaPlus, Brookhaven Instruments Corp., Holtsville, NY, USA) to determine the Debye screening length to be 5.55 nm. The permittivity of water was $80.2\varepsilon_0$ at 25 °C, where ε_0 is the permittivity of vacuum [38]. The gravitational force of 3 μm Janus particle was 0.23 pN, calculated with $\Delta\rho Vg$, where $\Delta\rho$ is the mass density difference of silica (2.65 g/cm^3) [39] and water (1.00 g/cm^3), V was the volume of particle, and g was gravity acceleration. Thus, the gap distance between the particle's bottom surface and the ITO surface was 47 ± 8 nm, which was the balance point of the particle at which the electric repulsive force and gravitational force were equal to each other.

In the absence of an electric field, Janus particles moved as Brownian particles in a 2-dimensional (2D) plane defined by the glass substrate. Video analysis of the particles' motion in 2D yielded a 2D mean-squared-displacement (MSD) shown in Figure 3. The MSD curve in the absence of the applied electric field was characteristic of a Brownian particle. Using $D_t = \frac{d(MSD)}{4dt}$ and the measured MSD curve, we determined the diffusivity to be 0.053 ± 0.008 μm^2/s in the absence of the applied electric field. With this diffusivity, we calculated the drag coefficient using the Stokes-Einstein relation [40]. The drag coefficient was determined to be $7.86 \pm 1.37 \times 10^{-8}$ Ns/m. Using Faxen's law [34], we determined the distance between the particle's bottom surface and the ITO surface to be 12.8 ± 0.2 *nm*; the gap size was about 4 times smaller compared to the distance estimated by the balance between the electrostatic repulsion and the weight of the particle. The discrepancy was probably caused by the overestimation of the zeta potentials of the particle surface, and the ITO surface since during the expriment, the pH value of the water could be ~5.6 instead of 7.0, which would reduce the thickness of the EDL of both surfaces.

When an AC electric field was applied, the Janus particles underwent induced-charge dielectrophoresis (ICEP). Modeling the ICEP driven particle as an active swimmer, its MSD vs. time curve was predicted by Marchetti [41] as Equation (19).

$$MSD = 4D_t t + 2v_0^2 \tau_r \left[t - \tau_r \left(1 - e^{-\frac{t}{\tau_r}} \right) \right] \tag{19}$$

According to the equation, a log-log plot of the MSD vs. time curve would yield the active ICEP speed from the intercept of the portion where the slope is 2. Equation (19) could be deducted to $MSD = v_0^2 t^2$, during the time interval between τ_r and $v_0^2/4D_t$. This deduction works when the $v_0^2/4D_t$ was larger than τ_r. The vertical axis intercept of the ballistic motion section of the curve was $2ln(v_0)$, and v_0 was calculated from the experimental intercept value.

We examined the size-dependent speed of the Janus particles under the same experimental conditions. The relationship between ICEP active velocity and applied voltage at 5 kHz was shown in Figure 5a. The speed of 3 μm, 10 μm, and 20 μm of metallic Janus particles were nearly proportional to the square of applied voltage as predicted theoretically by Squires and Bazant [15] for normal ICEP at low frequency. The metallic Janus particle driven by ICEP with movement towards the dielectric side had a speed of:

$$v(ICEP) = \frac{9}{64} \frac{\epsilon a}{\eta(1+\delta)} E^2 \tag{20}$$

where ϵ is the electric permittivity, η is the viscosity of the bulk solvent, a is the radius of the particle, E is the electric field, and δ is the ratio between the capacitances of the compact and diffuse layer in the electric double layer. With a uniform thickness in the experimental cell, the ICEP velocity should be proportional to the square of the voltage according to this theory, which agreed with Figure 5a, as well as Gangwal et al.'s experiment data [14]. The size dependence, which was predicted to be linear with particle diameter, however, was not in agreement with our experimental results. In this work, measured $v(ICEP)$ ratios of 3:3.9:6.6, shown in the Figure 5a insert, were lower than the theoretical prediction, which was 3:10:20. It is possible that the discrepancy in size dependence for the largest particles of the group might be due to increased hydrodynamic drag between the bottom of the particle and the glass substrate since gap sizes decrease as the particle sizes increase in our experiment. It is also possible that the discrepancies might be caused by the fact that the parameters δ in these three cases were different.

The frequency-dependent ICEP velocity was measured and shown in Figure 5b. There was a characteristic frequency f_f at which the ICEP velocity was at the maximum in the normal direction (from the metal side toward the dielectric side). As the frequency increases, we observed that the velocity first decreases gradually to zero, then the direction of motion reverses at a frequency f_c, the crossover frequency. The negative velocity (the reverse direction) increased as the frequency continuously increased until a peak was reached at a frequency f_r, after which the ICEP velocity reaches zero at a very high frequency.

According to Squires et al. [15], the ICEP movement is generated by slip flow around the particle which appears in a certain range of the driving frequency, $\omega_e \leq \omega \leq \omega_p$. The lower bound ω_e is the minimium charging frequency of the electrode. Here, $\omega_e = \frac{D}{\lambda L}$ and D is the ion diffusion coefficient, λ is the Debye length, and $2L$ is the electrode separation. The upper bound, ω_p is the maximum formation frequency of induced ionic cloud screening on the particle, with $\omega_p = \frac{D}{\lambda R}$, and R is the particle radius. For our experiment the $\omega_e = 1$ kHz, and for the 3 μm particle, $\omega_p = 30$ kHz. In Figure 5b, our experiment shows there was a peak for 3 μm particles in velocity between 1 kHz–5 kHz. The larger 10 μm and 20 μm particles archived maximum speed at even lower frequencies. The frequency range predicted by Squires does include these frequencies where maximum speeds occurred in our experiments. When testing the frequency response of 10 μm and 20 μm diameter particles at frequencies below 1 kHz, Joule heating was found to not be negligible; water vapor bubbles were observed to form at these low frequencies. For frequencies in the range of 100 kHz to 1 MHz, reversed ICEP was observed for the 3 μm particle, where the particle moved towards the metal side. The reversed ICEP had a much lower speed at the same applied voltage (2 V). The 3 μm Janus particle seemed to have a higher reversed ICEP speed in comparison to 10 μm and 20 μm particles. The crossover frequencies for these particles were difficult to determine by video imaging methods because video analysis was not sensitive enough for particles moving at low speeds and when displacements were small.

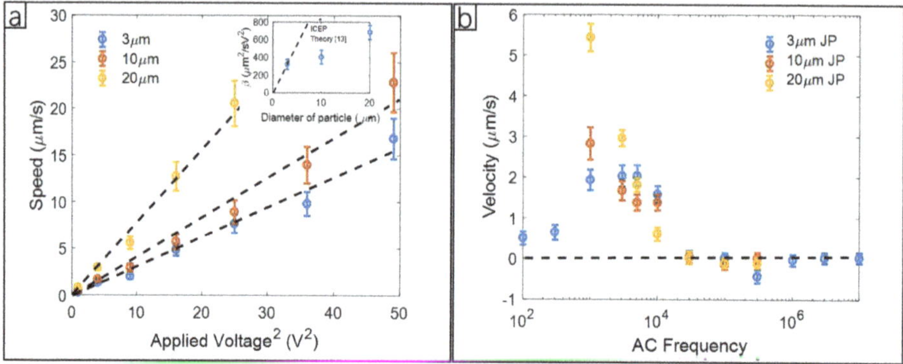

Figure 5. ICEP of a Janus particle without confinement. (**a**) ICEP speed of 3 μm, 10 μm, and 20 μm Janus particles under different applied voltage at 5 kHz. Insert: β vs. the diameter of the particle. The dashed line is the prediction using Equation (20). (**b**) The velocity of 3 μm, 10 μm, and 20 μm diameter particles under the potential difference 2 V over a 50 μm gap between the electrodes and with different frequencies from 100 Hz to 1 MHz. The positive velocity (normal direction) is defined as moving toward the dielectric side.

3.2. ICEP Movements in 1D of a Phoretic Particle Confined in a Quadratic Potential

The use of PFS to detect the phoretic motion requires the orientation of the Janus particle to be fixed by a magnetic field, and the amplitude of the AC electric field be modulated. Two permanent magnets applied the magnetic field in a direction perpendicular to the vertical AC electric field. Janus particles with a ferromagnetic Ni layer in the metal cap were used to align the direction of the particle's largest magnetic moment to the B field. The magnetic field fixed the direction of one axis, and the AC electric field fixed the direction of the other axis of the magnetic Janus particle, thus maintaining the orientation of particles in-plane. With the application of the AC electric field and the DC magnetic fields, the orientation of the particle was fixed in a direction perpendicular to both fields.

We used PFS to determine the crossover frequency and phoretic drag coefficient of the phoretic Janus particle which moved in a linear motion. According to Equation (3), particle motion had two harmonic components and an offset. The lock-in amplifier calculated the amplitude and the phase lag of the particle motion with a frequency component in the first harmonic of the amplitude modulation (AM) frequency. Using PFS, the crossover frequency was measured at the point when the phase lag signal shifted 180 degrees. Using Equations (8) and (9), we calculated the drag coefficient and mobility of the ICEP-driven motion under a range of base wave frequencies (1 kHz–1 MHz).

Figure 6a shows a plot of the phase shift vs. frequency. A 180-degree phase shift between 20 kHz to 30 kHz indicated the normal direction of motion changed into the reverse direction; the frequency at which the 180-degree phase shift was identified as the crossover frequency. Figure 6b shows the speed of the Janus particle as calculated by $v = \Omega_M A(\Omega_M)$, i.e., the maximum speed during the oscillating movement under amplitude-modulated ICEP. At frequencies lower than the crossover frequency (~27 kHz), the particle moved towards the dielectric side, defined as the positive velocity (normal direction). At frequencies higher than the crossover frequency of 27 kHz, the particle moved with the metal side in the front. It is informative to compare the crossover frequency of the 3 μm Janus particle with other experiments. Our crossover frequency for the 3 μm Janus particle was 30 kHz. Mano et al.'s experiment on a 3 μm rotational Janus particle showed a cross-over frequency at 22 kHz [16]. Suzuki et al.'s work, also using a 3 μm Janus particle, showed a cross-over frequency at around 30 kHz [42]. The reversal of ICEO often showed smaller crossover frequencies of around 5 kHz [17]. The frequency scales, in the range 5 kHz to 30 kHz, revealed a length scale of 0.2~0.6 μm using the length scale defined by $L = \sqrt{D_{ion}/f_c}$. The ionic diffusivity, D_{ion}, of K^+ or Cl^- is about 2×10^{-9} m^2/s [39]. The 30-microsecond time scale revealed by the crossover frequency was on the same order of magnitude of the diffusion time for ions to diffuse the perimeter of the Janus particle, suggesting that the relaxation of the electrokinetic flows in the electric double layer must play a significant role.

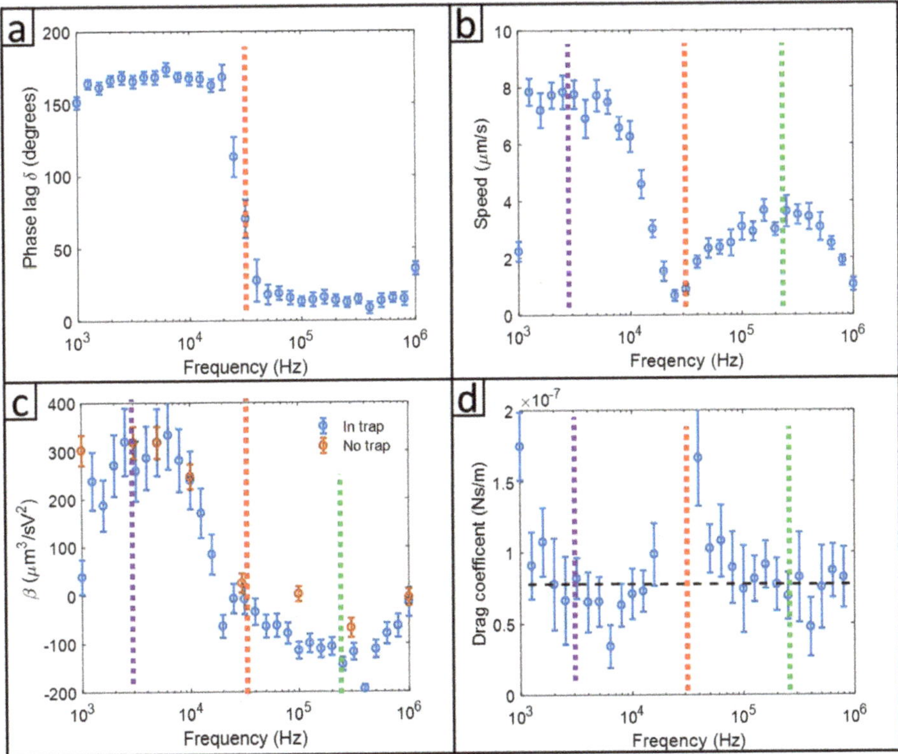

Figure 6. Experimental results of PFS at 0.1 V/μm of the carrier frequency of the electric field from 1 kHz to 1 MHz with 1 Hz amplitude modulation frequency at 30% modulation amplitude. The vertical dash rad lines (f_c) represent the crossover frequency at 27 kHz. The vertical dash green line (f_r) represents the maximum reversal velocity at 220 kHz. The vertical dash purple line $\left(f_f\right)$ represents the maximum forward velocity at 1–5 kHz. (**a**) Phase delays of a 3 μm Janus particle. (**b**) Speeds of the same 3 μm Janus particle. (**c**) Comparison of the Beta values of the Janus particles measured with trapping (Equation (3)) and without trapping. (**d**) Drag coefficients determined from the phase delay data.

Frequency-dependent ICEP mobility β calculated according to Equations (8) and (9), is shown in Figure 6c. The frequency when the speed approached zero in Figure 6c is in agreement with the frequency at which the phase delay changed by 180 degrees. The frequency at which the reversed ICEP speed occurred was the same as that observed in the frequency-dependent β or the frequency-dependent speed. The β vs. frequency curve looked different from that of the speed vs. frequency because of the $\sin \delta$ term in the denominator of the right side of Equation (8). Using amplitude-modulated ICEP, the oscillation of the driving force had a phase lag (δ) relative to the particle motion. The in-phase term of the driving force was $0.1275E_0^2 \sin\delta$, which was used to calculate the β. The out-off-phase term of the driving force was $0.1275E_0^2 \cos\delta$, which was used to balance the optical trap force. Mobility β obtained by the AM modulated, and the phoretic force spectroscopy agreed with the value of the 2D measurements for particles not confined by the optical trapping. The maximum speed in the normal direction was at a frequency of 1 kHz–5 kHz, and the peak frequency for the reverse direction was around 200 kHz. We don't know exactly why the reversed ICEP had much slower mobility than that of the forward motion. A few papers [14–20] that mentioned the frequency-dependent reversal of electroosmotic flows did not propose mechanisms that would explain the reversal.

We determined the phoretic drag coefficient ξ according to Equation (9), shown in Figure 6d. The drag coefficients were found to be independent of frequency except a few outliers obtained when the particle had very low moving speeds (near crossover frequency) with large error bars. Taking the average over all the frequency, we determined the ICEP drag to be $7.5 \pm 2.7 \times 10^{-8}$ N·s/m. The Stokes drag coefficient, measured by oscillating tweezers [29,41] for the same Janus particle in the absence of ICEP, was $7.9 \pm 1.4 \times 10^{-8}$ N·s/m, similar to the ICEP drag coefficient.

4. Conclusions

This paper reports a study that used phoretic force spectroscopy (PFS) to determine the crossover frequency, the maximum phoretic speeds in the normal and the reverse direction, ICEP mobility, and the ICEP drag coefficient. The direction of the ICEP-driven metallic-dielectric Janus particles was in the normal direction at low frequencies and the reversed direction at high frequencies. Phoretic mobility was proportional to the square of an applied electric field. Our data are in agreement with prior experimental studies. However, our PFS detection method provided better accuracy in determining the crossover frequencies.

The 30-microsecond time scale revealed by the crossover frequency is on the same order of magnitude as the diffusion time for ions to diffuse the perimeter of the Janus particle, suggesting the relaxation of the electrokinetic flows in the electric double layer must play a role. This study found the ICEP drag coefficients of the forward and the reversed motion were similar, and both are comparable to the Stokes drag. Unfortunately, large error bars in our data did not permit more detailed analysis to distinguish the ICEP drag from the Stokes drag. Microscopic interpretations of both the frequency at which ICEP mobility switched signs could benefit from the accurately determined crossover frequencies. The reason for the large difference between the magnitudes of ICEP mobility in the forward and reversed modes of ICEP requires further study.

With the demonstrated ability to determine the high-resolution frequency-dependent response function of the metallic Janus particle by a phase-sensitive detection method, this study can help broaden the application areas and enhance the performance of current applications of colloidal microrobots. For example, a more accurately determined crossover frequency can provide more precise sorting or separation of particles by exposing unsorted particles to judiciously controlled frequencies in a microfluidic channel setting. Metallic Janus particles coated with layers of different magnetic properties exposed in cleverly arranged magnetic fields could be used to transport different kinds of medicines to different locations. Combining the use of electric fields with different frequencies and/or an added magnetic field might also be useful to control the self-assembly of active colloids to achieve complex and time-varying microstructures.

Author Contributions: Conceptualization, C.S. and H.D.O.; methodology, C.S., J.F.G. and H.D.O.; software, C.S.; data curation, C.S., Z.J. and L.L.; writing—original draft preparation, C.S., Z.J. and H.D.O.-Y.; writing—review and editing, C.S., L.L., Z.J., J.F.G. and H.D.O.-Y.; supervision, H.D.O.-Y.; funding acquisition, H.D.O.-Y. All authors have read and agreed to the published version of the manuscript.

Funding: This research received no external funding.

Acknowledgments: This work was supported in part by the Emulsion Polymers Institute and the Department of Physics of Lehigh University. The authors thank Thitiporn Kaewpetch of the Department of Chemical and Biomolecular Engineering for assistance in preparing the monolayer particle arrays and Theadosia Kurniawan for critical reading of the manuscript.

Conflicts of Interest: The authors declare no conflict of interest.

References

1. Choi, J.; Zhao, Y.; Zhang, D.; Chien, S.; Lo, Y.H. Patterned Fluorescent Particles as Nanoprobes for the Investigation of Molecular Interactions. *Nano Lett.* **2003**, *3*, 995–1000. [CrossRef]

2. Yang, Y.; Bevan, M.A. Optimal Navigation of Self-Propelled Colloids. *ACS Nano* **2018**, *12*, 10712–10724. [CrossRef]

3. Solovev, A.A.; Sanchez, S.; Pumera, M.; Mei, Y.F.; Schmidt, O.C. Magnetic Control of Tubular Catalytic Microbots for the Transport, Assembly, and Delivery of Micro-Objects. *Adv. Funct. Mater.* **2010**, *20*, 2430–2435. [CrossRef]

4. Walther, A.; Müller, A.H.E. Janus Particles: Synthesis, Self-Assembly, Physical Properties, and Applications. *Chem. Rev.* **2013**, *113*, 5194–5261. [CrossRef] [PubMed]

5. Bogue, R. The Development of Medical Microrobots: A Review of Progress. *Ind. Rob.* **2008**, *35*, 294–299. [CrossRef]

6. Shields, C.W.; Velev, O.D. The Evolution of Active Particles: Toward Externally Powered Self-Propelling and Self-Reconfiguring Particle Systems. *Chem* **2017**, *3*, 539–559. [CrossRef]

7. Boymelgreen, A.M.; Balli, T.; Miloh, T.; Yossifon, G. Active Colloids as Mobile Microelectrodes for Unified Label-Free Selective Cargo Transport. *Nat. Commun.* **2018**, *9*, 1–8. [CrossRef]

8. Ebbens, S.J. Active Colloids: Progress and Challenges towards Realising Autonomous Applications. *Curr. Opin. Colloid Interface Sci.* **2016**, *21*, 14–23. [CrossRef]

9. Jiang, H.R.; Yoshinaga, N.; Sano, M. Active Motion of a Janus Particle by Self-Thermophoresis in a Defocused Laser Beam. *Phys. Rev. Lett.* **2010**, *105*, 268302. [CrossRef]

10. Uspal, W.E.; Popescu, M.N.; Dietrich, S.; Tasinkevych, M. Self-Propulsion of a Catalytically Active Particle near a Planar Wall: From Reflection to Sliding and Hovering. *Soft Matter* **2015**, *11*, 434–438. [CrossRef]

11. Kilic, M.S.; Bazant, M.Z. Induced-Charge Electrophoresis near a Wall. *Electrophoresis* **2011**, *32*, 614–628. [CrossRef] [PubMed]

12. Sanchez, S.; Solovev, A.A.; Harazim, S.M.; Schmidt, O.G. Microbots Swimming in the Flowing Streams of Microfluidic Channels. *J. Am. Chem. Soc.* **2011**, *133*, 701–703. [CrossRef] [PubMed]

13. Zhang, L.; Zhu, Y. Directed Assembly of Janus Particles under High Frequency Ac-Electric Fields: Effects of Medium Conductivity and Colloidal Surface Chemistry. *Langmuir* **2012**, *28*, 13201–13207. [CrossRef] [PubMed]

14. Gangwal, S.; Cayre, O.J.; Bazant, M.Z.; Velev, O.D. Induced-Charge Electrophoresis of Metallodielectric Particles. *Phys. Rev. Lett.* **2008**, *100*, 058302. [CrossRef]

15. Squires, T.M.; Bazant, M.Z. Breaking Symmetries in Induced-Charge Electro-Osmosis and Electrophoresis. *J. Fluid Mech.* **2006**, *560*, 65–101. [CrossRef]

16. Mano, T.; Delfau, J.B.; Iwasawa, J.; Sano, M. Optimal Run-And-Tumble-Based Transportation of a Janus Particle with Active Steering. *Proc. Natl. Acad. Sci. USA* **2017**, *114*, E2580–E2589. [CrossRef]

17. Storey, B.D.; Edwards, L.R.; Kilic, M.S.; Bazant, M.Z. Steric Effects on Ac Electro-Osmosis in Dilute Electrolytes. *Phys. Rev. E* **2008**, *77*, 036317. [CrossRef]

18. Bazant, M.Z.; Kilic, M.S.; Storey, B.D.; Ajdari, A. Towards an Understanding of Induced-Charge Electrokinetics at Large Applied Voltages in Concentrated Solutions. *Adv. Colloid Interface Sci.* **2009**, *152*, 48–88. [CrossRef]

19. Ng, W.Y.; Lam, Y.C.; Rodríguez, I. Experimental Verification of Faradaic Charging in Ac Electrokinetics. *Biomicrofluidics* **2009**, *3*, 022405. [CrossRef]

20. González, A.; Ramos, A.; García-Sánchez, P.; Castellanos, A. Effect of the Combined Action of Faradaic Currents and Mobility Differences in Ac Electro-Osmosis. *Phys. Rev. E.* **2010**, *81*, 016320. [CrossRef]
21. Ramos, A.; Morgan, H.; Green, N.G.; González, A.; Castellanos, A. Pumping of Liquids with Traveling-Wave Electroosmosis. *J. Appl. Phys.* **2005**, *97*, 084906. [CrossRef]
22. Yang, H.; Jiang, H.; Shang, D.; Ramos, A.; Garcia-Sanchez, P. Experiments on Traveling-Wave Electroosmosis: Efect of Electrolyte Conductivity. *IEEE Trans. Dielectr. Electr. Insul.* **2009**, *16*, 417–423. [CrossRef]
23. Wei, M.T.; Junio, J.; Ou-Yang, D.H. Direct Measurements of the Frequency-Dependent Dielectrophoresis Force. *Biomicrofluidics* **2009**, *3*, 012003. [CrossRef] [PubMed]
24. Liao, M.-J.; Wei, M.-T.; Xu, S.-X.; Daniel Ou-Yang, H.; Sheng, P. Chinese Physics B Non-Stokes Drag Coefficient in Single-Particle Electrophoresis: New Insights on a Classical Problem. *Chin. Phys. B* **2019**, *28*, 084701. [CrossRef]
25. Park, H.; Wei, M.T.; Ou-Yang, H.D. Dielectrophoresis Force Spectroscopy for Colloidal Clusters. *Electrophoresis* **2012**, *33*, 2491–2497. [CrossRef] [PubMed]
26. Muangnapoh, T.; Weldon, A.L.; Gilchrist, J.F. Enhanced Colloidal Monolayer Assembly via Vibration-Assisted Convective Deposition. *Appl. Phys. Lett.* **2013**, *103*, 181603. [CrossRef]
27. Lin, C.H.; Chen, Y.L.; Jiang, H.R. Orientation-Dependent Induced-Charge Electrophoresis of Magnetic Metal-Coated Janus Particles with Different Coating Thicknesses. *RSC Adv.* **2017**, *7*, 46118–46123. [CrossRef]
28. Han, M.; Yan, J.; Granick, S.; Luijten, E. Effective Temperature Concept Evaluated in an Active Colloid Mixture. *Proc. Natl. Acad. Sci. USA* **2017**, *114*, 7513–7518. [CrossRef]
29. Sbalzarini, I.F.; Koumoutsakos, P. Feature Point Tracking and Trajectory Analysis for Video Imaging in Cell Biology. *J. Struct. Biol.* **2005**, *151*, 182–195. [CrossRef]
30. Shen, C.; Ou-Yang, H.-C.D. The Far-from- Equilibrium Fluctuation of an Active Brownian Particle in an Optical Trap. In *Optical Trapping and Optical Micromanipulation XVI, Proceedings of the SPIE Nanoscience + Engineering, San Diego, CA, USA, 11–15 August 2019*; Dholakia, K., Spalding, G.C., Eds.; SPIE: Bellingham, WA, USA, 2019; Volume 11083, p. 63. [CrossRef]
31. Valentine, M.T.; Dewalt, L.E.; Ou-Yang, H.D. Forces on a Colloidal Particle in a Polymer Solution: A Study Using Optical Tweezers. *J. Phys. Condens. Matter* **1996**, *8*, 9477–9482. [CrossRef]
32. Wang, J.; Wei, M.T.; Cohen, J.A.; Ou-Yang, H.D. Mapping Alternating Current Electroosmotic Flow at the Dielectrophoresis Crossover Frequency of a Colloidal Probe. *Electrophoresis* **2013**, *34*, 1915–1921. [CrossRef] [PubMed]
33. Zia, R.N. Active and Passive Microrheology: Theory and Simulation. *Annu. Rev. Fluid Mech.* **2018**, *50*, 371–405. [CrossRef]
34. Ha, C.; Ou-Yang, H.D.; Pak, H.K. Direct Measurements of Colloidal Hydrodynamics near Flat Boundaries Using Oscillating Optical Tweezers. *Phys. A Stat. Mech. Appl.* **2013**, *392*, 3497–3504. [CrossRef]
35. BUTLER, J.A.V. Theory of the Stability of Lyophobic Colloids. *Nature* **1948**, *162*, 315–316. [CrossRef]
36. Behrens, S.H.; Plewa, J.; Grier, D.G. Measuring a Colloidal Particle's Interaction with a Flat Surface under Nonequilibrium Conditions Total Internal Reflection Microscopy with Absolute Position Information. *Eur. Phys. J. E* **2003**, *10*, 115–121. [CrossRef]
37. Kline, T.R.; Chen, G.; Walker, S.L. Colloidal Deposition on Remotely Controlled Charged Micropatterned Surfaces in a Parallel-Plate Flow Chamber. *Langmuir* **2008**, *24*, 9381–9385. [CrossRef]
38. Kaatze, U. Complex Permittivity of Water as a Function of Frequency and Temperature. *J. Chem. Eng. Data* **1989**, *34*, 371–374. [CrossRef]
39. Lide, D.R. *CRC Handbook of Chemistry and Physics*; CRC Press: Boca Raton, FL, USA, 2004; ISBN 0-8493-0487-3.
40. Kholodenko, A.L.; Douglas, J.F. Generalized Stokes-Einstein Equation for Spherical Particle Suspensions. *Phys. Rev. E* **1995**, *51*, 1081–1090. [CrossRef]
41. Marchetti, M.C.; Fily, Y.; Henkes, S.; Patch, A.; Yllanes, D. Minimal Model of Active Colloids Highlights the Role of Mechanical Interactions in Controlling the Emergent Behavior of Active Matter. *Curr. Opin. Colloid Interface Sci.* **2016**, *21*, 34–43. [CrossRef]
42. Suzuki, R.; Jiang, H.-R.; Sano, M. Validity of Fluctuation Theorem on Self-Propelling Particles. *arXiv* **2011**, arXiv:1104.5607.

 *micromachines*

Article

Selective Retrieval of Individual Cells from Microfluidic Arrays Combining Dielectrophoretic Force and Directed Hydrodynamic Flow

Pierre-Emmanuel Thiriet [1],*, **Joern Pezoldt** [2], **Gabriele Gambardella** [1], **Kevin Keim** [1], **Bart Deplancke** [2] and **Carlotta Guiducci** [1]

1 Laboratory of Life Sciences Electronics, École Polytechnique Fédérale de Lausanne, 1015 Lausanne, CH, Switzerland; gabriele.gambardella@epfl.ch (G.G.); kevin.keim@epfl.ch (K.K.); carlotta.guiducci@epfl.ch (C.G.)
2 Laboratory of Systems Biology and Genetics, École Polytechnique Fédérale de Lausanne, 1015 Lausanne, CH, Switzerland; jorn.pezoldt@epfl.ch (J.P.); bart.deplancke@epfl.ch (B.D.)
* Correspondence: pierre-emmanuel.thiriet@epfl.ch; Tel.: +41-216-931-345

Received: 14 February 2020; Accepted: 19 March 2020; Published: 20 March 2020

Abstract: Hydrodynamic-based microfluidic platforms enable single-cell arraying and analysis over time. Despite the advantages of established microfluidic systems, long-term analysis and proliferation of cells selected in such devices require off-chip recovery of cells as well as an investigation of on-chip analysis on cell phenotype, requirements still largely unmet. Here, we introduce a device for single-cell isolation, selective retrieval and off-chip recovery. To this end, singularly addressable three-dimensional electrodes are embedded within a microfluidic channel, allowing the selective release of single cells from their trapping site through application of a negative dielectrophoretic (DEP) force. Selective capture and release are carried out in standard culture medium and cells can be subsequently mitigated towards a recovery well using micro-engineered hybrid SU-8/PDMS pneumatic valves. Importantly, transcriptional analysis of recovered cells revealed only marginal alteration of their molecular profile upon DEP application, underscored by minor transcriptional changes induced upon injection into the microfluidic device. Therefore, the established microfluidic system combining targeted DEP manipulation with downstream hydrodynamic coordination of single cells provides a powerful means to handle and manipulate individual cells within one device.

Keywords: single-cell microfluidics; single-cell recovery; single-cell array; hydrodynamic trapping; electrokinetics; tridimensional electrodes; dielectrophoresis (DEP); mRNA sequencing; Drop-seq

1. Introduction

The analysis of single cells is routinely carried out by means of in-flow measurements such as optical marker detection in flow cytometry. Despite the high throughput that can be achieved with those techniques, they only provide a snapshot of cells' properties at a certain time point with limited possibility to observe their behavior over time [1,2]. Cell arraying, i.e., the separation and localization of individual cells or doublets on a surface, allows observation of cells over extended periods of time, collection of secreted biomolecules [3] and recording of the response to specific stimuli [4]. Observation systems based on cell arraying are unique tools used to unveil mechanisms of cell-to-cell interaction [5], polarized cell proliferation [6], in vitro fertilization [7], etc.

Parallel immobilization of cells can be obtained through various methods—for instance, cell sedimentation on microwells [8], localization of single cells by means of optical tweezers [9], trapping by dielectrophoretic cages [10,11], segregation into small chambers sealed by PDMS valves [12] or hydrodynamic trapping [13].

Hydrodynamic trapping is a method based on the immobilization of cells of a specific size range at various locations of a microfluidic channel. Trapping sites are defined by tight side-wall openings of low fluidic resistance where single cells are led to by the laminar flow. Cells captured with this method are continuously exposed to a flow of medium, allowing for delivery of nutrients and disposal of waste. Previous works characterizing hydrodynamic cell-trapping systems have reported a trapping efficiency—defined as the percentage of traps filled after injection of cells—between 75% and 99%. These methods have been used to capture multiple cell types [14] and, in some cases, to localize rare cells [15]. They also proved to be valuable tools for investigation of single-cell behaviors. For instance, Dura et al. [5] could conjointly place single dendritic cells and T-cells and measure heterogeneity in the activation of T-cells.

Furthermore, the association of the molecular state of a cell with its on-chip characterization could help unveil new cellular mechanisms. Kimmerling et al. [16] compared intra- and inter-lineage transcriptomes within a cell population by capturing multiple generations of a single starting cell in subsequent traps. In that study, the analysis of the transcriptome was performed upon the retrieval of the entire lineage from the chip. The release of cells from their hydrodynamic traps was carried through application of a backflow pushing the cell out of the trap. This approach was also used by Kim et al. [12], who trapped unicellular microalgae in PDMS chambers for on-chip culturing and selectively retrieved the content of one chamber with a sophisticated three-layer PDMS valving system. Yeo et al. [15] combined centrifugal and hydrodynamic forces to isolate circulating tumor cells from a mixed cell population; to enable release, each trap was connected to an independent backflow channel. This method is suitable for collection of extremely rare cells such as circulating tumor cells, but it suffers from very low throughput and poor scalability, as the number of cells that can be retrieved is limited by the number of backflow channels that can be placed on the chip. Tan et al. [17] could retrieve cells encapsulated in hydrogel beads, creating an air expansion on the trap site generated with laser heating. This platform is limited by the complexity of the setup and the damage to the cells that may be induced by heat.

These proof-of-concept technologies underscore the necessity and also the challenges in combining continuous observation of cells on chip with further off-chip investigations. Dielectrophoresis (DEP) forces localized on the trap site by implementing microelectrodes in its vicinity could be utilized to control cell trapping in very high-throughput arrays. In fact, electric signals can be easily multiplexed, while keeping a very small footprint of the DEP actuator. Zhu et al. [18] used such forces to selectively release single yeast cells from their trapping site. The yeasts, placed in synthetic low-conductivity medium [19], were not recovered from the chip. A possible cause for this could be the well-known challenge of fabricating valves combined with microfluidics fabricated with stiff materials [20].

It is important to mention that a potential issue of DEP electrokinetic actuators could be cellular stress resulting from the application of polarizing electric fields. The impact of DEP signals on a population of cells in a single large chamber has been investigated by Nerguizian et al. [21] and by Flanagan et al. [22], although no study has dissected transcriptional changes induced by exposing cells to DEP forces applied identically to the whole population and on single cells. Additionally, to date, no studies have addressed the impact of DEP application and general effects of off- and on-chip handling on the molecular state of the cell [15,16].

In this study, we introduce a Microfluidic Platform for Arraying and Release of single Cell (MiPARC). This DEP-based platform allows for selective trapping and off-chip recovery of individual mammalian cells in their native culture medium. Selective release of T-lymphocytes was achieved by three-dimensional DEP actuators, integrated for the first time in SU-8 channels with a width of 25 μm. In order to obtain a precise handle on the flow in the microfluidic branches of the chip, and thus to recover single cells, we developed a PDMS valve technology compatible with SU-8 microfluidics, overcoming one of the main limitation of devices using walls made of stiff materials. We investigated the stress induced on cells by our platform through analysis of the molecular phenotype via mRNA sequencing, revealing no impact of DEP application on the transcriptional signature of the cells, super-seeded by minor alterations of the cellular molecular state introduced by hydrodynamic forces within the microfluidic system.

2. Materials and Methods

2.1. Device Fabrication

Our device consists of sixteen microfluidic traps arrayed inside a tree-like structure, as shown in Figure 1. The two valves located upstream allow control of liquid injection in the chip while the two valves located downstream enable the recovery of single cells. The height of the microfluidic channel as well as of the electrodes is 15 µm.

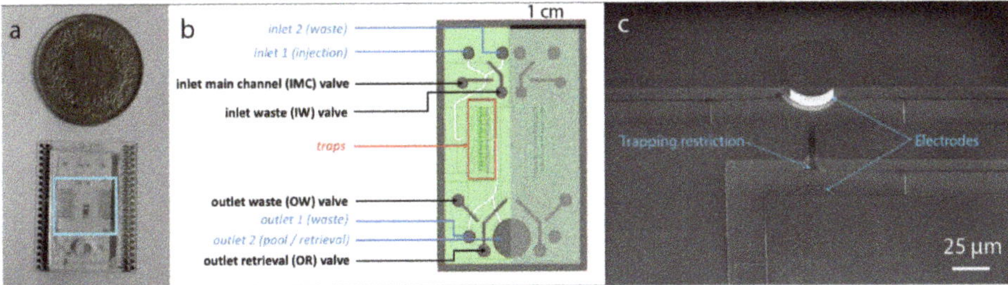

Figure 1. Device layout description and fabrication. (**a**) Picture of the microfluidic chip used for the experiments. The chip hosts two symmetrical parallel channels that share the outlet-well used for cell recovery. The blue box indicates the area where the cells are trapped in hydrodynamic sites. (**b**) Microfluidic layout of the chip. The SU-8 based microfluidic main channels are shown in green, while PDMS control channels are depicted in gray. When a cell enters the chip through inlet 1, it can be directed either towards the traps for immobilization and observation or to the inlet 2 if it needs to be discarded. After release for a trap, a cell can be recovered from the pool at outlet 2 or disposed through the outlet waste. (**c**) Magnified scanning electron microscopy image of a trap with electrodes embedded in the microfluidic channel. The aperture is 5 µm in width and the electrode extrusion from the microfluidic channel measures approximately 7 µm.

2.1.1. Fabrication of the Microfluidic Chip

Our system is based on a glass substrate. A detailed picture of the process flow is reported in Supplementary Material Figure S1. After sputtering of a layer of Ti/Pt/Ti (20/200/20 nm) on the wafer (Pfeiffer Spider 600, Pfeiffer Vacuum, Asslar, Germany), planar metal lines are patterned through standard photolithography and ion beam etching (Veeco Nexus IBE 350, Veeco, Plainview, NY, USA). In order to isolate electrical lines from the liquid, a 300 nm layer of oxide is sputtered on the wafer. This layer is then etched (SPTS APS dielectric etcher, SPTS Technologies, Newport, UK) in the region where the pillars will be connected to the planar metal lines. Cylindrical vertical pillars of SU-8 (Microchem 3025, Microresist Technologies, Berlin, Germany) are patterned on the exposed planar electrodes. Then, a thin layer of Ti/Pt (20/200 nm) is sputtered to cover the entire wafer with metal. This metal layer is then removed everywhere except on the pillar walls through vertical ion beam etching. In a successive process step, microfluidic channels of the same height of the electrodes are patterned in SU-8, hence allowing a micrometer-scale alignment of the electrodes inside the microfluidic system (Figure 1c). The most critical step of this process flow is the development of this second SU-8 layer that creates the microfluidic structure. In fact, an appropriate development time is crucial to avoid SU-8 residues clogging the fluidic restriction (insufficient development) and excessive development of the SU-8 that would result in apertures larger than designed and a delamination. With the chosen parameters (development time of 4′30″ in PGMEA: Propylene glycol methyl ether acetate), we recorded a 95% yield.

2.1.2. Fabrication of the PDMS Top Layer and Valves

The PDMS valving system placed on the top of the microfluidic channel is composed of two layers: a control layer made of a thick PDMS layer and a thin membrane that will be deflected in order to close or open the microfluidic channel (for more details, please refer to Supplementary Material Figure S2. The fabrication of the molds for the control layer is carried out with standard photolithography and dry etching (AMS 200 Alcatel) on a silicon wafer. PDMS prepared at a ratio of 5:1 is poured over the silicon mold which has previously been treated with trimethylchlorosilane (TMCS). The membrane is fabricated by spin-coating of PDMS at a ratio of 5:1 on a TMCS-treated silicon wafer, the speed of rotation is 2500 rpm and results in a thickness of 15 µm. Both this PDMS membrane layer and the lid PDMS layer are partially cured at 80 °C for 30 min and then aligned and bonded at 80 °C for 1.5 h. Finally, the PDMS coverslip is treated with 3-aminopropyl triethoxysilane (APTES) and irreversibly bonded by incubating the system at 150 °C for 2 h. This step presented some experimental challenges, since the alignment of both the fluidic and control channels is carried out manually. Additionally, pressing the chip with the PDMS coverslip with too much pressure might lead to the adhesion of the membrane with the bottom of the channel before the baking step, resulting in irreversible bonding between the membrane and the microfluidic channel after curing, preventing the valve form operating properly. Nonetheless, a yield above 80% could be obtained for the whole fabrication and packaging process.

2.2. Finite Element Simulations

Finite element simulations of the device were carried out using COMSOL Multiphysics 5.3 (Comsol Inc., Burlington, USA). The main objectives of these simulations were to optimize the dimensions of the microfluidic channels in order to achieve efficient hydrodynamic trapping of cells and to describe the electrical field in the trapping region to evaluate the best geometry to maximize the DEP force obtained. The simulation results reported in Figure 2 were all conducted on a 3D model. The flow velocity in the channels was simulated using the Laminar Flow module and a normal mesh, the pressure difference between inlet and outlet was set at 0.1 mbar and a no slip condition was applied on the fluidic walls (Figure 2a,b). The electric field was simulated using the Electrostatics module and a fine mesh, the voltage difference between both electrodes was set at 10V, which corresponds to the amplitude that will be later used for cell actuation (Figure 2c). All computations were computationally inexpensive and could be conducted in less than 10 min.

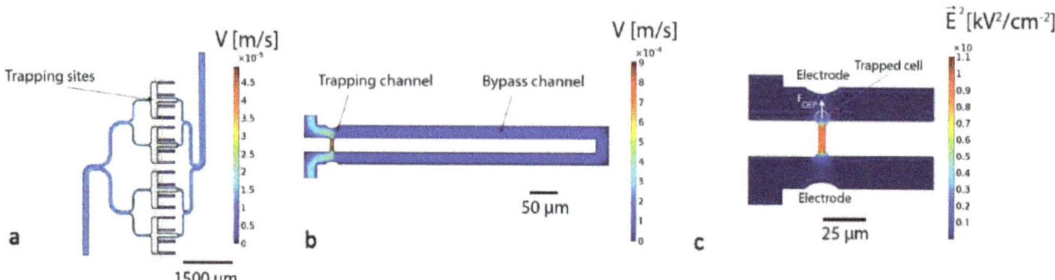

Figure 2. Simulations of the device layout and working principle. (**a**) Finite element simulation of the fluid velocity in the microfluidic channel (Inlet pressure: 0.1 mbar). (**b**) Finite element simulation of the fluid velocity in the microfluidic channel used to determine the ideal length of the channel ensuring the trapping of a single cell (Inlet pressure: 0.1 mbar). (**c**) Finite element simulation of the electric field in the channel. The field gradient is higher in the trapping region of the cell, generating a negative dielectrophoretic (DEP) force sufficient for the release of the cell from the trap (voltage difference between electrodes: 10 V).

2.3. Cell Preparation

The human T-lymphocytes Jurkat cell line was cultured in suspension in a Roswell Park Memorial Institute (RPMI) medium supplemented with 10% fetal bovine serum and 1% antibiotics (L-Glutamine-penicillin-streptomycin). Prior to application of cells to the microfluidics chip, cells were washed two times in phosphate-buffered saline (PBS) and resuspended in RPMI medium at a concentration of 300,000 cells/mL.

2.4. Cell Injection and Recovery

Prior to injection, the chip was primed with RPMI medium to coat the internal channel surfaces with albumin. All injections were carried out using a pressure system (OB-1, Elveflow (FR)). Hybrid SU-8/PDMS valves were filled with ultrapure water and closed applying a pressure of 3500 mbar. Cells were injected at a pressure of 0.5 mbar into the channel. Trapping of a cell can be monitored with an inverted microscope (Leica DMIL) equipped with a CMOS camera (Leica DFC295). Once a cell is trapped hydrodynamically, it can be released by applying an electrical signal via a signal generator (Agilent 33220A) with a frequency ranging from 1 to 20 MHz and an amplitude ranging from 8 to 10 Vpp. The signal was addressed specifically to each single electrode through a multiplexing system implemented on a custom made PCB and controlled by a Python interface (Supplementary Material Figure S3). After release, the valves were configured to allow guided release of cells to the recovery chamber that was priorly filled with 5 µl of RPMI medium.

2.5. Sequencing Methodology

2.5.1. Library Preparation

Recovered cells (approximately 400) were centrifuged at 600 g for 5 min and the supernatant was replaced with 25 µl of PBS containing 0.01% bovine serum albumin (Sigma, Kanagawa, Japan). Subsequently, 25 µL of lysis buffer was added, containing 0.1% Sarkosyl (Sigma), 5 mM EDTA (Life Technologies, Carlsbad, CA, USA), 0.1 M Tris (pH 7.5, Sigma) and 25 mM 1,4-Dithioreitol (DTT, Sigma), 800 units/mL RNase inhibitor (NEB, Ipswich, MA, USA) and 500 Drop-seq beads (Beads, Lot 120817, ChemGenes, Wilmington, MA, USA). All bead pelleting steps were carried out at 1000 g for 1 min in 1.5 mL microtubes (Axygen, Union City, CA, USA). Reverse transcription (RT), exonuclease I (ExoI) treatment and PCR were performed as described by Macosko et al. with minor adaptations [23]. Lysed cells were incubated at 1400 rpm for 5 min at room temperature and subsequently washed twice with 1 mL of 6x SSC buffer (Sigma). Reverse transcription was performed for 90 min at 42 °C in 50 µl of 1 mM dNTPs (Clontech, Mountain View, CA, USA), 2.5 µM template switch oligo (see Table 1), 1250 units/mL RNase inhibitor, 1x Maxima RT buffer and 10,000 units/mL Maxima H minus reverse transcriptase (Thermo Fisher Scientific, Waltham, MA, USA). Drop-seq beads were washed twice with 0.5% SDS (Applichem, Omaha, NE, USA) in 10 mM Tris (TE-SDS), twice with 0.01% Tween-20 (Sigma) in 10 mM Tris (TE-TW) and once with 10 mM Tris pH 7.5. The Drop-seq bead pellet was then incubated with 50 µl of exonuclease mix containing 1x Exonuclease I Buffer and 1000 units/mL Exonuclease I (NEB) and incubated at 37 °C for 45 min at 1400 rpm). Drop-seq beads were washed twice with TE-SDS, twice with TE-TW and once with double-distilled H_2O. Beads were amplified by PCR in 25 µL of 1x Hifi HotStart Readymix (Kapa Biosystems, Wilmington, MA, USA) and 0.8 µM TSO-PCR primer (Table 1) at 95 °C for 3 min; 4 cycles of: 98 °C for 20 sec, 65 °C for 45 sec, 72 °C for 3 min; then, 16 cycles of: 98 °C for 20 sec, 67 °C for 20 sec, 72 °C for 3 min and an extension step of 5 min. Libraries were purified using Ampure XP beads (at a ratio of 0.6x to remove small fragments), cDNA was quantified using a Qubit HS kit (Thermo Fisher Scientific) and integrity was analyzed on a Fragment Analyzer (Agilent). Libraries were prepared using in house-produced Tn5 loaded with adapters, as described [24]. Size selected and purified libraries were sequenced paired-end on a NextSeq 500 system (Illumina, San Diego, CA, USA) in High Output mode following recommendations from the original protocol (read 1—20 bp and read 2—50 bp) [23].

Table 1. Primers used for reverse transcription and library preparation.

ID	Sequence
TSO	AAGCAGTGGTATCAACGCAGAGTGAATrGrGrG
TSO-PCR	AAGCAGTGGTATCAACGCAGAGT

2.5.2. Data Analysis and Availability

The data analysis was performed using the Drop-seq tools package on the EPFL SCITAS HPC platform. After trimming and sequence tagging, reads were aligned to the human reference genome (hg38) using STAR (version 2.7.0.e) [25]. Following the alignment, the gene annotation was added, bead synthesis errors were corrected, and cell barcodes extracted. Subsequently, the BAM files containing the processed data were used to obtain digital gene expression matrices. Only cell barcodes with at least 50 UMI (Unique Molecular Identifier) were retained. Downstream data analysis was carried out using R (version 3.5.0), with *DESeq2* (version 1.22.2) [26] for identifying differentially expressed genes in pair-wise comparison. Plots were generated using the R package *ggplot2* (version 3.0.0). The Gene Expression Omnibus (GEO) accession number for the RNA-seq data reported in this paper is GSE143190.

3. Results and Discussion

3.1. Cell Trapping and Release

Sequential injection of individual cells at the trap location is an important feature of our platform, and it can be ensured by setting the width of the channel in the vicinity of the traps at 25 µm. Single-cell hydrodynamic traps are placed along this channel. Hydrodynamic trapping is a technique based on the use of mechanical restrictions to segregate particles from a main channel. The separation can be carried out efficiently if the flow going through the restriction channel is slightly higher than the flow in the main channel, however, the flow in the restriction should not be too high to avoid trapping of multiple cells. The traps are arranged on the branches of a tree-like fluidic structure shown in Figure 2a. Parallel-channel design is used to restrain possible clogging due to contamination to single branches. Figure 2b shows the finite element simulation of a single trap (COMSOL Multiphysics 5.3) that is composed of two elements: a fluidic bypass along the channel and a fluid path through the trap. The fluidic resistance of the bypass is 1.2-fold larger than that through the empty trap, which leads the cell towards the constriction (Figure 3a,c). Assuming an average diameter of lymphocyte of 10 µm, the height of the channel was set at 15 µm to avoid multiple stacking of cells in the trapping sites. The width of the main channel is 25 µm and the diameter of a cell is approximately 10 µm, resulting in a distance between the electrode extruding from the SU-8 wall and the cell in the trap of 15 µm. This distance has been chosen to allow a cell flowing in the channel after a trapping event to be guided in the bypass channel without risking clogging the whole channel. The number of traps which are filled with single cells upon injection is typically 90%, in agreement with the trapping efficiency values reported in literature [14]. We also measured the probability of a cell to be trapped by an empty trap as 75% in case of T-lymphocytes.

Having achieved targeted single-cell localization in the traps, we next aimed to use electrodes embedded in close proximity to the microfluidic channel to selectively release one specific single cell by means of DEP. Dielectrophoresis phenomena results in the displacement of polarizable particles in a non-uniform electric field. The particle experiences the formation of a dipole—the orientation of which depends on the relative permittivity of both the particle and its surrounding medium. If the particle is more polarizable than the medium, the induced dipole is oriented along the electric field. Reciprocally, the induced dipole is oriented against the electric field if the medium is more polarizable than the particle. Additionally, if the electrical field applied is non-uniform, the particle will move due to a higher field strength on one side of the particle. The electric field generated by the electrodes is constrained in the trap aperture as shown in the COMSOL simulation in Figure 2c. The field gradient is highest in the region of cell trapping. As the cell polarizability is lower than the polarizability of the surrounding medium, the cell is pushed toward the regions of the weaker field, i.e., out of the trap. This repulsive force—called negative dielectrophoretic (DEP) force—permits the targeted release of the cell (Figure 3b,d, Supplementary Material Video S1). The optimization of the operation parameters was carried out both experimentally and theoretically. The frequency was maintained above 1 MHz to avoid electrolysis that could be observed below that threshold and could lead to gaseous species formation. The force applied to the cell due to DEP is proportional to the real part of the Clausius–Mossotti factor according to the following equation:

$$F_{DEP} = 2\pi R^3 \varepsilon_{medium} Re(CM)\nabla E^2 \tag{1}$$

where R is the radius of the cell, ε_{medium} the dielectric permittivity of the medium, E the electric field in the channel and CM the Clausius–Mossotti factor.

Importantly, in the case of T-lymphocytes, the real part of the Clausius–Mossotti factor will drastically decrease at frequencies higher than 20 MHz [27]. Consequently, we set a working frequency range between 1 and 20 MHz. The minimal voltage value experimentally-identified to trigger efficient release in our configuration is 8 Vpp. Hence, in order to minimize the impact of application of electrical fields on cells, we maintained the voltage applied to cells between 8 and 10 Vpp throughout this study.

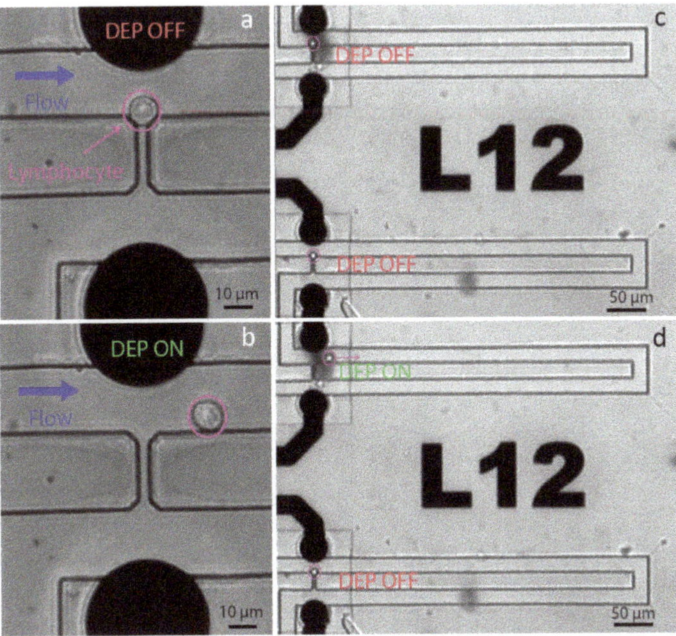

Figure 3. Selective single-cell retrieval. (**a,b**) A single lymphocyte can be trapped in the hydrodynamic constriction (**a**) and gently released (b) through application of negative DEP force activated by 10 Vpp voltage at 10 MHz. (**c,d**) The cell at the top is released, while the cell at the bottom is kept in the trap. The release is carried out with a 10 Vpp voltage at 10 MHz. A custom-made printed circuit board (PCB) enables the selective release of a single T lymphocyte. For more details, please refer to the Supplementary Material Video S3.

Each electrode is singularly addressable through a printed circuit board (PCB) directing the electrical signals toward the chosen electrodes, allowing a selective release of the cell of interest, as depicted in Figure 3c,d. Furthermore, the negative DEP force can be used in order to prevent trapping of unwanted cells (Supplementary Material Videos S2 and S3).

The most immediate approach to generate an electric field in a microfluidic system is to place two electrodes at the inlet and exit reservoirs, respectively [28]. The large distance between the electrodes and the active regions where the electrical gradient is required necessitates high voltages [28], which has many drawbacks—for instance, (i) the need to generate high-amplitude AC signals, (ii) the generation of heat and (iii) the induction of water electrolysis phenomena at the electrodes and consequent bubble generation, affecting cell viability.

The integration of electrodes near the trapping regions reduces those issues, since the amplitude of the signal can be drastically reduced to achieve the desired dielectrophoretic force. Planar microelectrodes and on-chip connections have been used to drive electrical signal to the site of interest [18]. However, this approach results in non-uniform electric fields over the channel height, decreasing towards the top of the channel. The employment of side-wall microelectrodes, instead, would generate the repulsive force more efficiently and homogeneously from the bottom to the top of the aperture.

One of the novel features of our approach resides in the replacement of standard planar electrodes by vertical electrodes according to a process previously developed by our group [29], consisting of a conformal coating of SU-8 pillars with metal, to obtain singularly addressable side-wall electrodes with a width of 40 µm and height of 15 µm (Figure 1c). In each cell trap, two vertical electrodes are placed across the hydrodynamic aperture, with the closest to the cell being only 15 µm away from it. This configuration can be successfully obtained only thanks to the high precision of the technology we developed to fabricate vertical electrodes in microfluidic channels. Vertical electrode fabrication has previously been reported through pyrolysis of photoresist structures [30] or metal ion implantation in PDMS [31] but none of those approaches enable such precision in the position of electrodes to ensure their integration inside a microfluidic channel as narrow as the one presented here.

The use of vertical electrodes improves the efficiency of force generation. In fact, we could set the amplitude of the DEP signals to lower values with respect to previous works that employ planar electrodes (10 Vpp of this work versus 20 Vpp [18]). Furthermore, we could afford to maintain the cells in their native medium (RPMI, conductivity 1300 mS/m) for all our experiments, a substantial advantage considering the requirements of cell biologists. Since DEP forces are weakened by high ionic strengths, most designs employing DEP to apply electrokinetic forces substitute the native medium with a synthetic one with lower conductivity [18,21]. Hence, by being able to retain cells in their native culture medium, cellular stress is reduced, washing and centrifugation steps limited and general compatibility with conventional cell-based assays achieved.

3.2. Single-Cell Handling for Accurate Retrieval

Our microfluidic system allows the recovery from the chip of a given cell that has been released from its trap. We designed and implemented a novel valve-based design to drive the released cell to a recovery line ending in an open well on the chip.

Microfluidic channels are made of SU-8, an epoxy-based negative photoresist that will not deform upon application of pressure. In order to create valves in the chip, we had to overpass the classical PDMS microfluidic valves, widely known as Quake's valves [32], which are based on elastomeric PDMS microfluidic structures. Previous studies have proposed solutions to integrate valves within microfluidics that were fabricated with hard material, as in our case. Huang et al. [33] designed valves by means of a PDMS layer sandwiched between two poly(methyl methacrylate (PMMA) channels. Similarly, Lee et al. [20] integrated a PDMS membrane between two polycarbonate (PC) layers.

In this study, we introduce for the first time PDMS-based valves combined with SU-8 microfluidics, a solution that opens new opportunities for lab-on-chip devices offered by the combination of hard photoresist microfluidics and on-chip PDMS valves. Four valves are placed on the microfluidic branches of the structure—the two inlet valves allow switching of solutions to be injected in the main channel, while the two outlet valves enable to switch between the recovery and the disposal channel, as previously mentioned (See Section 2.1.2 and Supplementary Material Figure S2 for more details). Figure 4a,b show the operation of one of our valves when the application of a high pressure on the control channel closes the valve sufficiently to stop the passage of cells. However, as the SU-8 fluidic cannot deform, the valve is only partially sealed, and a subcellular-sized aperture remains between the SU-8 walls and the PDMS membrane (See Supplementary Material Figure S2). The residual flow through the valve is used to drive the cells to the recovery well. Figure 4c,d show the recovery of a single cell leaving the chip and entering the recovery well from which it can be simply pipetted out.

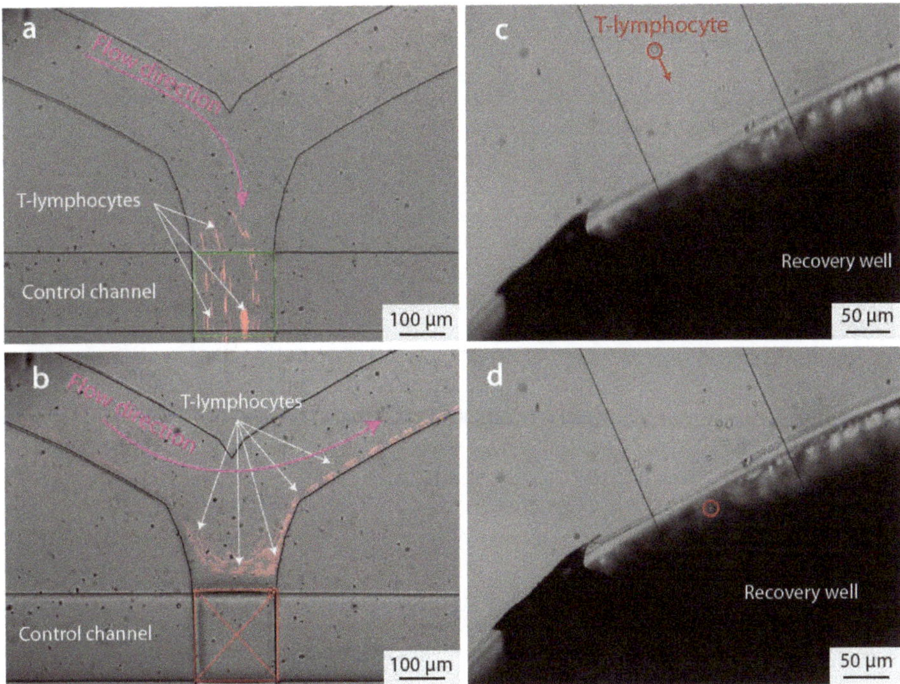

Figure 4. Single-cell recovery. (**a,b**) Operation of a hybrid SU-8/PDMS valve used to control the flow in the microfluidic channel. Trajectories of the lymphocytes in the channel is depicted in red. As SU-8 is stiff, high pressure (3500 mbar) must be exerted on the above PDMS layer in order to close the valve. When the valve is open, all cells are directed towards the main microfluidic channel (**a**), while closing the valve will prevent them from entering the channel (**b**). Combining four of those valves enables control of the injection and recovery of cells from the chip. (**c,d**) A single lymphocyte exiting the chip and entering the recovery channel, where it can be pipetted and further analyzed. This direct recovery in a well prevents the cells from sticking and being lost in outlet pipes. A video showing the trajectory of a cell from the trapping area to the recovery well is available as Supplementary Material (Video S4).

3.3. Transcriptional Profiling of MiPARC Processed Cells Reveals Negligible Impact of DEP Application on the Cellular Molecular State

The established MiPARC platform traps single cells in microfluidic constrictions, allowing for their release in the main channel and their selective extraction off chip. As those cells could be further analyzed or potentially expanded for adoptive transfer, the impact of the microfluidics operation and the applied electric field on cell function needed to be assessed. To globally inspect the molecular changes to the cell, we thus analyzed their transcriptome using RNA-sequencing in the framework of DEP manipulation.

In order to determine whether the microfluidics setup or the applied DEP impact the molecular properties, Jurkat T-cells were injected and either passed through the microfluidics setup for an average duration of three minutes in absence or in the presence of DEP forces. After retrieval, cells were collected and cultured off chip for three hours to permit potential alterations by the DEP field or the fluidic forces to be represented transcriptionally. Input cells that were solely cultured served as negative control (Input), whilst cells cultured for three hours under Phorbol-12-myristate 13-acetate and Ionomycin activation (PMA/Iono), globally activating transcription based on protein kinase C (PKC) activation and calcium ion influx [34], served as positive control for global T-cell. Due to the low number of maximally 400 cells per sample and the high volume of up to 100 μl of cell culture medium, we employed mRNA capture beads to obtain the mRNA of the lysed cells and detected over 7600 genes across the experimental conditions (Figure 5a). Importantly, cDNA quality was similar across the conditions with marginally higher yield for "PMA/Iono"-activated cells (Figure S4a). Additionally, hierarchical clustering of the detected mitochondrial genes revealed that cells processed with MiPARC showed similar expression intensities as compared to the "Input", indicating that cells were viable to a similar extent as unprocessed cells (Figure S4b). Differential expression analysis identified 114 genes across all pair-wise comparisons. Principal component (PC) analysis on all differentially expressed genes (DEGs) revealed that the first PC1 segregated "PMA/Iono"-activated cells from all other experimental conditions, whereas PC2 separated untreated cells from those subjected to the microfluidics chip. Importantly, there was no defined separation between cells subjected to DEP (Chip-DEP) or solely injected in the chip (Chip-Ctrl) (Figure 5b). The high concordance between "Chip-DEP" and "Chip-Ctrl" was underscored by the observation that no DEG could be detected, whereas 74 or 59 DEGs were identified when comparing "PMA/Iono" cells to "Chip-Ctrl" or "Chip-DEP", respectively (Figure 5c up). Importantly, the number of DEGs when comparing "Chip-Ctrl" or "Chip-DEP" to the "Input" conditions was substantially lower with 14 or 24, respectively (Figure 5c down). Importantly, genes associated with stress responses such as heat-shock proteins, *HSPA6* and *HSP90AA1*, chaperons like *CLU* and genes involved in stress recovery responses including *DNAJB1* and Ubiquitin (*UBC*) were significantly upregulated only under "PMA/Iono" conditions (Figure 5d left). Interestingly, a minor proportion of genes was upregulated for "PMA/Iono" and both Chip conditions, encompassing predominantly genes involved in cell proliferation including *EGR1*, *SMC2*, *FOS* and *FOSB* or activation *TRAC* and *H3F3B* (Figure 5d left). When comparing the genes consistently differentially expressed between PMA/Iono-activated cells and cells injected into the chip, the vast majority of activation and stress response genes was upregulated solely under "PMA/Iono" condition, whereas CXCR4 expression was only upregulated in cells that were flown through the chip (Figure 5d right).

Based on these transcriptional profiling results, we conclude that the impact of injection into and extraction from the microfluidics chip outweighs the changes wrought by the electric field onto the transcriptional landscape of the cell under the implemented culture conditions. This could be in part due to the very confined DEP, permitting the utilization of a low voltage of 8 Vpp at frequencies of 20 MHz, thereby limiting the extent of transcriptional changes observed in a previous study [21]. Furthermore, it was previously shown that the application of negative DEP, as utilized in the MiPARC system, does not alter the viability or differentiation capacity of neuronal embryonic stem cells, even at long DEP exposure times of up to 30 min [22], which exceeds the pulsed approach used within MiPARC by far. Although the transcriptional alterations instigated within the MiPARC system are minor, as compared to global activation of the cell, and unavoidable when implementing microfluidic cell handling, special care should be taken to minimize stressors such as high pressure or long retention times within the microfluidic devices. Regardless, application of short-term DEP for accurate retrieval of cells does only minorly impinge on the transcriptome at the obtained resolution.

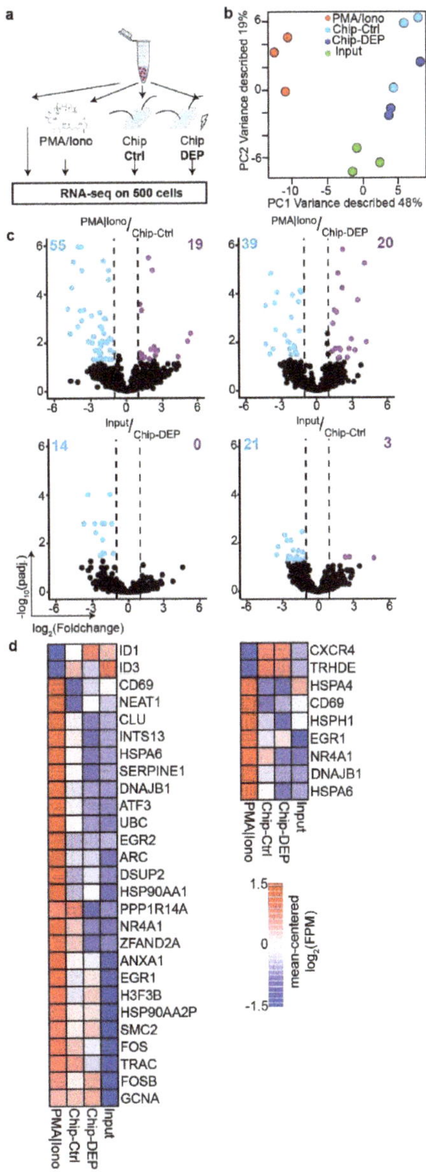

Figure 5. The electric field has minor molecular impact on Jurkat T-cells. (**a**) Jurkat cells were either injected into the microfluidics chip (Chip-Ctrl) or additionally subjected to the electric field used for accurate capture and retrieval of cells (Chip-DEP). Controls were either the input cells (Input) or cells activated for three hours under Phorbol-12-myristate 13-acetate and Ionomycin activation (PMA/Iono). Cells from all conditions were cultured for three hours to permit transcriptional changes to take place subsequent to treatment. (**b**) Principal component analysis on all differentially expressed genes (number of DEGs: 117). (**c**) Volcano plots of mean RNA-seq FPM (Fragment Per Million) comparing indicated samples. Number of DEGs is indicated. (**d**) Heatmaps represent expression of selected DEGs. Left: DEGs between PMA/Iono-stimulated and Input cells. Right: DEGs common on comparing Chip-Ctrl and Chip-DEP to PMA/Iono-stimulated cells. Experiments were performed in three independent biological replicates. DEG, differentially expressed gene (absolute(log2[foldchange] >= 1 and padj. <= 0.05).

4. Conclusions

Here, we presented a fully integrated system for single-cell isolation and off-chip recovery. The device is based on trapping constriction with dimensions that were designed to obtain efficient single-cell arraying. Moreover, hybrid SU-8/PDMS valve fabrication is described and implemented here for the first time to allow control of different flow sources and outputs. We showed for the first time the integration of vertical SU-8 electrodes in a narrow microfluidic channel. SU-8 electrodes were used in this work to enable selective release of single cells from specific traps by means of a negative DEP force generated by low voltage application at the electrodes and achieved in the cell's native culture medium. After release, cells were successfully recovered off chip and the phenotypic effect of injection in the microfluidic channel and exposure to DEP force were investigated through mRNA sequencing. For the first time, phenotypical effects induced by microfluidics and DEP were characterized separately, showing that the changes triggered by handling the cell in the microfluidic device outweigh the changes caused by the electrical field.

Currently, our device is composed of sixteen traps that can be singularly addressed. The MiPARC can be scaled up to permit the trapping, selective release and recovery of hundreds of cells, thus increasing the throughput of our platform. Indeed, hydrodynamic trap-based systems for single-cell arraying have already been achieved with parallelism of 1000 trapping locations on one chip [12]. Our approach for specific release by means of electrical actuators is compatible with this and even higher scales.

Supplementary Materials: The following are available online at http://www.mdpi.com/2072-666X/11/3/322/s1, Figure S1: Microfabrication process flow. Figure S2: Fabrication of PDMS coverslips. Figure S3: (A) CAD (Computer Aided Design) design of the PCB layout and picture of the assembled PCB. (B) Python interface developed to control the platform. Figure S4: Cell viability. Video S1: Cell trapping and release. Video S2: Forbidding cell trapping. Video S3: Selective cell release. Video S4: Cell off-chip recovery. Folders: Arduino DEP release, Masks DEP release, PCB DEP release, and Python interface DEP release.

Author Contributions: Conceptualization, P.-E.T., J.P. and C.G.; methodology, P.-E.T., J.P. and G.G.; software, P.-E.T. validation, P.-E.T. and J.P.; formal analysis, J.P.; investigation, P.-E.T. and J.P.; data curation, J.P.; writing—original draft preparation, P.-E.T.; writing—review and editing, J.P., K.K., B.D. and C.G.; project administration, C.G. All authors have read and agreed to the published version of the manuscript.

Funding: This research was funded by the Swiss National Science Foundation (205321_179086 and 310030_182655) and by the Precision Health & Related Technologies Grant (PHRT-502).

Acknowledgments: We are thankful for the support of the Center for Nano- and Microtechnology at EPFL, where we fabricated the chips. We would like to thank Raphael Genolet from the Ludwig Cancer center for supplying the human immortalized T lymphocytes, Marjan Biočanin and Johannes Bues for their valuable help while carrying out the sequencing part, Enrico Tenaglia for valuable discussions regarding cell biology and Paul Ery for his help in the fabrication of the chips.

References

1. Carlo, D.; Lee, L.P. Dynamic Single-Cell Analysis for Quantitative Biology. *Anal. Chem.* **2006**, *78*, 7918–7925. [CrossRef]

2. Armbrecht, L.; Dittrich, P.S. Recent Advances in the Analysis of Single Cells. *Anal. Chem.* **2016**, *89*, 2–21. [CrossRef]

3. Ma, C.; Fan, R.; Ahmad, H.; Shi, Q.; Begonya, C.A.; Chodon, T.; Koya, R.C.; Liu, C.C.; Kwong, G.A.; Radu, C.G.; et al. A clinical microchip for evaluation of single immune cells reveals high functional heterogeneity in phenotypically similar T cells. *Nat. Med.* **2011**, *17*, 738–743. [CrossRef]

4. Song, J.; Ryu, H.; Chung, M.; Kim, Y.; Blum, Y.; Lee, S.; Pertz, O.; Jeon, N. Microfluidic platform for single cell analysis under dynamic spatial and temporal stimulation. *Biosens. Bioelectron.* **2018**, *104*, 58–64. [CrossRef]

5. Dura, B.; Dougan, S.K.; Barisa, M.; Hoehl, M.M.; Lo, C.T.; Ploegh, H.L.; Voldman, J. Profiling lymphocyte interactions at the single-cell level by microfluidic cell pairing. *Nat. Commun.* **2015**, *6*, 5940. [CrossRef]

6. Geng, T.; Bredeweg, E.L.; Szymanski, C.J.; Liu, B.; Baker, S.E.; Orr, G.; Evans, J.E.; Kelly, R.T. Compartmentalized microchannel array for high-throughput analysis of single cell polarized growth and dynamics. *Sci. Rep.* **2015**, *5*, 16111. [CrossRef]

7. Han, C.; Zhang, Q.; Ma, R.; Xie, L.; Qiu, T.; Wang, L.; Mitchelson, K.; Wang, J.; Huang, G.; Qiao, J.; et al. Integration of single oocyte trapping, in vitro fertilization and embryo culture in a microwell-structured microfluidic device. *Lab Chip* **2010**, *10*, 2848–2854. [CrossRef]

8. Park, M.; Hur, J.; Cho, H.; Park, S.H.; Suh, K.Y. High-throughput single-cell quantification using simple microwell-based cell docking and programmable time-course live-cell imaging. *Lab Chip* **2010**, *11*, 79–86. [CrossRef]

9. Gao, D.; Ding, W.; Manuel, N.V.; Ding, X.; Rahman, M.; Zhang, T.; Lim, C.; Qiu, C.W. Optical manipulation from the microscale to the nanoscale: Fundamentals, advances and prospects. *Light Sci. Appl.* **2017**, *6*, e17039. [CrossRef]

10. Boral, D.; Vishnoi, M.; Liu, H.N.; Yin, W.; Sprouse, M.L.; Scamardo, A.; Hong, D.S.; Tan, T.Z.; Thiery, J.P.; Chang, J.C.; et al. Molecular characterization of breast cancer CTCs associated with brain metastasis. *Nat. Commun.* **2017**, *8*, 196. [CrossRef]

11. Keim, K.; Rashed, M.Z.; Kilchenmann, S.C.; Delattre, A.; Gonçalves, A.F.; Éry, P.; Guiducci, C. On-chip technology for single-cell arraying, electrorotation-based analysis and selective release. *Electrophoresis* **2019**, *40*, 1830–1838. [CrossRef] [PubMed]

12. Kim, H.; Devarenne, T.P.; Han, A. A high-throughput microfluidic single-cell screening platform capable of selective cell extraction. *Lab Chip* **2015**, *15*, 2467–2475. [CrossRef] [PubMed]

13. Sauzade, M.; Brouzes, E. Deterministic trapping, encapsulation and retrieval of single-cells. *Lab Chip* **2017**, *17*, 2186–2192. [CrossRef] [PubMed]

14. Kim, H.; Lee, S.; Lee, J.; Kim, J. Integration of a microfluidic chip with a size-based cell bandpass filter for reliable isolation of single cells. *Lab Chip* **2015**, *15*, 4128–4132. [CrossRef] [PubMed]

15. Yeo, T.; Tan, S.; Lim, C.; Lau, D.; Chua, Y.; Krisna, S.; Iyer, G.; Tan, G.; Lim, T.; Tan, D.; et al. Microfluidic enrichment for the single cell analysis of circulating tumor cells. *Sci. Rep.* **2016**, *6*, 22076. [CrossRef]

16. Kimmerling, R.J.; Szeto, G.; Li, J.W.; Genshaft, A.S.; Kazer, S.W.; Payer, K.R.; de Borrajo, J.; Blainey, P.C.; Irvine, D.J.; Shalek, A.K.; et al. A microfluidic platform enabling single-cell RNA-seq of multigenerational lineages. *Nat. Commun.* **2016**, *7*, 10220. [CrossRef]

17. Tan, W.H.; Takeuchi, S. Dynamic microarray system with gentle retrieval mechanism for cell-encapsulating hydrogel beads. *Lab Chip* **2007**, *8*, 259–266. [CrossRef]

18. Zhu, Z.; Frey, O.; Ottoz, D.; Rudolf, F.; Hierlemann, A. Microfluidic single-cell cultivation chip with controllable immobilization and selective release of yeast cells. *Lab Chip* **2011**, *12*, 906–915. [CrossRef]

19. Pucihar, G.; Kotnik, T.; Kandušer, M.; Miklavčič, D. The influence of medium conductivity on electropermeabilization and survival of cells in vitro. *Bioelectrochemistry* **2001**, *54*, 107–115. [CrossRef]

20. Lee, K.S.; Ram, R.J. Plastic–PDMS bonding for high pressure hydrolytically stable active microfluidics. *Lab Chip* **2009**, *9*, 1618–1624. [CrossRef]

21. Nerguizian, V.; Stiharu, I.; Nosayba, A.A.; Bader, Y.D.; Alazzam, A. The effect of dielectrophoresis on living cells: Crossover frequencies and deregulation in gene expression. *Analyst* **2019**, *144*, 3853–3860. [CrossRef]

22. Lu, J.; Barrios, C.A.; Dickson, A.R.; Nourse, J.L.; Lee, A.P.; Flanagan, L.A. Advancing practical usage of microtechnology: A study of the functional consequences of dielectrophoresis on neural stem cells. *Integr. Biol.* **2012**, *4*, 1223–1236. [CrossRef]

23. Macosko, E.Z.; Basu, A.; Satija, R.; Nemesh, J.; Shekhar, K.; Goldman, M.; Tirosh, I.; Bialas, A.R.; Kamitaki, N.; Martersteck, E.M.; et al. Highly Parallel Genome-wide Expression Profiling of Individual Cells Using Nanoliter Droplets. *Cell* **2015**, *161*, 1202–1214. [CrossRef]

24. Picelli, S.; Björklund, A.K.; Reinius, B.; Sagasser, S.; Winberg, G.; Sandberg, R. Tn5 transposase and tagmentation procedures for massively scaled sequencing projects. *Genome Res.* **2014**, *24*, 2033–2040. [CrossRef]

25. Biočanin, M.; Bues, J.; Dainese, R.; Amstad, E.; Deplancke, B. Simplified Drop-seq workflow with minimized bead loss using a bead capture and processing microfluidic chip. *Lab Chip* **2019**, *19*, 1610–1620. [CrossRef]

26. Love, M.I.; Huber, W.; Anders, S. Moderated estimation of fold change and dispersion for RNA-seq data with DESeq2. *Genome Biol.* **2014**, *15*, 550. [CrossRef]

27. Keim, K.; Gonçalves, A.; Guiducci, C. Trapping of Single-Cells Within 3D Electrokinetic Cages. In Proceedings of the 2018 COMSOL Conference, Lausanne, Switzerland, 22–24 October 2018.

28. Moncada-Hernández, H.; Lapizco-Encinas, B.H. Blanca Simultaneous concentration and separation of microorganisms: Insulator-based dielectrophoretic approach. *Anal. Bioanal. Chem.* **2010**, *396*, 1805–1816. [CrossRef]

29. Kilchenmann, S.C.; Rollo, E.; Maoddi, P.; Guiducci, C. Metal-Coated SU-8 Structures for High-Density 3-D Microelectrode Arrays. *J. Microelectromech. Syst.* **2016**, *25*, 425–431. [CrossRef]

30. Rodrigo, M.D.; Renaud, P.; Madou, M.J. A novel approach to dielectrophoresis using carbon electrodes. *Electrophoresis* **2011**, *32*, 2385–2392.

31. Choi, J.W.; Rosset, S.; Niklaus, M.; Adleman, J.R.; Shea, H.; Psaltis, D. 3-dimensional electrode patterning within a microfluidic channel using metal ion implantation. *Lab Chip* **2010**, *10*, 783–788. [CrossRef]

32. Thorsen, T.; Maerkl, S.J.; Quake, S.R. Microfluidic Large-Scale Integration. *Science* **2002**, *298*, 580–584. [CrossRef] [PubMed]

33. Huang, S.; He, Q.; Hu, X.; Chen, H. Fabrication of micro pneumatic valves with double-layer elastic poly(dimethylsiloxane) membranes in rigid poly(methyl methacrylate) microfluidic chips. *J. Micromech. Microeng.* **2012**, *22*, 085008. [CrossRef]

34. Brignall, R.; Cauchy, P.; Bevington, S.L.; Gorman, B.; Pisco, A.O.; Bagnall, J.; Boddington, C.; Rowe, W.; England, H.; Rich, K.; et al. Integration of Kinase and Calcium Signaling at the Level of Chromatin Underlies Inducible Gene Activation in T Cells. *J. Immunol.* **2017**, *199*, 2652–2667. [CrossRef]

Article

Dielectrophoretic Immobilization of Yeast Cells Using CMOS Integrated Microfluidics

Honeyeh Matbaechi Ettehad [1,*], **Pouya Soltani Zarrin** [1], **Ralph Hölzel** [2]
and **Christian Wenger** [1,3]

1. IHP–Leibniz-Institut für innovative Mikroelektronik, Im Technologiepark 25,
   15236 Frankfurt/Oder, Germany; soltani@ihp-microelectronics.com (P.S.Z.);
   wenger@ihp-microelectronics.com (C.W.)
2. Fraunhofer Institute for Cell Therapy and Immunology, Branch Bioanalytics and Bioprocesses (IZI-BB),
   14476 Potsdam-Golm, Germany; Ralph.Hoelzel@izi-bb.fraunhofer.de
3. BTU Cottbus-Senftenberg, 03046 Cottbus, Germany
* Correspondence: matbaechi@ihp-microelectronics.com; Tel.: +49-335-5625 (ext. 663)

Received: 15 April 2020; Accepted: 13 May 2020; Published: 15 May 2020

Abstract: This paper presents a dielectrophoretic system for the immobilization and separation of live and dead cells. Dielectrophoresis (DEP) is a promising and efficient investigation technique for the development of novel lab-on-a-chip devices, which characterizes cells or particles based on their intrinsic and physical properties. Using this method, specific cells can be isolated from their medium carrier or the mixture of cell suspensions (e.g., separation of viable cells from non-viable cells). Main advantages of this method, which makes it favorable for disease (blood) analysis and diagnostic applications are, the preservation of the cell properties during measurements, label-free cell identification, and low set up cost. In this study, we validated the capability of complementary metal-oxide-semiconductor (CMOS) integrated microfluidic devices for the manipulation and characterization of live and dead yeast cells using dielectrophoretic forces. This approach successfully trapped live yeast cells and purified them from dead cells. Numerical simulations based on a two-layer model for yeast cells flowing in the channel were used to predict the trajectories of the cells with respect to their dielectric properties, varying excitation voltage, and frequency.

Keywords: dielectrophoresis; cell immobilization; cell separation; interdigitated electrodes; microfluidics; lab-on-a-chip

---

1. Introduction

Cell characterization and manipulation are critical when it comes to clinical and diagnostic applications [1]. Immobilization and isolation of specific cells as a way to detect diseases [2–4], separation of live and dead cells as a means for early-stage disease diagnosis [5,6], as well as filtering and purification of cells, viruses, proteins, and micro/nanoparticles [7–10] are essential examples in a variety of biological and biomedical applications. Development of lab-on-a-chip (LOC) devices such as microfluidic platforms has simplified the handling of complex and costly laboratory-based sample preparations and analyses, using a single device in the scale of a few centimeters [8–10]. Performing various tasks on a single device not only increases the precision of analysis but also improves the accuracy, reliability, and reproducibility of sample preparation procedure.

Among various cell manipulation techniques for LOC devices, dielectrophoresis (DEP) has been utilized widely for biomedical applications [11,12]. DEP is a non-invasive, label-free, and low-cost method which provides high accuracy and efficiency analyses [11]. Since this method exploits the intrinsic dielectric properties (relative permittivity and electrical conductivity) of the cells and their surrounding medium [12], it can be used selectively for the characterization and manipulation of cells. When polarizable cells subject to a non-uniform AC electric field, DEP force is induced as a result of the interaction between the cells' induced dipole and the electric field [13]. A non-uniform electric field can be generated as a result of imposing an AC signal to an electrode. Variation in the frequency of the applied signal can generate DEP forces in two opposite directions, resulting in either positive DEP (pDEP) or negative DEP (nDEP). Based on the selective DEP forces, specific cells can be trapped and detected [2,4,14], collected for further analyses (e.g., viability test) [15,16], or isolated from a mixture of cell suspension in blood for purifying processes [17]. Furthermore, dead cells, which cause bias during experimental measurements, can be removed from live cells [18].

Planar [18–20] and three-dimensional (3D) electrode structures [21,22] are commonly used for these applications. 3D electrodes are fabricated on the top and bottom, or sidewalls of microfluidic channels, whereas planar electrodes are commonly embedded on the bottom of microfluidic channels [23]. Prominent examples of planar electrodes are, interdigitated electrode arrays (IDEs), castellated [24], quadrupole [25], curved [26], spiral [1], oblique [27] and matrix [28]. Among these planar electrodes, IDEs are the convenient form of electrode geometry for dielectrophoretic immobilization [29] and separation [8] of certain cell population. IDEs have been previously used to immobilize biological entities [6], proteins [30], and to detect the dielectric constant of organic fluids [31], etc.

Over recent decades, many studies have been conducted on characterization and detection of the biological species on a single chip. Flanagan et al. [32] explored the use of DEP for characterization and identification of stem cells and their differentiated progeny. To create DEP force, IDEs were fabricated on glass wafers and were placed at the bottom of a polymer-based microfluidic channel. DEP showed that stem cells and their differentiated deviations develop different dielectric properties. Although this approach presented a platform to distinguish specific cells, it employs a large-scale setup. Lyu et al. [33] presented a numerical model using COMSOL simulations on the development of an electroporation technology for simultaneously calculating the DEP forces and electroporation of yeast and E.coli cells in the fluid flowing on a non-electrolytic micro/nano electroporation (NEME) electrode surface. Although this advancement could lead to new medical applications such as cell separation and destruction of unwanted cells, the applicability of this method has not been validated and confirmed experimentally. Ning et al. [34] described a test system for simultaneous microwave measurement and visual validation of cytoplasm resistance of a live Jurkat using broadband electrical detection technique. The setup is based on a homemade probe station mounted on top of an inverted microscope. This system included gold-based coplanar waveguide (CPW) placed between a quartz substrate and a PDMS cover. A 150 µm wide and 50 µm height channel etched underside of the PDMS cover. In other work, Li et al. [35] proposed a similar microfluidic setup to differentiate between the small number of live and heat-killed Escherichia coli cells suspended in culture media using microwave measurement. The differentiation principle between live and dead cells is based on the comparison of the transmitted and reflected microwave signals. The off-chip analysis showed that the difference is due to the decrease of cytoplasmic dielectric properties over cell death. A bipolar complementary metal-oxide-semiconductor (BiCMOS) based LOC platform was proposed by Manczak et al. [36] for discrimination of Glioblastoma (GBM), undifferentiated from differentiated cells, using ultra-high frequency (UHF) DEP technique to characterize cancer stem cells. Using this technique, characterization and detection of cells were achieved based on the intracellular dielectric properties of individual cells. To avoid the contact of cell population suspended in a liquid carrier with electronics of the chip, a polydimethylsiloxane (PDMS) microfluidic channel fabricated above the sensors on top of the BiCMOS device. All of these published methods offer opportunities to characterize and detect bio-particles on the same chip. However, they utilized relatively large-scale setups with polymer-based microfluidic channels that are not compatible with complementary metal-oxide-semiconductor (CMOS) process flows. The bulky polymer-based LOC setups limit the device performance by introducing parasites to the system. Moreover, PDMS microfluidics is more convenient for lab-based researches than for industrial applications because of the limited reproducibility of the fabrication process [37].

Among many alternative materials, such as polymethylmethacrylate (PMMA), amorphous polymers, thermoplastics and epoxy photoresist SU-8 [38], silicon is a reliable substitute for polymer in microfluidic applications due to its high integration robustness with electronics. Furthermore, silicon can be used in conjunction with fluidic applications which requires, high temperature resilience, very high precision channel alignments and high aspect ratio structures [38]. One of the most outstanding advantageous of silicon is the possibility of fabricating thin membranes which improves the stability of the device to high temperature ramp-rates by reducing thermal mass [38,39]. High thermal conductivity of the silicon warrant a uniform temperature distribution [40]. Thus, on the one hand, the interest in using silicon-based microfluidic LOC is increasing. On the other hand, the need for physically interfacing the fluidic samples with electrodes and sensors for analyzing biological and nonbiological samples is increasing the demand for combing the capabilities of microfluidics and CMOS integrated circuits. Integrating these technologies provides remarkable opportunities in the biomedical field for point-of-care diagnostics, high throughput screening, and implantable devices [41]. The hetero-integrated CMOS technology allows the fabrication of microfluidic channel, sensors, and circuitry as monolithic devices. Due to the fact, that trapping, sensing, detecting and analyzing can be achieved on a single chip, the hetero-integrated CMOS approach is very beneficial for future applications, while polymeric-based microfluidic channel approaches lacks from sensor and circuitry integration on a single chip solution. Combination of CMOS and microfluidics on the same die allows highly miniaturized LOC fabrication. Moreover, the high alignment accuracy of CMOS processing enables smaller distances between the isolated fluidic and electrical interfaces. Integrating microfluidics process steps into CMOS fabrication for miniaturized microsystems not only facilitate the LOC portability, but also enable fast diagnostic results even under non-laboratory conditions. However, lack of promising integration methods remains a big a challenge and realizing a fully functional device is under research.

In this paper, we investigated a 5×5 mm^2 CMOS integrated silicon microfluidic device utilizing six various IDEs, with different geometrical ratios, for the immobilization and separation of live and dead yeast cells using dielectrophoresis. The idea of combining the fluidic solution and electrical components improves the functionality and precision of this highly miniaturize LOC by using separate interfaces for electrical connections and microfluidics. This approach provides a low voltage DEP technique and an operational simplicity that enables the portability of the LOC device. The hetero-integration technology which allows the fabrication of microfluidic channels, sensors, and circuitry on a single chip, is replaced by the costly multi-step fabrication processes of various chips. The high alignment accuracy of the microfluidic channel on CMOS electronics ensures a reproducible and reliable integration process compared to relatively large size polymeric-based microfluidic LOC systems. The opportunity of immobilizing, sensing, and detecting cells on the same chip increases the reproducibility of the measurements by using less complex setups. Contamination-free fabrication process of CMOS integrated microfluidic offers reliable measurements. Using silicon instead of polymer for the fabrication of the microfluidic channel benefits the high integration level of circuitry and sensors on a single chip. State-of-the-art of this CMOS technology offers the opportunity of immobilizing, sensing, and detecting the particles on the same chip. However, sensing and detecting of the cells are not in the scope of this article and will be described in details elsewhere. The IDE structures used in this study followed the Guha. et al. approach, which used a similar structure for sensing and detecting biological cells on a single chip [42–45]. To optimize and adapt this IDE to our application, a systematic simulation study was conducted using COMSOL Multiphysics (version 5.3) [46,47]. A wide range of different electrodes with varying electrode width and spacing between fingers were modeled. To confirm the simulation results, some promising structures were selected and proposed for fabrication. Two main concepts have been scrutinized and demonstrated throughout this paper, which include the applicability of dielectrophoresis for cell immobilization and the impact of voltage, frequency, flow rate, and geometry ratio (spacing to width) of various IDEs on the time-dependent DEP behavior of live yeast cells suspended in deionized (DI) water. The choice of yeast cell as a model organism and DI-water as a model liquid carrier was done to keep the first model as simple as possible and reduce the number of complex parameters to obtain a trustable comparison between simulation and experimental results. However, this device can also be used for the analysis of cells suspended in more complex mediums. The cells motion were investigated optically and compared with the simulation results. Moreover, this paper proposes the adaption of the developed LOC device for the isolation and separation of viable and non-viable yeast cells in a mixture.

2. Materials and Methods

2.1. Microsystem

The LOC platform introduced in this work [48] combines a microfluidic channel with high-performance CMOS electronics. Based on this technique, separate microfluidic and electrical interfaces can be achieved. The developed silicon microfluidic channel is integrated into a CMOS device and encapsulated with transparent glass for simultaneous electrical and optical measurements. The combination of microfluidics and CMOS technologies offers great benefits in terms of high throughput integration level and cost reduction, thus making the approach favorable for biomedical applications. By taking advantage of the system miniaturization, designing small-sized channels and integrating sensors near the fluidic interface are possible, which ultimately leads to a higher sensitivity of the LOC system.

2.2. Microfluidics

The state-of-the-art of our LOC device is due to the hetero-integration of the microfluidics and compatibility of the in-house CMOS technology with the standard processing technology. It is noteworthy that the fabrication cost reduction, reproducibility and reliability are the main benefit of this approach. CMOS electronics, Si channels, and the glass wafer are integrated (a three-wafer-stack approach) on a single chip using 200 mm wafer bonding techniques [48]. Figure 1 illustrates the fabrication process of the microfluidics LOC device. In this process, the first wafer used to fabricate the CMOS device, including active circuitry and sensors. Next, the inlet and outlet for the microfluidic channel were opened by Localize backside etching (LBE) from the backside of this wafer. The second bare Si wafer is patterned to structure the channel by etching. Using plasma-activated oxide-oxide fusion bonding, these wafers are bonded together from their front sides at 300 °C. This step is followed by grinding the backside of the microfluidic channel to achieve the desired channel height. Finally, to seal the microfluidic channel, the third wafer, which is a glass wafer, is adhesively bonded to the top of the channel at 200 °C.

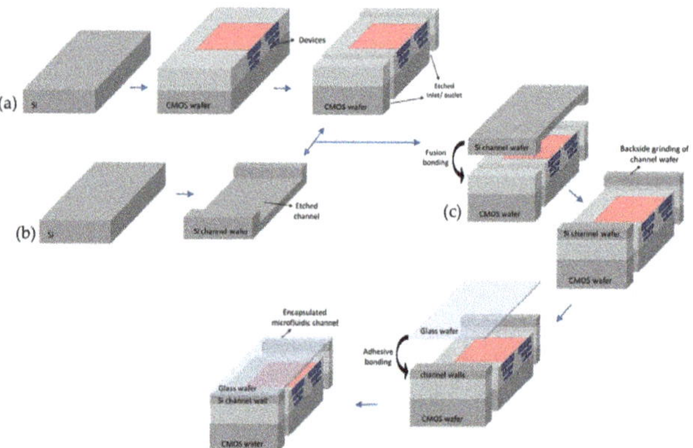

Figure 1. Schematic cross-sectional view of the microfluidics lab-on-a-chip (three-wafer stack) packaging process [48]: (**a**) CMOS fabrication; (**b**) Formation of the microfluidic channel; (**c**) Three-wafer-stack approach bonding process.

The reduced silicon-based channel dimensions (low channel height) compared to the polymeric-based fluidic channel with relatively larger sizes, increase the chance of bringing cells closer to the fringing electric field created by the IDEs. Larger channels increase the probability of cell tracing from above the effective distance of fringing field over IDEs which results in the discard of cells from the channel. Development of the current device satisfy the need of a more reliable analyzation of small sample amounts.

2.3. Interdigitated Electrodes

Arrays of microfabricated IDEs are the convenient form of electrode geometry for the dielectrophoretic characterization of biological particles (e.g. cells and viruses), through microfluidic biochips. In this work, a multi-fingered planar IDE is used as electrodes (Figure 2a) and embedded in the microfluidic channel (Figure 2b).

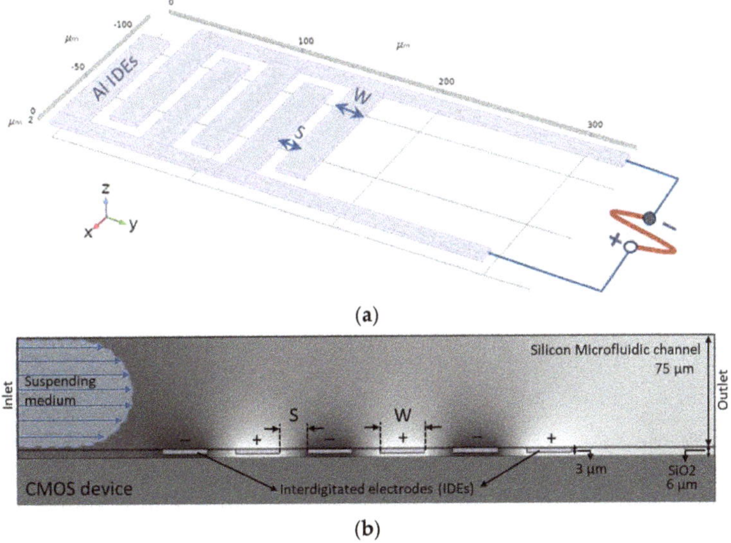

Figure 2. (**a**) Multi-fingered planar IDEs, (**b**) Cross-sectional view of IDEs embedded in the microfluidic channel.

These electrodes are used for the separation of particles or purification of the live cells from dead cells. In this work, we initially used the same IDE structures which were previously established for high-frequency CMOS dielectric sensors [49]. However, these IDEs were then geometrically optimized to enhance the DEP performance, using COMSOL simulations [47]. Various IDEs with different geometrical parameters were simulated. To confirm and validate the simulation results experimentally, various IDEs were fabricated. For the first prototype of the CMOS integrated microfluidic channel, due to design limitations for electrical contacts, these IDE structures were fabricated perpendicularly to the microfluidic channel. Table 1 represents the geometrical parameters of the manufactured IDE structures [47]. IDE structures were fabricated in the standard 0.25 µm CMOS technology of IHP. Figure 3 illustrates the device chip. The commonly used CMOS compatible material chosen to fabricate the IDEs are known to be long term stable in CMOS based products. The reliability issues of the same material used for biochip fabrication have to be evaluated in the future.

Table 1. Parametrical geometries of the IDEs, chosen based on the results reported in [49].

IDE Structure	IDE 1	IDE 2	IDE 3	IDE 4	IDE 5	IDE 6
S/W ratio	0.1	0.25	0.4	0.6	1	1.3
Spacing between finger (S)	5 (μm)	10 (μm)	15 (μm)	20 (μm)	20 (μm)	20 (μm)
IDE finger width (W)	45 (μm)	40 (μm)	35 (μm)	30 (μm)	20 (μm)	15 (μm)

Figure 3. CMOS integrated microfluidic lab-on-a-chip devices with zoom-in of the chip and embedded IDEs in the channel.

2.4. Cells under Test

Yeast cells (Saccharomyces cerevisiae RXII) were used for DEP studies as the model particles. Live yeasts were diluted in 40 mL deionized (DI) water at a concentration of 15×10^2 μg mL^{-1} and incubated at room temperature for 15 min and were stirred every 5 min. Dead cells obtained by heating live cell suspension in DI-water with the same concentration, at 100 °C for 20 min, and mixed with live ones for the separation experiment. The average diameters of the live and dead cells were measured as 8 μm and 6 μm, respectively. Sample suspensions were introduced into the microfluidic chip using a syringe pump.

2.5. Experimental Setup

Figure 4 presents our experimental setup, which consists of an AC signal generator (Agilent-33220A, Agilent Technologies/Keysight Technologies, Santa Clara, CA, USA) to generate a fringing electric field between the IDE fingers, the programmable syringe pump (NEMESYS, CETONI GmbH, Korbußen, Germany) to flow the cells which are suspended in DI-water, a tabletop, and an upright microscope (Nikon Eclipse-LV100ND, Nikon GmbH, Tokyo, Japan) equipped with a CCD video camera (Nikon-DS-Fi2, Nikon GmbH, Tokyo, Japan) connected to a computer for simultaneous optical measurement and analysis of the acquired videos and images.

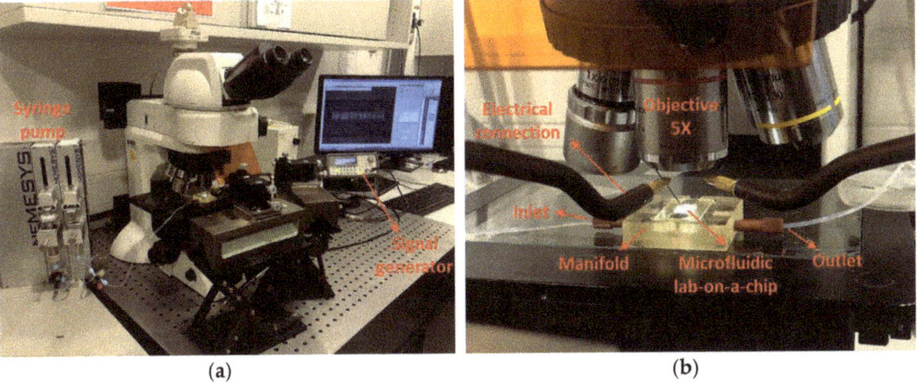

(a)   (b)

Figure 4. Experimental setup: (**a**) equipment used for dielectrophoresis characterization of yeast cells; (**b**) lab-on-a-chip with electrical connections under the objective.

To provide an interface to control the fluid flow and sample injection through the microfluidic channel, an external macrofluidic technology was employed. To this end, a fluidic manifold was designed, Figure 5a, and fabricated out of transparent hard polymer (PMMA) using a commercial 3D printer (Keyence Agilista-3200W, Keyence Co., Osaka, Japan), Figure 5b, interfacing the micro-device to the macroscale fluidic connections. A cavity with the same size as the chip, 5×5 mm^2, is created in the manifold. The square-shaped inlet and outlet (with the dimension of 150 μm) of the microfluidics were aligned directly on the manifold channels from the bottom side of the CMOS chip. At the interface of the chip and manifold, two O-rings with an inner diameter of 0.5 mm were used to seal the fluidic connections between the manifold and the chip to prevent leakage. The chip is clamped between the fluidic manifold base and cap by two screws. The external tubing connections were made via thread connectors, which are screwed into the inlet/outlet ports of the manifold.

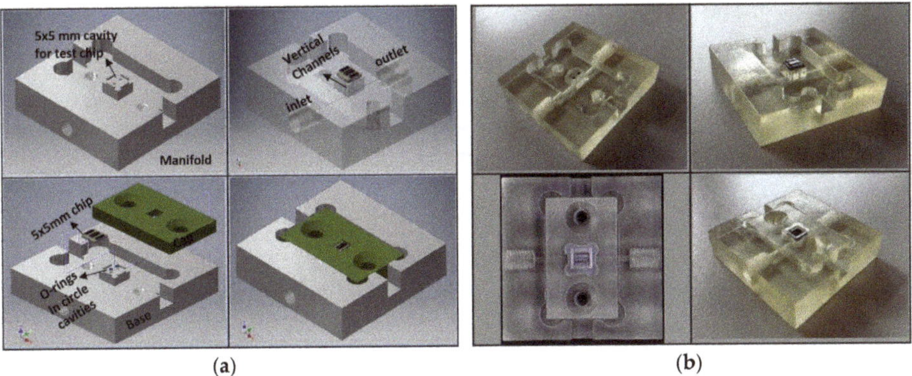

(a)   (b)

Figure 5. Manifold technology development using 3D printing for holding the lab-on-a-chip: (**a**) 3D schematic of the manifold design; (**b**) Fabricated manifold using 3D printing including a test chip.

The dielectrophoretic characterization and immobilization of yeast cells in the microfluidic channel was attained by imposing the AC signal (electric field) across the IDEs. Six IDE structures with varied ratios of spacing to width were used separately to conduct DEP characterization studies on yeast cells. A signal generator supplied the electric field. The AC signal frequency was varied from 1 kHz to 20 MHz (limited by the signal generator with the maximum output voltage of 20 V). The syringe pump loaded the cell suspension into the chip. Yeast suspensions were driven through the channel with a flow rate of 50 µm s^{-1}. When the cells reached the region of the IDEs, the flow was stopped, and when the cells were settled (after 15 s), then the flow rate was increased to 1 µm s^{-1} and was kept constant during the DEP immobilization. Initiation of the high flow rate fluid through the channel results in discards of the cells from the channel rather than immobilization and entrapment of the cells to the IDEs. Disconnecting the AC signal from the electrodes after cell immobilization results in desorption of the entire entrapped cells from the IDEs and removing the cells from the channel. Cells trajectory were observed under the microscope for various voltages, frequencies, and flow rates. A CCD camera with 5× objective was used during experiments to capture and record videos and images of the microfluidic channel. An AC signal with 20 Vpp (peak-to-peak) and 1 MHz, where yeast cells experience pDEP, was applied as the initial input signal.

2.6. Finite Element Simulation

When an external electric field is imposed on the fluidic medium and suspending cells, medium and cells are being polarized. As shown in Figure 6, a net DEP force is induced in the direction of the high electric field intensity as a result of the non-uniform electric field distribution.

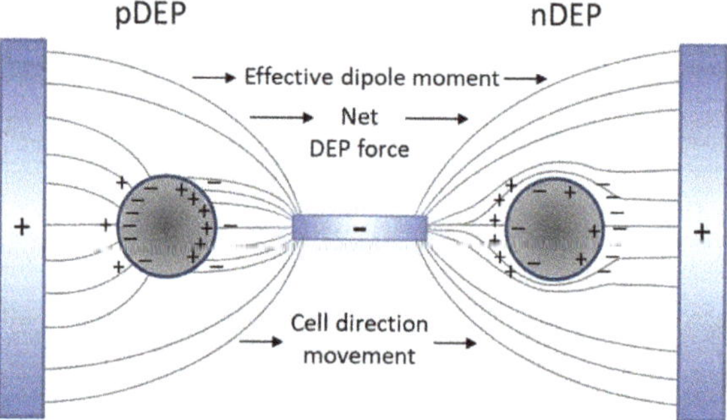

Figure 6. The induced dipole of the cell and medium with the presence of the non-uniform electric field and generation of pDEP and nDEP.

Since this force is unique for every biological or nonbiological particle and exploits the differences in their dielectric properties, it can be used for characterization and manipulation of the cells in a fluidic medium. The time-dependent DEP on a cell in an inhomogeneous and time-varying electric field is proportional to the volume of the cell, as shown in the following equation [50]:

$$F_{DEP}(t) = 2\pi\varepsilon_m r^3\, Re[f_{CM}]\nabla E^2_{rms} \tag{1}$$

where ε_m, is the fluidic medium permittivity, r is particle radius, E_{rms} is the root-mean-square of the electric field strength and $Re[f_{CM}]$ is the real part of the Clausius-Mossotti (CM) factor as defined in the equation below [50]:

$$f_{CM} = \frac{\varepsilon^*_c - \varepsilon^*_m}{\varepsilon^*_c + 2\varepsilon^*_m} \tag{2}$$

$$\varepsilon^* = \varepsilon - i\frac{\sigma}{\omega} \tag{3}$$

where ε^*_c and ε^*_m are the complex permittivity of the cell and the suspending medium, respectively. Complex permittivity is a function of conductivity (σ) and angular frequency of the applied electric field (ω). The f_{CM} of biological cells, such as yeast cells in this study, can be evaluated by modeling concentric layers with different dielectric properties [50]. Based on Equations (2) and (3), f_{CM} is a frequency dependent parameter and a function of relative magnitude of the cell with respect to its medium [50]. When the cell is more polarizable than the medium ($Re[f_{CM}] > 0$), positive DEP (pDEP) moves the cells towards the maximum electric field intensity locations. When the cells are less polarizable than the medium ($Re[f_{CM}] < 0$), they experience negative DEP (nDEP) which pushes them towards the zones of minimum electric field intensity.

The f_{CM}, as a function of the electric field frequency, for both live and dead yeast cells suspended in DI-water was numerically calculated using MATLAB and myDEP software [51], based on the two-shell model [50], where cells are assumed to possess two concentric layers of various electric and dielectric properties, as shown in Figure 7. Table 2 represents the dielectric values of yeast cells [50], and DI-water used for simulations. For live cells, the real part of the CM factor is bounded between 0.9 and ~−0.2. For dead cells this value is bounded between ~0.6 and ~−0.2. Variation of the applied signal frequency to the electrodes gives rise to DEP force in two opposite directions, which results in pDEP and nDEP. For the frequency ranges below crossover frequency (fc), $Re[f_{CM}]$ is positive, while for higher frequencies $Re[f_{CM}]$ is negative. The transition from the pDEP (top half) to nDEP (bottom half) which occurs at around 45 MHz and 1.45 MHz for live and dead yeast, respectively, is called crossover frequency (fc). This is a specific frequency at which the intrinsic properties of cells can be defined.

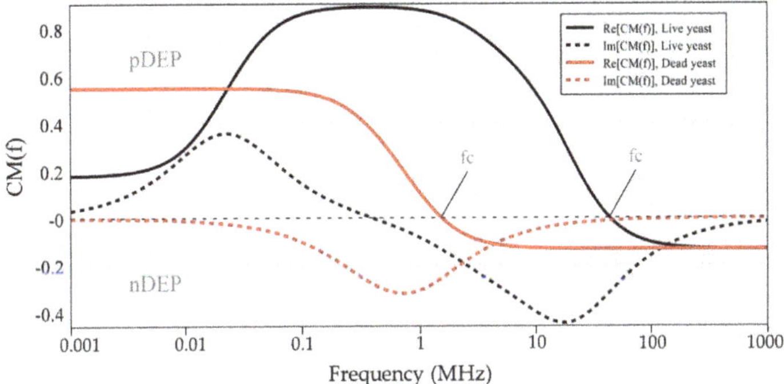

Figure 7. The Clausius-Mossotti factor of live and dead yeast cells suspended in DI-water as a function of frequency, using a two-shell model with the yeast parameters listed in Table 2.

Table 2. Yeast cell [50] and DI-water dielectric properties.

MUT [1]	Permittivity			Conductivity (S/m)		
Di-water	78			1×10^{-3}		
Yeast	cp [2]	cm [3]	cw [4]	cp [2]	cm [3]	cw [4]
Live yeast cell	50	6	60	0.2	2.5×10^{-7}	1.4×10^{-2}
Dead yeast cell	50	6	60	7×10^{-3}	1.6×10^{-3}	1.5×10^{-3}

[1] Material under test. [2] Cytoplasm. [3] Cell membrane. [4] Cell wall.

Several IDE structures were modeled in COMSOL 5.5 using a 2D model [47]. Using this model, the trajectory of the live and dead yeast cells through the microfluidic channel was simulated and the capabilities of different IDEs for cell immobilization were evaluated. Predictions from the developed simulations were compared with the experimental results. Figure 8 shows the electric potential contours (lines) imposed on the IDEs and the electric field distribution (arrows) over the electrodes in the channel. The electric field is intensified between IDE fingers and maximized at the rectangular corners of the electrodes [47]. This results in the non-uniform distribution of the electric field. The magnitude of the electric field over the IDEs decays with the distance over the IDEs towards the top of the microfluidic channel.

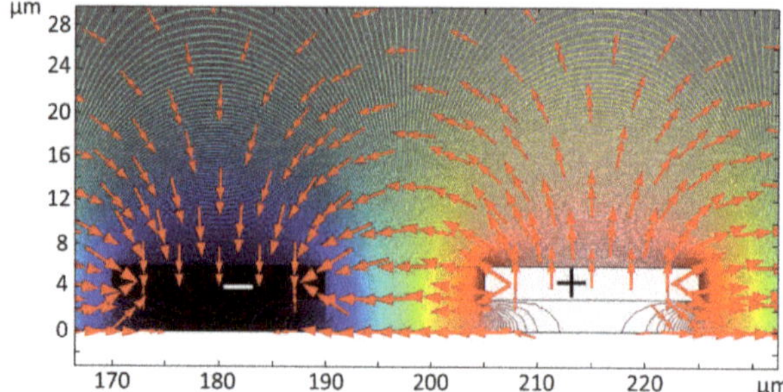

Figure 8. Numerical simulation results for the electric potential applied (line contour) to the IDEs and the electric field distribution (red arrows).

3. Results

3.1. Dielectrophoretic Immobilization of Living Yeast Cells

After applying 20 Vpp at 1 MHz, obvious cell immobilization was observed. Cell entrapment started a few seconds after cell suspension had reached the IDEs. Trapping started at the electrode edges, where the electric field gradient intensity was increased. Figure 9 demonstrates the cell trapping performance of different IDE geometries at three time intervals. Cell polarization effect in a non-uniform electric field led to dipole-dipole interaction and forming of pearl chains of cells [21]. As shown in Figure 9, the number of immobilized yeast cells is strongly dependent on the geometrical ratio of the IDEs. The number of immobilized yeast cells is reduced with increasing IDEs geometrical ratio. Cell entrapment reaches its highest efficiency by using IDEs with the largest finger width (45 µm) and smallest gap spacing (5 µm) between adjacent fingers. Furthermore, immobilization of cells using IDEs with higher geometrical ratios is challenging. This is due to the fact that the entrapped cells desorb from the IDEs with greater S/W ratios throughout the immobilization process. In addition, it is observed that increasing an electrode width, with a constant spacing size, expand the number of entrapped cells. These optical observations are in line with the simulation results presented in Figure 10. Finite element modeling (FEM) simulations support the impact of geometrical parameters on the DEP effect. The probability of immobilizing cells increases with reduced geometrical (spacing to width) ratios of the IDEs. Figure 10 illustrates the impact of various geometry ratios on the immobilization probability (IP) of yeast cells. The immobilization probability is defined by the number of trapped cells to the total number of cells suspended in the fluidic medium.

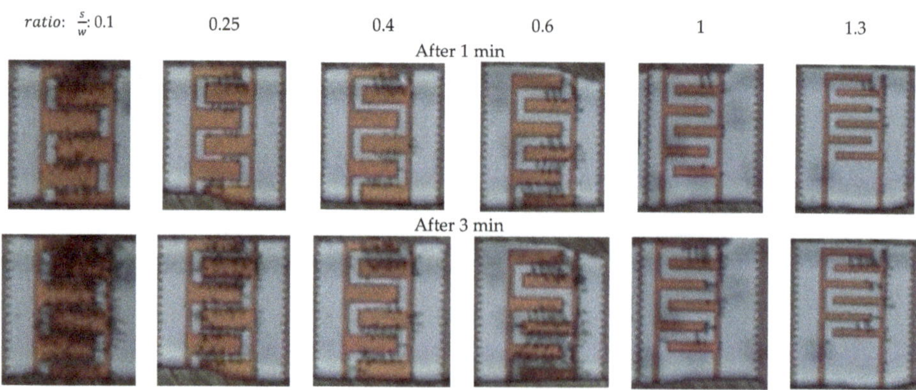

Figure 9. *Cont.*

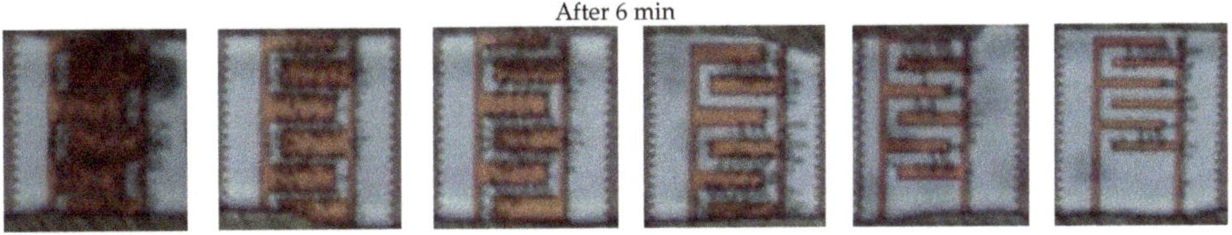

Figure 9. Micrographs of the immobilized yeast cells as a function of IDEs geometry (spacing to width ratios) at three time intervals of 1 min, 3 min, and 6 min. Immobilization conditions: 20 Vpp, 1 MHz, 1 μm·s^{-1} flow rate.

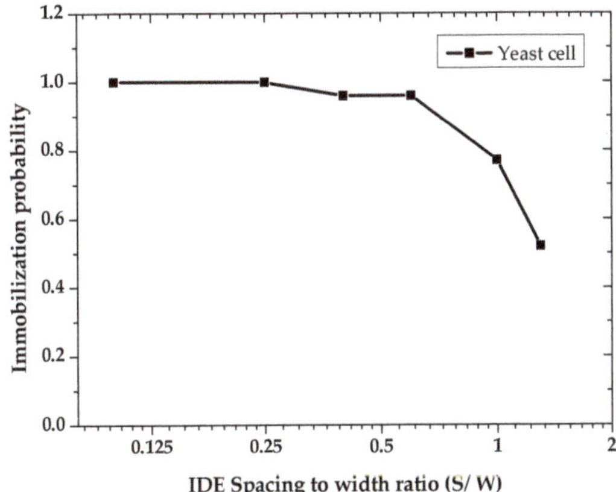

Figure 10. Calculated impact of IDE's spacing to width ratio on the immobilization of yeast cells. Finite element modeling (FEM) for the dielectrophoretic immobilization of yeast cells was performed at 20 Vpp and 1 MHz.

Increasing the IDEs ratio reversely impacts the IP. Furthermore, it can be seen that by keeping the spacing constant at 20 μm, there has been a steady decline in IP with decreasing width (at ratios of 0.6, 1, and 1.3). Therefore, it can be concluded that the DEP efficiency is highly influenced by IDE's dimensional ratio (S/W). As illustrated in Figure 11, experimental results indicate that smaller IDE ratios reduce the required peak voltages for DEP immobilization.

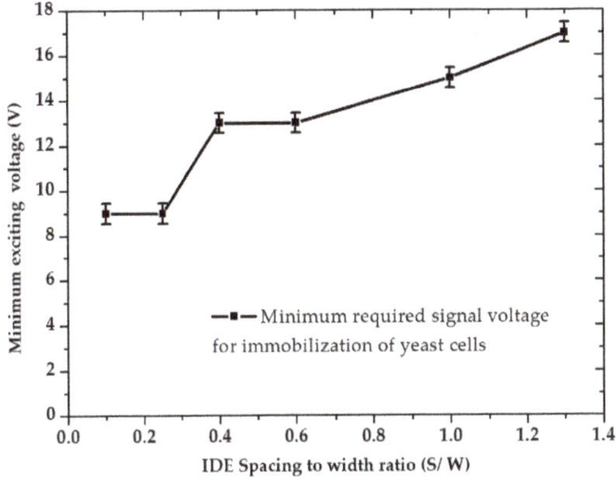

Figure 11. Minimum DEP driving voltage for the immobilization of yeast cells as function of spacing to width ratio, according to experimental data.

For the largest IDE dimensional ratio, the required voltage values were roughly twice as much as the required value for the smallest IDE ratio. In addition, cell trapping was increased drastically with an increasing voltage trend. By increasing the gradient of the electric potential, the DEP force was raised and thus more cells were attracted to the electrodes.

According to the experimental results, live cells can be trapped in a frequency range between 700 kHz and 9 MHz. Using frequencies above 10 MHz and below 300 kHz cells experience a repulsive force, which results in significant desorption of the immobilized cells from the electrodes. Figure 12 shows the frequency dependency of cell entrapment at 20 Vpp. In the frequency range between 900 kHz and 6 MHz, desorption rates are very low. Desorption rates increase drastically at lower frequencies ($f_o \leq 300$ kHz) due to weak pDEP and at higher frequencies (10 MHz $\leq f_o$) due to nDEP.

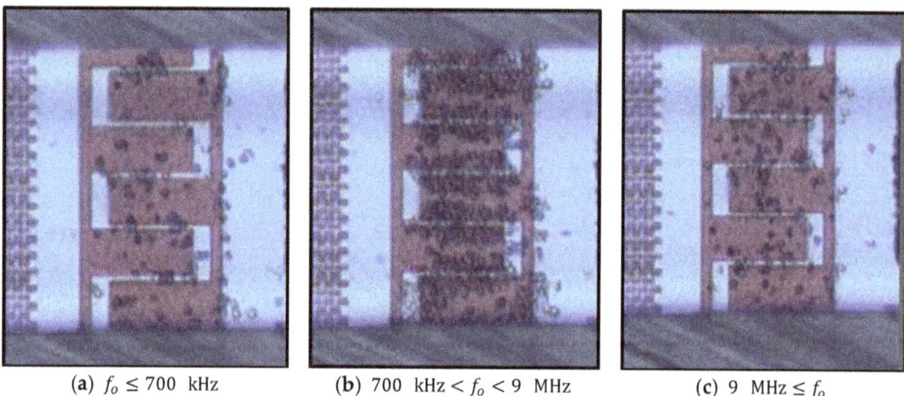

(a) $f_o \leq 700$ kHz   (b) 700 kHz $< f_o < 9$ MHz   (c) 9 MHz $\leq f_o$

Figure 12. Frequency dependency of cell immobilization at 20 Vpp and a flow rate of 1 μm s^{-1}. (**a**) High desorption rate of immobilized yeast cells; (**b**) Very high cell immobilization rate and very low desorption rate of immobilized cells; (**c**) High desorption rate of immobilized cells.

At frequencies between 700 kHz and 900 kHz and between 7 MHz and 9 MHz, the immobilization stability of cells mainly dropped, and with the passage of the time immobilized cells gradually tend to desorb from the electrodes. Figure 13 illustrates the weak immobilization as a result of imposing an AC voltage of 9 Vpp at 8 MHz.

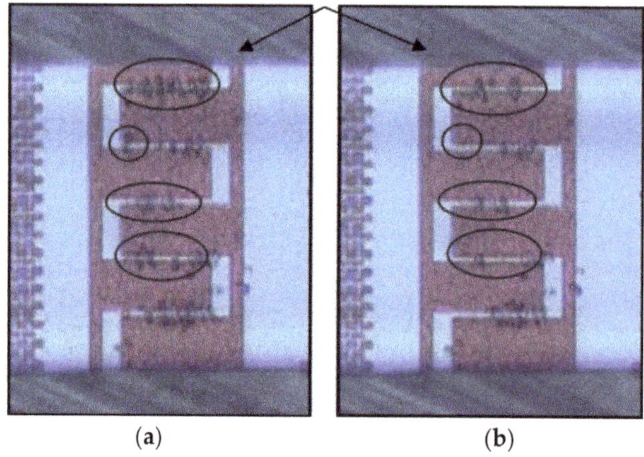

(**a**)   (**b**)

Figure 13. An example of weak immobilization at 8 MHz (9 V_{pp}, and 1 μm·s^{-1}): (**a**) Immobilization of cells after 2 min, and (**b**) 3 min, partial desorption of trapped cells from the IDEs.

The experimental results are in agreement with the simulation (see Figure 7), the yeast cells are forced by pDEP at frequencies lower than 10 MHz. However, fc was found to be 10 MHz, whereas the simulated prediction of fc is about 45 MHz. The large difference between the simulated and the experimentally evaluated values of fc could be caused by the simplicity of the used model in terms of cell wall, membrane, cytoplasm, and nucleus size.

3.2. Dielectrophoretic Separation of Live and Dead Yeast Cells

In order to differentiate between live and dead yeast cells, the impact of the AC frequency on the trajectory of live and dead cells was investigated experimentally and simulated using COMSOL. On the basis of the simulation shown in Figure 7, the DEP response of live and dead cells is significantly different at high frequencies [16]. Yeast cells subject to some modifications when they expose to heat shock. Such modifications could include a reduction in size and alterations to the dielectric properties of the cell. Due to the heat shock, intercellular water is reduced, and yeast cell experience water stress. This results in the wrinkling of the cell membrane and reduction in the cell diameter, which is associated with the shrinkage of the entire yeast cell [52–54]. The fc of dead yeasts occurs at ~1.45 MHz because dead cells lose their viability due to an impaired membrane. Their cytoplasmic conductivity is decreased while their membrane conductivity is increased. An impaired membrane of a dead cell polarized differently when it is exposed to an electric field. Thus, due to these dielectric discrepancies, responses of live and dead cells to the fringing electric field are dissimilar [16,21,55]. Our experimental results demonstrate that dead cells experienced an attractive force between 40 kHz and 1.45 MHz and can be trapped at the IDEs between 60 kHz and 1.45 MHz. At lower frequencies (<40 kHz), no DEP response was observed for dead yeasts.

Taking into account the distinct DEP responses of live and dead yeast cells at specific frequency ranges, preliminary demonstrations of the separation were performed. The concept was also simulated using COMSOL Multiphysics. Our simulation results were reasonably consistent with experimental results. Figure 14a shows a snapshot image for the separation of dead cells from live cells, where live cells immobilized at the IDEs at 3 MHz and 20 Vpp. In contrast, separation of live cells from dead ones, when dead cells immobilized at the electrodes at 90 kHz and 20 Vpp, is shown in Figure 14b. Therefore, separation of live and dead cells using the proposed method is achievable. Furthermore, the desorbed dead or live cells can be collected at the outlet of the microfluidic channel for further investigations and analyses.

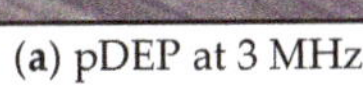

(a) pDEP at 3 MHz   **(b) pDEP at 90 kHz**

Figure 14. Separation of live and dead cells based on DEP characteristics. (**a**) Immobilize of live cells at the IDEs, while dead ones repel from the IDEs and leave the microfluidic channel at 20 Vpp, and (**b**) Immobilization of dead yeast cells at the IDEs).

Figure 15 demonstrates how DEP can be used diversely to isolate specific cells from a mixture of cells using the distinct DEP behavior of live and dead yeast cells. Keeping the signal frequency constant at 3 MHz, dead cells were separated from live cells, which were simultaneously immobilizing at the IDEs (Figure 15a). The opposite situation happens when a signal frequency in the range of 70 kHz $\leq f_0 \leq$ 100 kHz is applied to the system (Figure 15b).

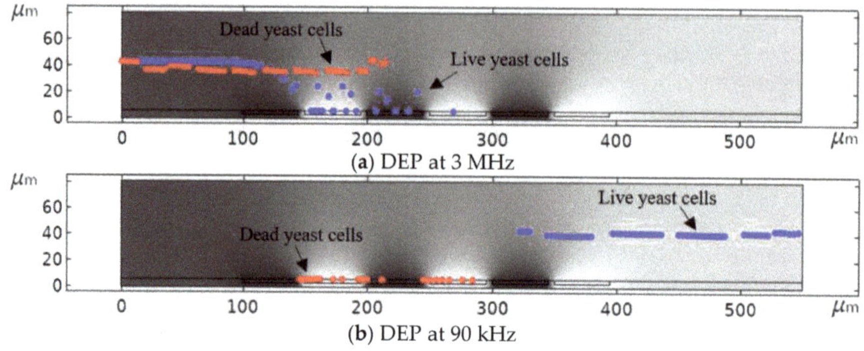

Figure 15. Simulated DEP separation of live and dead yeast cells: (**a**) Separation of dead cells from live cells, and (**b**) vice versa as a function of frequency.

4. Conclusions

A silicon-based CMOS integrated microfluidic device for immobilization of live and dead yeast cells via DEP was investigated. The device has been used to differentiate between live and dead yeast cells based on the selective DEP forces, pDEP and nDEP. IDEs with various geometrical parameters were studied. The effect of DEP force on the trajectory of yeast cells as functions of voltage, frequency, flow-rate, and IDE geometry was studied experimentally. Besides, finite element modeling was used to predict the trajectories of the cells. Experimental and simulation results demonstrate that based on the specific properties of cells, the microfluidic device can be used to immobilize and separate specific cells by varying the AC frequency. It was found that the experimental results are in agreement with the simulation.

Author Contributions: Methodology, H.M.E.; Validation, H.M.E.; Software, H.M.E., and P.S.Z.; Formal analysis, H.M.E., R.H., and C.W.; Data curation, H.M.E.; Writing—Original draft preparation, H.M.E.; Writing—Review and editing, H.M.E., P.S.Z., R.H. and C.W.; Supervision, C.W.; Funding acquisition, C.W. All authors have read and agreed to the published version of the manuscript.

Funding: This research was funded by the Brandenburg Ministry of Science, Research, and Cultural Affairs for the project within the StaF program in Germany. The publication of this article was funded by the Open Access Fund of the Leibniz Association.

Acknowledgments: The authors would like to thank the technology department of IHP for the fabrication of the sensor chip. The authors would like to thank Rita Winkler for the fabrication of the manifold.

Conflicts of Interest: The authors declare no conflict of interest.

References

1. Wang, X.-B.; Huang, Y.; Gascoyne, P.R.C.; Becker, F.F. Dielectrophoretic manipulation of particles. *IEEE Trans. Ind. Appl.* **1997**, *33*, 660–669. [CrossRef]
2. Altomare, L.; Leonardi, A.; Medoro, G.; Guerrieri, R.; Tartagni, M.; Manaresi, N. A lab-on-a-chip for cell detection and manipulation. *IEEE Sens. J.* **2003**, *3*, 317–325.
3. Henslee, E.A.; Sano, M.B.; Rojas, A.D.; Schmelz, E.M.; Davalos, R.V. Selective concentration of human cancer cells using contactless dielectrophoresis. *Electrophoresis* **2011**, *32*, 2523–2529. [CrossRef] [PubMed]
4. Jen, C.P.; Chen, T.W. Selective trapping of live and dead mammalian cells using insulator-based dielectrophoresis within open-top microstructures. *Biomed. Microdevices* **2009**, *11*, 597. [CrossRef] [PubMed]

5. Adekanmbi, E.O.; Srivastava, S.K. Dielectrophoretic applications for disease diagnostics using lab-on-a-chip platforms. *Lab Chip* **2016**, *16*, 2148–2167. [CrossRef] [PubMed]

6. Li, H.; Zheng, Y.; Akin, D.; Bashir, R. Characterization and modeling of a microfluidic dielectrophoresis filter for biological species. *J. Microelectromechanical Syst.* **2005**, *14*, 103–112. [CrossRef]

7. West, J.; Becker, M.; Tombrink, S.; Manz, A. Micro Total Analysis Systems: Latest Achievements. *Anal. Chem.* **2008**, *80*, 4403–4419. [CrossRef]

8. Zhang, H.; Chang, H.; Neuzil, P. DEP-on-a-Chip: Dielectrophoresis Applied to Microfluidic Platforms. *Micromachines* **2019**, *10*, 423. [CrossRef]

9. Pesch, G.R.; Lorenz, M.; Sachdev, S.; Salameh, S.; Du, F.; Baune, M.; Boukany, P.E.; Thöming, J. Bridging the scales in high-throughput dielectrophoretic (bio-)particle separation in porous media. *Sci. Rep.* **2018**, *8*, 10480. [CrossRef]

10. Suehiro, J.; Zhou, G.; Imamura, M.; Hara, M. Dielectrophoretic filter for separation and recovery of biological cells in water. *IEEE Trans. Ind. Appl.* **2003**, *39*, 1514–1521. [CrossRef]

11. Abd Rahman, N.; Ibrahim, F.; Yafouz, B. Dielectrophoresis for Biomedical Sciences Applications: A Review. *Sensors* **2017**, *17*, 449. [CrossRef] [PubMed]

12. Yang, J.; Huang, Y.; Wang, X.; Wang, X.-B.; Becker, F.F.; Gascoyne, P.R.C. Dielectric Properties of Human Leukocyte Subpopulations Determined by Electrorotation as a Cell Separation Criterion. *Biophys. J.* **1999**, *76*, 3307–3314. [CrossRef]

13. Pohl, H.A. The Motion and Precipitation of Suspensoids in Divergent Electric Fields. *J. Appl. Phys.* **1951**, *22*, 869–871. [CrossRef]

14. Park, K.; Kabiri, S.; Sonkusale, S. CMOS dielectrophoretic Lab-on-Chip platform for manipulation and monitoring of cells. In Proceedings of the 2015 37th Annual International Conference of the IEEE Engineering in Medicine and Biology Society (EMBC), Milan, Italy, 25–29 August 2015; Volume 2015, pp. 7530–7533.

15. Huan, Z.; Chu, H.K.; Yang, J.; Sun, D. Characterization of a honeycomb-like scaffold with dielectrophoresis-based patterning for tissue engineering. *IEEE Trans. Biomed. Eng.* **2017**, *64*, 755–764. [CrossRef]

16. Patel, S.; Showers, D.; Vedantam, P.; Tzeng, T.-R.; Qian, S.; Xuan, X. Microfluidic separation of live and dead yeast cells using reservoir-based dielectrophoresis. *Biomicrofluidics* **2012**, *6*, 034102. [CrossRef]

17. Piacentini, N.; Mernier, G.; Tornay, R.; Renaud, P. Separation of platelets from other blood cells in continuous-flow by dielectrophoresis field-flow-fractionation. *Biomicrofluidics* **2011**, *5*, 1–8. [CrossRef]

18. Cheng, I.-F.; Chang, H.-C.; Hou, D.; Chang, H.-C. An integrated dielectrophoretic chip for continuous bioparticle filtering, focusing, sorting, trapping, and detecting. *Biomicrofluidics* **2007**, *1*, 021503. [CrossRef]

19. Zeinali, S.; Cetin, B.; Oliaei, S.N.B.; Karpat, Y. Fabrication of continuous flow microfluidics device with 3D electrode structures for high throughput DEP applications using mechanical machining. *Electrophoresis* **2015**, *36*, 1432–1442. [CrossRef]

20. García-Sánchez, P.; Ramos, A.; Green, N.G.; Morgan, H. Experiments on AC electrokinetic pumping of liquids using arrays of microelectrodes. *IEEE Trans. Dielectr. Electr. Insul.* **2006**, *13*, 670–677. [CrossRef]

21. Yildizhan, Y.; Erdem, N.; Islam, M.; Martinez-Duarte, R.; Elitas, M. Dielectrophoretic separation of live and dead monocytes using 3D carbon-electrodes. *Sensors* **2017**, *17*, 2619. [CrossRef]

22. Yafouz, B.; Kadri, N.; Ibrahim, F. Microarray Dot Electrodes Utilizing Dielectrophoresis for Cell Characterization. *Sensors* **2013**, *13*, 9029–9046. [CrossRef] [PubMed]

23. Yafouz, B.; Kadri, N.; Ibrahim, F. Dielectrophoretic Manipulation and Separation of Microparticles Using Microarray Dot Electrodes. *Sensors* **2014**, *14*, 6356–6369. [CrossRef] [PubMed]

24. Becker, F.F.; Wang, X.B.; Huang, Y.; Pethig, R.; Vykoukal, J.; Gascoyne, P.R. Separation of human breast cancer cells from blood by differential dielectric affinity. *Proc. Natl. Acad. Sci.* **1995**, *92*, 860–864. [CrossRef] [PubMed]

25. Jang, L.-S.; Huang, P.-H.; Lan, K.-C. Single-cell trapping utilizing negative dielectrophoretic quadrupole and microwell electrodes. *Biosens. Bioelectron.* **2009**, *24*, 3637–3644. [CrossRef]

26. Khoshmanesh, K.; Zhang, C.; Tovar-Lopez, F.J.; Nahavandi, S.; Baratchi, S.; Kalantar-zadeh, K.; Mitchell, A. Dielectrophoretic manipulation and separation of microparticles using curved microelectrodes. *Electrophoresis* **2009**, *30*, 3707–3717. [CrossRef]

27. Pommer, M.S.; Zhang, Y.; Keerthi, N.; Chen, D.; Thomson, J.A.; Meinhart, C.D.; Soh, H.T. Dielectrophoretic separation of platelets from diluted whole blood in microfluidic channels. *Electrophoresis* **2008**, *29*, 1213–1218. [CrossRef]

28. Fatoyinbo, H.O.; Hoettges, K.F.; Hughes, M.P. Rapid-on-chip determination of dielectric properties of biological cells using imaging techniques in a dielectrophoresis dot microsystem. *Electrophoresis* **2008**, *29*, 3–10. [CrossRef]

29. Otto, S.; Kaletta, U.; Bier, F.F.; Wenger, C.; Hölzel, R. Dielectrophoretic immobilisation of antibodies on microelectrode arrays. *Lab Chip* **2014**, *14*, 998. [CrossRef]

30. Laux, E.-M.; Knigge, X.; Bier, F.F.; Wenger, C.; Hölzel, R. Aligned Immobilization of Proteins Using AC Electric Fields. *Small* **2016**, *12*, 1514–1520. [CrossRef]

31. Guha, S.; Jamal, F.I.; Schmalz, K.; Wenger, C.; Meliani, C. CMOS lab on a chip device for dielectric characterization of cell suspensions based on a 6 GHz Oscillator. In Proceedings of the 43rd European Microwave Conference, Nuremberg, Germany, 6–10 October 2013; pp. 471–474.

32. Flanagan, L.A.; Lu, J.; Wang, L.; Marchenko, S.A.; Jeon, N.L.; Lee, A.P.; Monuki, E.S. Unique Dielectric Properties Distinguish Stem Cells and Their Differentiated Progeny. *Stem Cells* **2008**, *26*, 656–665. [CrossRef]

33. Lyu, C.; Wang, J.; Powell-Palm, M.; Rubinsky, B. Simultaneous electroporation and dielectrophoresis in non-electrolytic micro/nano-electroporation. *Sci. Rep.* **2018**, *8*, 2481. [CrossRef] [PubMed]

34. Ning, Y.; Ma, X.; Multari, C.R.; Luo, X.; Gholizadeh, V.; Palego, C.; Cheng, X.; Hwang, J.C.M. Improved broadband electrical detection of individual biological cells. In Proceedings of the 2015 IEEE MTT-S International Microwave Symposium, Phoenix, AZ, USA, 17–22 May 2015; pp. 1–3.

35. Li, H.; Multari, C.; Palego, C.; Ma, X.; Du, X.; Ning, Y.; Buceta, J.; Hwang, J.C.M.; Cheng, X. Differentiation of live and heat-killed E. coli by microwave impedance spectroscopy. *Sens. Actuators B Chem.* **2018**, *255*, 1614–1622. [CrossRef]

36. Manczak, R.; Baristiran Kaynak, C.; Kaynak, M.; Palego, C.; Lalloue, F.; Pothier, A.; Saada, S.; Provent, T.; Dalmay, C.; Bessette, B.; et al. UHF-Dielectrophoresis Crossover Frequency as a New Marker for Discrimination of Glioblastoma Undifferentiated Cells. *IEEE J. Electromagn. RF Microw. Med. Biol.* **2019**, *3*, 191–198. [CrossRef]

37. Becker, H. Mind the gap! *Lab Chip* **2010**, *10*, 271–273. [CrossRef] [PubMed]

38. Iliescu, C.; Taylor, H.; Avram, M.; Miao, J.; Franssila, S. A practical guide for the fabrication of microfluidic devices using glass and silicon. *Biomicrofluidics* **2012**, *6*, 016505. [CrossRef] [PubMed]

39. Tiggelaar, R.M.; Van Male, P.; Berenschot, J.W.; Gardeniers, J.G.E.; Oosterbroek, R.E.; De Croon, M.H.J.M.; Schouten, J.C.; van den Berg, A.; Elwenspoek, M.C. Fabrication of a high-temperature microreactor with integrated heater and sensor patterns on an ultrathin silicon membrane. *Sens. Actuators A Phys.* **2005**, *119*, 196–205. [CrossRef]

40. Pipper, J.; Inoue, M.; Ng, L.F.-P.; Neuzil, P.; Zhang, Y.; Novak, L. Catching bird flu in a droplet. *Nat. Med.* **2007**, *13*, 1259–1263. [CrossRef]

41. Huang, Y.; Mason, A.J. Lab-on-CMOS integration of microfluidics and electrochemical sensors. *Lab Chip* **2013**, *13*, 3929. [CrossRef]

42. Guha, S.; Schumann, U.; Jamal, F.I.; Wagner, D.; Meliani, C.; Schmidt, B.; Wenger, C.; Wessel, J.; Detert, M. Integrated high-frequency sensors in catheters for minimally invasive plaque characterization. In Proceedings of the 20th European Microelectronics and Packaging Conference and Exhibition: Enabling Technologies for a Better Life and Future, EMPC, Friedrichshafen, Germany, 14–16 September 2015; pp. 1–6.

43. Guha, S.; Schmalz, K.; Meliani, C.; Krautschneider , W. CMOS MEMS based Microfluidic System for Cytometry at 5GHz. In Proceedings of the MFHS, Microfluidic Handling System, Entschede, The Netherlands, 10–12 October 2012.

44. Guha, S.; Wenger, C. Radio Frequency CMOS Chem-bio Viscosity Sensors based on Dielectric Spectroscopy. In *10th International Joint Conference on Biomedical Engineering Systems and Technologies*; SCITEPRESS—Science and Technology Publications: Porto, Portugal, 2017; pp. 142–148.

45. Guha, S.; Schmalz, K.; Wenger, C.; Herzel, F. Self-calibrating highly sensitive dynamic capacitance sensor: Towards rapid sensing and counting of particles in laminar flow systems. *Analyst* **2015**, *140*, 3262–3272. [CrossRef]

46. Matbaechi Ettehad, H.; Guha, S.; Wenger, C. Simulation of CMOS compatible sensor structures for dielectrophoretic biomolecule immobilization. In Proceedings of the COMSOL Conference—Bioscience and Bioengineering, Rotterdam, The Netherlands, 19 October 2017; p. 6.

47. Matbaechi Ettehad, H.; Yadav, R.K.; Guha, S.; Wenger, C. Towards CMOS Integrated Microfluidics Using Dielectrophoretic Immobilization. *Biosensors* **2019**, *9*, 77. [CrossRef]

48. Inac, M.; Wietstruck, M.; Goritz, A.; Cetindogan, B.; Baristiran-Kaynak, C.; Marschmeyer, S.; Fraschke, M.; Voss, T.; Mai, A.; Palego, C.; et al. BiCMOS Integrated Microfluidic Packaging by Wafer Bonding for Lab-on-Chip Applications. In Proceedings of the 2017 IEEE 67th Electronic Components and Technology Conference (ECTC), Orlando, FL, USA, 30 May–2 June 2017; pp. 786–791.

49. Guha, S.; Lisker, M.; Trusch, A.; Wolf, A.; Meliani, C.; Wenger, C. 12 GHz CMOS MEMS Lab-on-chip System for Detection of Concentration of Suspended Particles in Bio-suspensions. In *International Conference on Biomedical Electronics and Devices*; SCITEPRESS—Science and Technology Publications: Lisbon, Portugal, 2015; pp. 49–57.

50. Huang, Y.; Holzel, R.; Pethig, R.; Xiao-Bo, W. Differences in the AC electrodynamics of viable and non-viable yeast cells determined through combined dielectrophoresis and electrorotation studies. *Phys. Med. Biol.* **1992**, *37*, 1499–1517. [CrossRef] [PubMed]

51. Cottet, J.; Fabregue, O.; Berger, C.; Buret, F.; Renaud, P.; Frénéa-Robin, M. MyDEP: A new computational tool for dielectric modeling of particles and cells. *Biophys. J.* **2019**, *116*, 12–18. [CrossRef] [PubMed]

52. Piper, P.W. The heat shock and ethanol stress responses of yeast exhibit extensive similarity and functional overlap. *FEMS Microbiol. Lett.* **1995**, *134*, 121–127. [CrossRef] [PubMed]

53. Hallsworth, J.E. Ethanol-induced water stress in yeast. *J. Ferment. Bioeng.* **1998**, *85*, 125–137. [CrossRef]

54. Munoz, A.J.; Wanichthanarak, K.; Meza, E.; Petranovic, D. Systems biology of yeast cell death. *FEMS Yeast Res.* **2012**, *12*, 249–265. [CrossRef]

55. Schwan, H.P. Electrical Properties of Tissue and Cell Suspensions. In *Advances in Biological and Medical Physics*; Elsevier. Amsterdam, The Netherlands, 1957; Volume 5, pp. 147–209.

 *micromachines*

Article

Driving Waveform Design of Electrowetting Displays Based on an Exponential Function for a Stable Grayscale and a Short Driving Time

Zichuan Yi [1], **Zhenyu Huang [2,3,*]**, **Shufa Lai [3]**, **Wenyao He [2]**, **Li Wang [2,3]**, **Feng Chi [1]**, **Chongfu Zhang [1]**, **Lingling Shui [1,2]** and **Guofu Zhou [2,3]**

[1] College of Electron and Information, University of Electronic Science and Technology of China, Zhongshan Institute, Zhongshan 528402, China; yizichuan@163.com (Z.Y.); chifeng@semi.ac.cn (F.C.); cfzhang@uestc.edu.cn (C.Z.); Shuill@m.scnu.edu.cn (L.S.)

[2] Institute of Electronic Paper Displays, South China Academy of Advanced Optoelectronics, South China Normal University, Guangzhou 510006, China; hwy956005579@outlook.com (W.H.); creekxi@163.com (L.W.); guofu.zhou@m.scnu.edu.cn (G.Z.)

[3] Shenzhen Guohua Optoelectronics Tech. Co., Ltd., Shenzhen 518110, China; laishufa518@163.com

* Correspondence: 2018023170@m.scnu.edu.cn; Tel.: +86-0755-2941-5855

Received: 8 February 2020; Accepted: 12 March 2020; Published: 16 March 2020

Abstract: The traditional driving waveform of the electrowetting display (EWD) has many disadvantages, such as the large oscillation of the target grayscale aperture ratio and a long time for achieving grayscale. Therefore, a driving waveform based on the exponential function was proposed in this study. First, the maximum driving voltage value of 30 V was obtained by testing the hysteresis curve of the EWD pixel unit. Secondly, the influence of the time constant on the driving waveform was analyzed, and the optimal time constant of the exponential function was designed by testing the performance of the aperture ratio. Lastly, an EWD panel was used to test the driving effect of the exponential-function-driving waveform. The experimental results showed that a stable grayscale and a short driving time could be realized when the appropriate time constant value was designed for driving EWDs. The aperture ratio oscillation range of the gray scale could be reduced within 0.95%, and the driving time of a stable grayscale was reduced by 30% compared with the traditional driving waveform.

Keywords: electrowetting display; driving waveform; aperture ratio; exponential function; time constant

---

1. Introduction

In recent years, display technology has been widely used in all aspects of daily life [1,2]. In 2003, Hayes et al. proposed the EWD structure based on the principle of ink electrowetting [3], which has the advantages of low power consumption, high reflectivity, high contrast, and fast response speed. As one of new display technologies [4], increasingly more attention has been paid to EWDs. However, EWD technology has not been industrialized, and its driving technology is one of the limiting factors.

Grayscale is displayed in EWDs by applying a voltage sequence, called as driving waveform, which is used to control the form of colored ink in a pixel [5]. The ratio of a pixel area not covered by the ink to a whole pixel area is the aperture ratio, which can reflect the performance of EWDs [6]. Because the ink is divided into several parts, called the ink dispersion [7], the ink coverage area is constantly changing in the process of providing a fixed grayscale display. Therefore, an oscillation of the aperture ratio can be formed, which can reduce the number of grayscale levels and the visual experience of users. At present, some problems with EWDs are being solved, such as ink dispersion [8], charge capture [9,10], hysteresis [11], ink reflow [12], and large oscillations of the aperture ratio [13]. We have used industrial electrophoretic electronic display driver chips to drive TFT (thin-film transistor) EWDs and a four-level gray scale was realized using a PWM (pulse-width modulation) driving waveform [14]. However, the ink in the EWD pixel has obvious shrinkage and the oscillation range of the aperture ratio is relatively large due to a high switching frequency of the voltage in PWM, which has a bad effect when aiming to display stable grayscales [15]. In addition, by comparing the duty cycle of different PWMs, we found that the smaller the duty cycle of the PWM, the larger the oscillation range of the aperture ratio in the EWDs was [13]; in contrast, the larger the duty ratio, the smaller the oscillation range, which provides a reference for the driving waveform design of stable grayscales. Based on a PWM driving waveform, where the starting point of a driving waveform is set to 10 V as a maintained voltage, the optimized design of the maintained voltage and its timing can reduce the ink breakage, which can reduce the oscillation range of aperture ratio effectively; however, its driving time is extended [16]. At the same time, by optimizing the slope of the driving waveforms to control movement state of the ink in EWDs, the ink dispersion can be improved and the reflectivity can effectively be increased. However, the driving waveform increases the oscillation range of the aperture ratio and lengthens the driving time [17]. However, the damped oscillation of an electrowetting liquid drop can be optimized by changing the rising speed of the driving waveform, and the transition between the under-damping state and the over-damping state in the electrowetting lens is realized [18], which provides a reference direction for optimizing the driving waveform of the EWDs.

In order to display a stable grayscale and reduce the driving time in EWDs, an exponential-function-driving waveform was proposed in this study. By testing the impact of different time-constant values, we set parameters of the exponential function when designing the driving waveform. Compared with traditional driving waveforms, the exponential-function-driving waveform had a better driving performance.

2. Driving Principle of the EWD

2.1. Model of the EWD

The EWD was mainly composed of a glass substrate, indium tin oxide glass (ITO), hydrophobic insulation layer, pixel wall, conductive liquid (NaCl solution), colored ink, and so on [19,20], as shown in Figure 1.

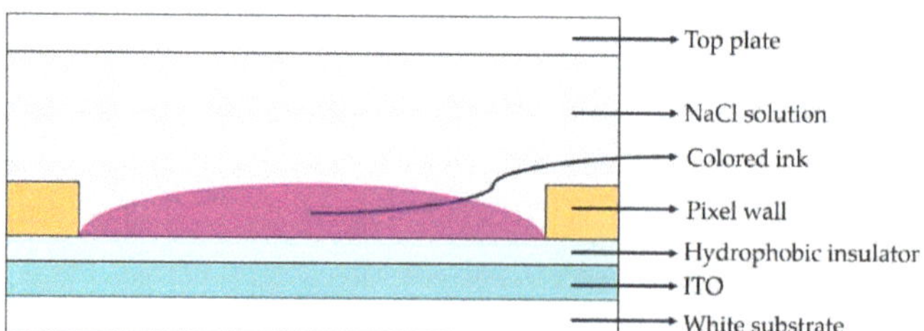

Figure 1. Structure of a pixel in electrowetting displays (EWDs). ITO: Indium tin oxide.

The ink is laid between a NaCl solution and a hydrophobic insulation layer when no voltage is applied, and the NaCl solution does not contact the hydrophobic insulation layer directly. Then, the color of ink is displayed, and the pixel is in the "off" state, as shown in Figure 2a. The interfacial tension between the NaCl solution and the hydrophobic insulation layer is changed by an electric field force when a certain amplitude of voltage is applied between the upper and lower electrodes. At this time, the NaCl solution contacts with the hydrophobic insulating layer directly, and the ink is pushed away.

Then, the pixel shows the white substrate, and the pixel is in the "on" state, as shown in Figure 2b. The ink can be spread on the hydrophobic insulating layer again when the voltage is removed such that the pixel can be continuously switched between "on" and "off " [21].

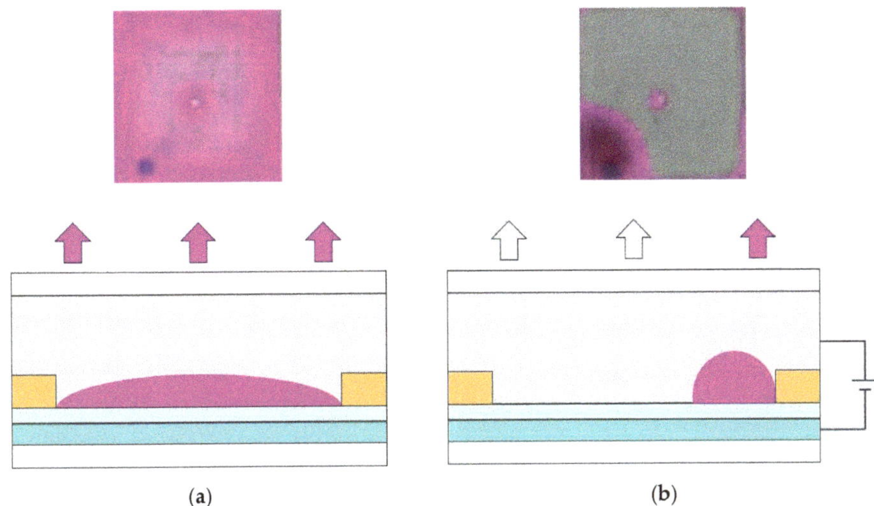

Figure 2. Ink state in an EWD pixel. (**a**) The pixel is "off" and the whole pixel is covered by ink such that the color of the ink is reflected. (**b**) The pixel is "on" and the ink is pushed into a corner of the pixel such that the white substrate is reflected.

For typical submillimeter-scale EWD pixels, the main driving forces of the ink movement are the interface tension and electrostatic force. The interface boundary is usually described using the Young–Lippmann equation [22,23], as shown in Equation (1):

$$\cos \theta = \cos \theta_0 + \frac{1}{2} \frac{\varepsilon_0 \varepsilon_{FP}}{d \gamma_{OW}} U^2 \tag{1}$$

In Equation (1), θ is the contact angle of the Lippmann, and θ_0 is the equilibrium contact angle between the NaCl solution and the hydrophobic insulating layer. ε_0 represents the dielectric constant of the vacuum, ε_{FP} represents dielectric constant of the hydrophobic insulating layer, d represents thickness of the hydrophobic insulating layer, U represents the driving voltage of the pixel, and γ_{OW} represents the interfacial tension between the ink and the NaCl solution.

It can be seen from Equation (1) that the contact angle of the NaCl solution on the surface of the hydrophobic insulating layer can be controlled by changing the electric potential. That is to say, due to the change of electric potential energy, the surface performance of the hydrophobic insulating layer is changed, which produces a change of the conductive liquid contact angle on the hydrophobic insulating layer surface [24]. With an increase of the electric potential, the interface tension between the hydrophobic insulation layer and the NaCl solution becomes larger, the contact angle can also become larger, and the voltage difference breaks the balanced state. At this time, the NaCl solution is in direct contact with the hydrophobic insulation layer and the colored ink is pushed to a corner of the pixel to expose the white substrate. The area ratio of the exposed white substrate is called the aperture ratio, which can directly represent the reflectivity of EWDs [25], where the expression is shown in Equation (2):

$$A = \left[1 - \left(\frac{S_{oil}}{S_{pix}} \right) \right] \times 100\% \tag{2}$$

where A is the aperture ratio, S_{oil} is the area of ink when it is pushed to a corner of a pixel, and the area of a whole pixel is S_{pix}.

2.2. Design Principle of an Exponential-Function-Driving Waveform

The exponential-function-driving waveform expression is shown in Equation (3):

$$U = U_0 \times \left(1 - e^{-\frac{t}{\tau}}\right) \tag{3}$$

where U represents the real-time voltage of a driving waveform; U_0 represents the maximum voltage in a driving waveform, which is a fixed value; t represents driving time; and τ represents the time constant of the exponential function. We can analyze the influence of τ on the exponential-function-driving waveform by changing its value.

We take the derivative of Equation (3) to obtain Equation (4):

$$U'(t) = \frac{U_0}{\tau} e^{-\frac{t}{\tau}}. \tag{4}$$

When $\tau = 1$ ms:

$$\begin{cases} U'(t) = U_0 e^{-t} \\ U'(1) = U_0 e^{-1}, \quad U'(2) = U_0 e^{-2} \\ U'(3) = U_0 e^{-3}, \quad U'(4) = U_0 e^{-4} \end{cases} \tag{5}$$

When $\tau = 6$ ms:

$$\begin{cases} U'(t) = U_0 e^{-\frac{t}{6}} \\ U'(1) = U_0 e^{-\frac{1}{6}}, \quad U'(2) = U_0 e^{-\frac{1}{3}} \\ U'(3) = U_0 e^{-\frac{1}{2}}, \quad U'(4) = U_0 e^{-\frac{2}{3}} \end{cases} \tag{6}$$

In Equations (5) and (6), U_0 is a fixed value. Therefore, the change rate of the derivative in an exponential function is determined by τ, which is used to change the rising speed of driving waveforms.

3. Design of the Exponential-Function-Driving Waveform

3.1. Maximum Voltage of the Driving Waveform

EWD has the characteristic of displaying hysteresis in the pixel [26,27]. In the rising-voltage stage, the ink area could be reduced slightly between 9 V and 16 V, but its aperture ratio was too small, and the pixel was not opened. However, the aperture ratio could be increased to about 50% when the voltage reached 17 V. At this time, the pixel was in an "on" state. Therefore, 17 V was the threshold voltage of a pixel. Then, the aperture ratio could reach the maximum value when the voltage was increased to 30 V. In the falling-voltage stage, the aperture ratio could be closed to 0 when the voltage was reduced to 5 V. Hence, the aperture ratio of the pixels was different between the rising-voltage stage and the falling-voltage stage at the same voltage value, as shown in Figure 3. In addition, it is easy to break through the hydrophobic insulation layer if the voltage is too high, which can cause damage to the EWD pixels [28]. Therefore, the maximum voltage of the driving waveform was set as 30 V.

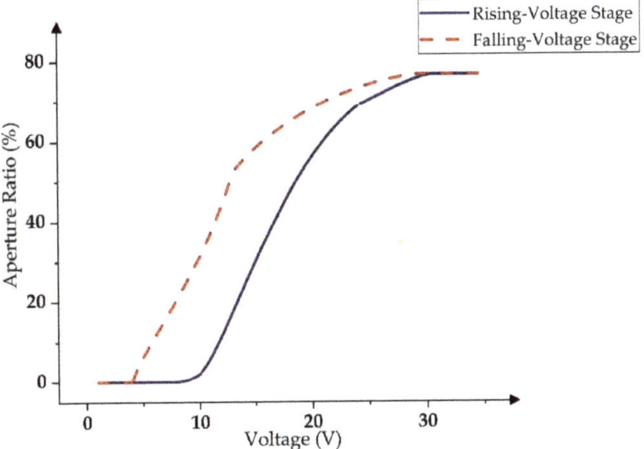

Figure 3. Rising-voltage stage and falling-voltage stage of the hysteresis curve in EWDs.

3.2. Time Constant of the Exponential Function

By taking the derivative of the exponential function, the derivative value at each time point can be obtained. Furthermore, it can be changed by using different time constants τ. The smaller the value of τ, the greater the variation of the exponential function derivative when the frequency and the maximum voltage U_0 are fixed. On the contrary, the larger the value of τ, the smaller the exponential function derivative change is. Therefore, the voltage-rising speed can be controlled by changing the derivative, which can be used to control the oscillation range of the aperture ratio for stable grayscales and reduce the driving time. As shown in Figure 4, the exponential function with $\tau = 1$ ms was the steepest one, whereas the exponential function with $\tau = 6$ ms was the smoothest one. The exponential function was close to PWM due to its fast voltage-rising speed when $\tau < 1$ ms, and it could not reach the maximum driving voltage in a driving cycle due to its slow voltage-rising speed if $\tau > 6$ ms at the same time. Therefore, the value of τ was in the range of 1 ms $\leq \tau \leq$ 6 ms.

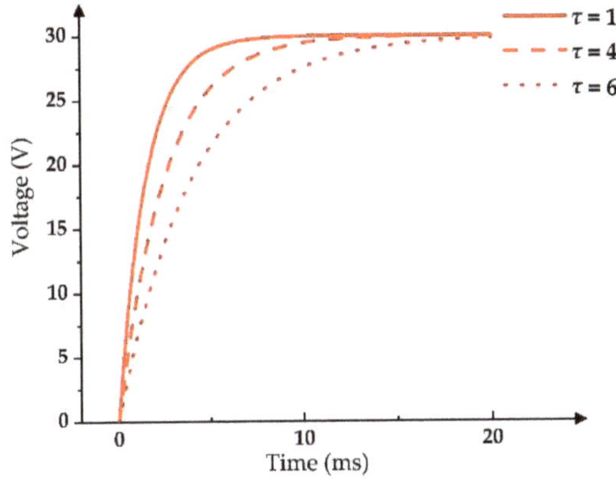

Figure 4. Exponential functions with different τ values.

Ink in the pixel can be pushed toward corners using a conductive NaCl solution by applying a driving voltage to EWDs when a grayscale needs to be displayed. However, the ink may be divided into several parts in this process; this phenomenon is called ink dispersion [29]. With the increase of the driving time, some smaller inks can gradually move closer to the larger one, as shown in Figure 5. In this process, the grayscale is unstable until all the ink is in one corner of a pixel.

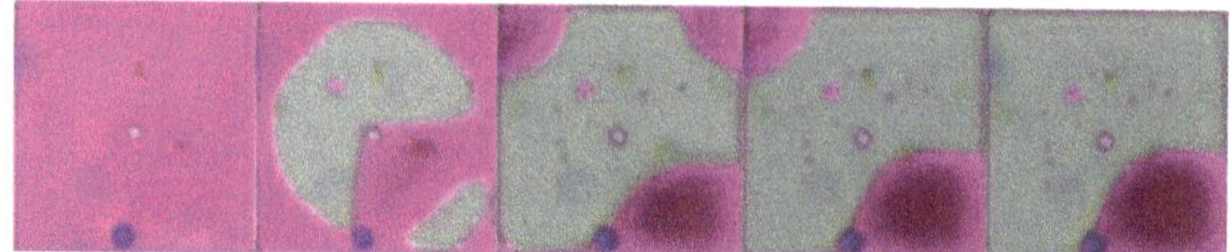

Figure 5. The ink formation change process in the driving process of a pixel.

Therefore, there is an oscillation phenomenon of the aperture ratio in all driving processes, which has a bad effect when aiming to display stable grayscales. The aperture ratio of the pixel decreased from 77% to 76%, and then increased to 77%, where the oscillation range of the aperture ratio is expressed as ΔR in this paper. In addition, the driving time for a stable grayscale is expressed as ST, where a complete driving process is shown in Figure 6.

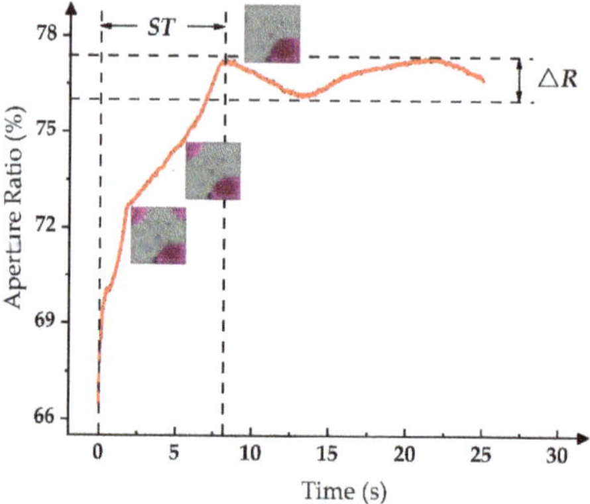

Figure 6. The aperture ratio change process when a driving waveform is applied to the EWD.

Then, the relationship between τ and ΔR can be tested by using the exponential-function-driving waveform to drive the EWD, and the same method can be used to test the relationship between τ and ST. In the testing, the value range of τ was from 1 ms to 6 ms, where the relationship curves are shown in Figure 7.

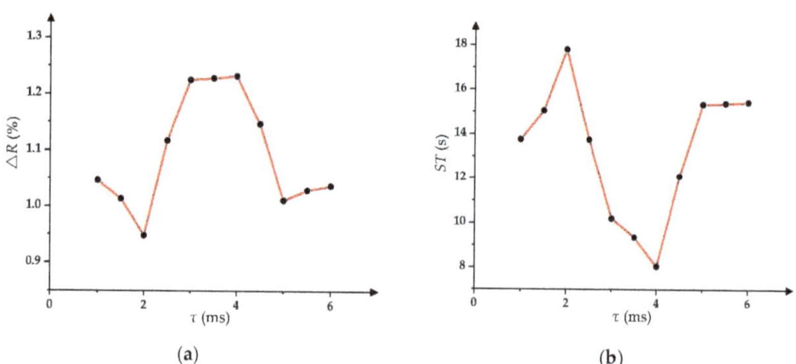

(a)

(b)

Figure 7. The relationship among parameters of the exponential-function-driving waveform. (**a**) The relationship between τ and ΔR. (**b**) The relationship between τ and the stable time.

As shown in Figure 7a, with the increase of τ, ΔR reached the minimum value of 0.95% when $\tau = 2$ ms. Furthermore, ΔR reached the maximum value of 1.25% when $\tau = 4$ ms. As shown in Figure 7b, with the increase of τ, the ST reached the minimum value of 8 s when $\tau = 4$ ms, where the maximum ST was 18 s when $\tau = 2$ ms.

4. Experimental Results and Discussion

4.1. Testing System

In order to test the effect of driving waveforms, we developed an experimental platform to record the aperture ratio and the driving process to measure the oscillation range of the aperture ratio and the driving time of the grayscales. The driving system consisted of a computer, a waveform generator and a high-voltage amplifier, which were used to edit the driving waveform for the EWD. The testing system consisted of a microscope and a high-speed camera. Relevant equipment information is shown in Table 1. Furthermore, an EWD panel was used in the experiment, where its parameters are shown in Table 2.

Table 1. Experimental platform for testing EWDs.

Category	Computer	Waveform Generator	High Voltage Amplifier	Microscope	Industrial Camera
Manufacturer	Lenovo	Tektronix	Agitek	Cossim	Koppace
Model	M425	AFG-3052C	ATA-2022H	SZ760T2LED	KP-AF200

Table 2. Parameters of the EWD panel.

Panel Size (cm)	Ink Color	Ink Thickness (μm)	Resolution	Pixel Size (μm²)	Pixel Wall Height (μm)	Pixel Wall Size (μm²)	Hydrophobic Layer Thickness (μm)
3.5×3.5	Purple	5	50×50	400×400	6	15×15	1

In the EWD panel, the hydrophobic insulation layer was not resistant to a high voltage, and the withstand voltage value was under 40 V. The hydrophobic layer material was Teflon AF1600, and its solution was FC-43, which is a fluorocarbon solvent from the 3M company (Maplewood, MN, USA); the pixel wall grid material was transparent polyimide; the ITO glass substrate came from Shenzhen Laibao Hi Tech Co., Ltd. (Shenzhen, China); the glass thickness was 0.7 mm and the impedance is 100 Ω/cm^2; and the electrolyte solution was NaCl with a concentration of 1×10^{-4} mol/L. Deionized water was obtained using an ultra-pure ultraviolet water purification system.

In the testing process, the driving waveform was edited using MATLAB (Mathworks, Natick, MA, USA) and ArbExpress software (V3.4, Tektronix, Beaverton, OR, USA), and then it was sent to the signal generator via a serial port. Since the maximum output voltage of the signal generator was 5 V, the EWD could not be driven by the signal generator directly. Therefore, the signal generator should be connected with a high-voltage amplifier for outputting the driving voltage. In addition, an industrial camera was used to record the ink state of the EWD in real time using a microscope. The microscope could magnify the pixel 200 times and the resolution of the industrial camera was 1920×1080. The real-time picture of a pixel could be captured every 50 ms and the aperture ratio of the pixels could be calculated for each picture using binary processing. Therefore, the relationship between the aperture ratio and the time could be obtained. During the experiment, the surrounding temperature was 25 °C and the humidity was 60%. Next, two comparative experiments are shown as follows.

4.2. Oscillation Range of the Aperture Ratio

Two traditional driving waveforms were used to analyze and compare the oscillation range ΔR. In the design of the traditional PWM driving waveform, the maximum voltage was still set to 30 V, and its duty cycle was changed from 50% to 90%, as shown in Figure 8a. The smaller the duty cycle, the smaller the ratio of the maximum voltage in an entire driving cycle.

The testing results of ΔR in PWM are shown in Figure 8b; as the duty cycle increased, ΔR decreased. The ΔR was as high as 6.1% when the duty cycle was at 50%, and the ΔR was reduced to 1.4% when the duty cycle was at 90%, which met the requirement for stable grayscales. However, the ink in a pixel reflowed when the duty cycle was larger than 90% and the pixel could be closed at this time [14,30]. Therefore, this situation was meaningless for the experiment.

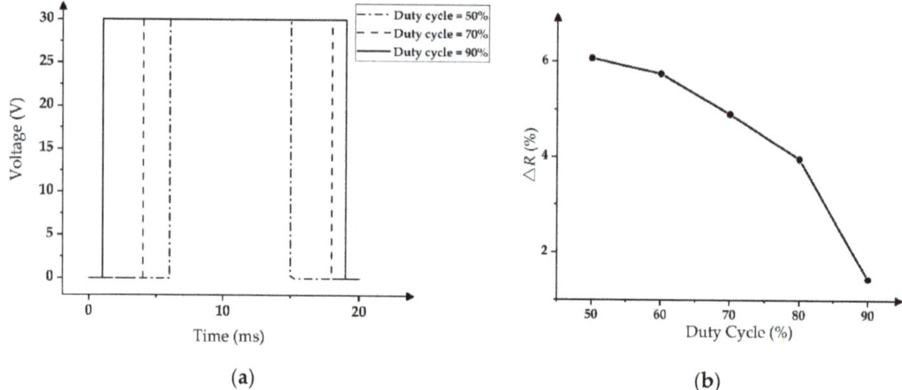

Figure 8. The pulse width modulation (PWM) driving waveform and its aperture ratio performance [14,30]. (**a**) The PWM driving waveform with different duty cycles. (**b**) Relationship between the duty cycle and ΔR.

In the contrast experiment of slope-driving waveforms, the maximum driving voltage was set to 30 V and the slope value range of driving waveforms was from 3 V/ms to 10 V/ms, as shown in Figure 9a, where K is the slope [17,31]. Furthermore, the relationship between K and ΔR is shown in Figure 9b, where the maximum value of ΔR was 2.2% when K was 3 V/ms; the minimum value of ΔR was 0.3% when K was 10 V/ms.

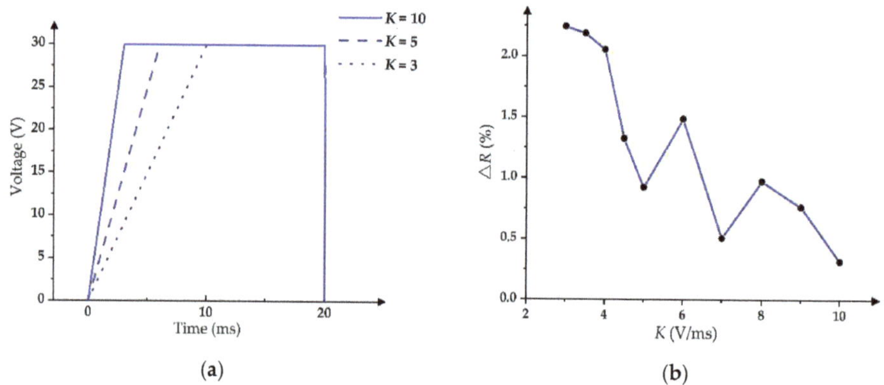

Figure 9. Slope-driving waveform and its aperture ratio performance [17,31]. (**a**) Slope-driving waveform with different K values. (**b**) Relationship between K and ΔR.

In order to compare ΔR among three kinds of driving waveforms, the parameters of three driving waveforms were optimized to obtain the minimum value of ΔR. The minimum ΔR of the PWM was 1.4% when the duty cycle was 90%, the minimum ΔR of the slope-driving waveforms was 0.3% when K was 10 V/ms, and the minimum ΔR of the exponential-function-driving waveform was 0.95 when τ was 2 ms. As shown in Figure 10, the PWM could cause a large grayscale oscillation but the slope-driving waveform and the exponential-function-driving waveform could solve the problem. As a result, more levels of the grayscale display could be achieved.

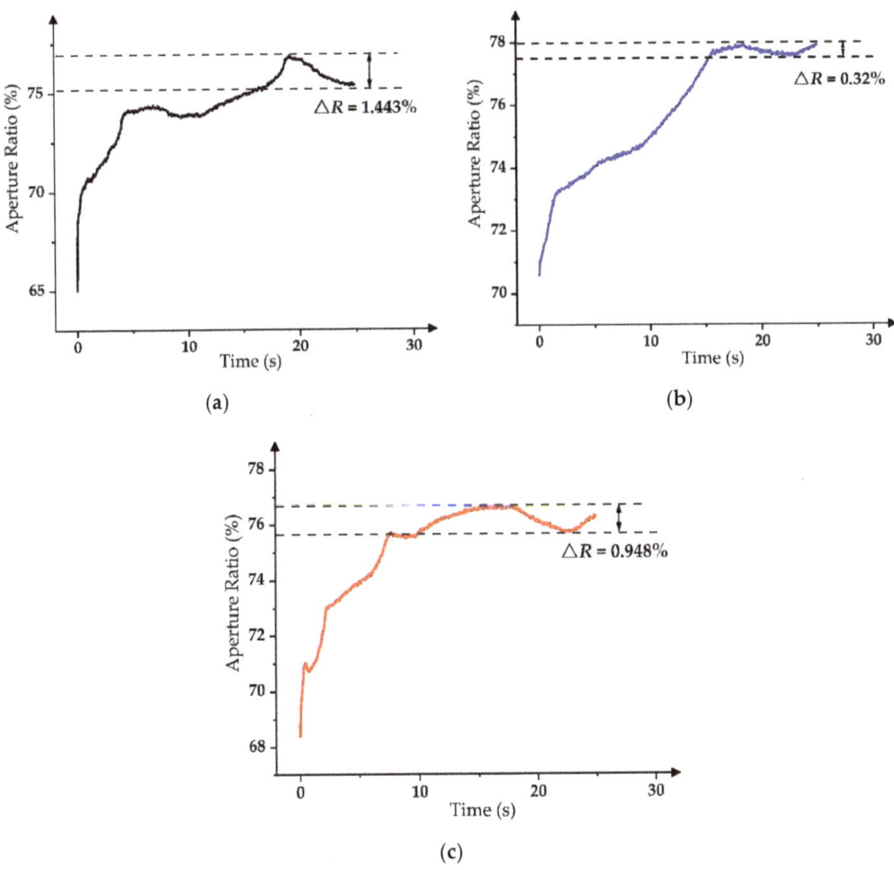

Figure 10. Aperture ratio response of different driving waveforms. (**a**) PWM with a duty cycle of 90% [14,30]. (**b**) Slope-driving waveform with K = 10 V/ms [17,31]. (**c**) Exponential function with $\tau = 2$ ms.

4.3. Driving Time of a Stable Grayscale

In the PWM driving waveform, ΔR was 1.4% when the duty cycle was 90%, and its ST for stable grayscales was 19 s. Furthermore, ΔR values of other duty cycle PWM waveforms were greater than 3%; therefore, the PWM driving waveform with a 90% duty cycle was the best one. In the slope-driving waveform, the ST could reach the minimum value when K was 5 V/ms, as shown in Figure 11.

In the exponential-function-driving waveform, ST was the shortest when τ was 4 ms. As shown in Figure 12, the ST of the PWM driving waveform was 19 s. Furthermore, the ST of the slope-driving waveform could be shortened to 10.5 s. However, the ST of the exponential-function-driving waveform was the shortest with 8 s when τ was 4 ms. Therefore, the driving time for stable grayscales could be reduced by using the exponential-function-driving waveform, and the oscillation of the aperture ratio could be restrained in a certain range at the same time. Then, it could be used to effectively improve the static display ability of EWDs.

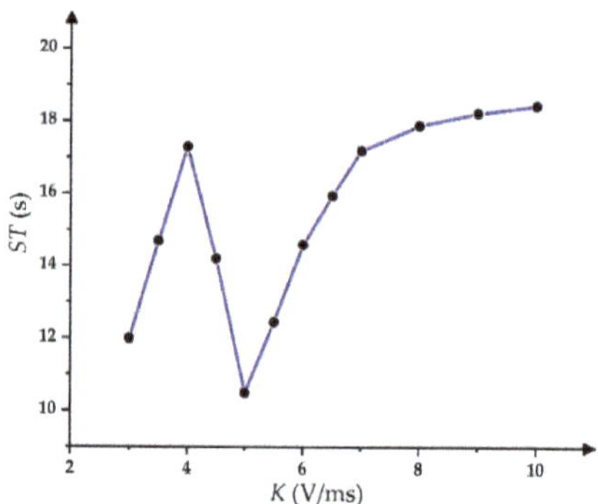

Figure 11. The relationship between *ST* and the slope of the driving waveform.

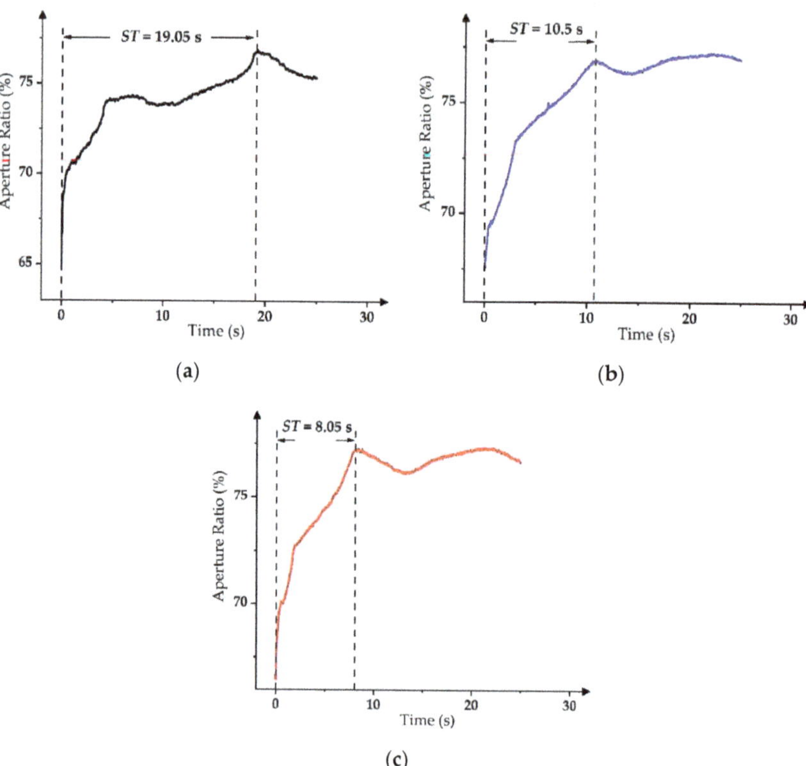

Figure 12. Driving time for the stable grayscale display of different driving waveforms. (**a**) The PWM with a duty cycle of 90% [14,30]. (**b**) The slope-driving waveform with K = 5 V/ms [17,31]. (**c**) The exponential function with τ = 4 ms.

5. Conclusions

In order to reduce the oscillation of the aperture ratio and the driving time for stable grayscales in pixels, an exponential-function-driving waveform was proposed for improving the performance of EWDs. Then, the time constant of the exponential function was optimized by testing the driving process of the aperture ratio. The results showed that the grayscale oscillation could be controlled in a certain range by using the exponential-function-driving waveform when its time constant was 2 ms, which could be used to reduce the flicker of EWDs. In addition, the shortest driving time of the stable grayscale could be obtained by using the proposed driving waveform when its time constant was 4 ms, which could improve the static display performance of EWDs.

Author Contributions: Z.Y. and S.L. designed this project. Z.H. and Z.Y. carried out most of the experiments and data analysis. W.H. and L.W. performed part of the experiments and helped with discussions during manuscript preparation. F.C. and L.S. contributed to the data analysis and correction. C.Z. and G.Z. gave suggestions on project management and provided helpful discussions on the experimental results. All authors have read and agreed to the published version of the manuscript.

Funding: This research was funded by the Guangdong Basic and Applied Basic Research Foundation (no. 2020A1515010420), the Key Research Platforms and Research Projects in Universities and Colleges of Guangdong Provincial Department of Education (no. 2018KQNCX334), the Zhongshan Innovative Research Team Program (no. 180809162197886), the Guangdong Government Funding (no. 2014A010103024), the Zhongshan Institute high-level talent scientific research startup fund project (no. 416YKQ04), the Project for Innovation Team of Guangdong University (no. 2018KCXTD033), and the National Key R&D Program of China (nos. 2018YFB0407100-02 and 2016YFB0401502).

Conflicts of Interest: The authors declare no conflict of interest.

References

1. Zhang, N.; Gu, W.; Xu, F.Q.; Yang, X.F. Electronic Paper Display Technology and its Application Development. *Adv. Mater. Res.* **2014**, *926*, 2333–2336. [CrossRef]
2. Riahi, M.; Brakke, K.A.; Alizadeh, E.; Shahroosvand, H. Fabrication and characterization of an electrowetting display based on the wetting–dewetting in a cubic structure. *Optik* **2016**, *127*, 2703–2707. [CrossRef]
3. Hayes, R.A.; Feenstra, B.J. Video-speed electronic paper based on electrowetting. *Nature* **2003**, *425*, 383–385. [CrossRef] [PubMed]
4. Mugele, F.; Baret, J.-C. Electrowetting: From basics to applications. *J. Phys. Condens. Matter* **2005**, *17*, R705–R774. [CrossRef]
5. Luo, Z.; Zhao, W.; Cao, Y.; Lin, W.; Zhou, G. A high-resolution and intelligent dead pixel detection scheme for an electrowetting display screen. *Opt. Rev.* **2017**, *25*, 18–26. [CrossRef]
6. Zhou, M.; Zhao, Q.; Tang, B.; Groenewold, J.; Hayes, R.; Zhou, G. Simplified dynamical model for optical response of electrofluidic displays. *Displays* **2017**, *49*, 26–34. [CrossRef]
7. Fan, M.; Zhou, R.; Jiang, H.; Zhou, G. Effect of liquid conductivity on optical and electric performances of the electrowetting display system with a thick dielectric layer. *Results Phys.* **2020**, *16*, 102904. [CrossRef]
8. Hsieh, W.-L.; Lin, C.-H.; Lo, K.-L.; Lee, K.-C.; Cheng, W.-Y.; Chen, K.-C. 3D electrohydrodynamic simulation of electrowetting displays. *J. Micromech. Microeng.* **2014**, *24*, 125024. [CrossRef]
9. Chen, Y.-C.; Chiu, Y.-H.; Lee, W.-Y.; Liang, C.-C. 56.3: A Charge Trapping Suppression Method for Quick Response Electrowetting Displays. *SID Symp. Dig. Tech. Pap.* **2010**, *41*, 842–845. [CrossRef]
10. Wu, H.; Dey, R.; Siretanu, I.; Ende, D.V.D.; Shui, L.; Zhou, G.; Mugele, F. Electrically Controlled Localized Charge Trapping at Amorphous Fluoropolymer-Electrolyte Interfaces. *Small* **2019**, *16*, e1905726. [CrossRef]
11. Gao, J.; Mendel, N.; Dey, R.; Baratian, D.; Mugele, F. Contact angle hysteresis and oil film lubrication in electrowetting with two immiscible liquids. *Appl. Phys. Lett.* **2018**, *112*, 203703. [CrossRef]
12. Giraldo, A.; Massard, R.; Mans, J.; Derckx, E.; Aubert, J.; Mennen, J. 10.3: Ultra low-power Electrowetting-based Displays Using Dynamic Frame Rate Driving. *SID Symp. Dig. Tech. Pap.* **2011**, *42*, 114–117. [CrossRef]
13. Yi, Z.; Liu, L.; Wang, L.; Li, W.; Shui, L.; Zhou, G. A Driving System for Fast and Precise Gray-Scale Response Based on Amplitude-Frequency Mixed Modulation in TFT Electrowetting Displays. *Micromachines* **2019**, *10*, 732. [CrossRef] [PubMed]

14. Yi, Z.; Shui, L.; Wang, L.; Jin, M.; Hayes, R.; Zhou, G. A novel driver for active matrix electrowetting displays. *Displays* **2015**, *37*, 86–93. [CrossRef]

15. Yang, G.; Liu, L.; Zheng, Z.; Henzen, A.; Xi, K.; Bai, P.; Zhou, G. A portable driving system for high-resolution active matrix electrowetting display based on FPGA. *J. Soc. Inf. Disp.* **2019**, *28*, 287–296. [CrossRef]

16. Zhao, Q.; Tang, B.; Dong, B.; Li, H.; Zhou, R.; Guo, Y.; Dou, Y.; Deng, Y.; Groenewold, J.; Henzen, A.V.; et al. Electrowetting on dielectric: Experimental and model study of oil conductivity on rupture voltage. *J. Phys. D Appl. Phys.* **2018**, *51*, 195102. [CrossRef]

17. Zhang, X.-M.; Bai, P.-F.; Hayes, R.A.; Shui, L.-L.; Jin, M.-L.; Tang, B.; Zhou, G.-F. Novel Driving Methods for Manipulating Oil Motion in Electrofluidic Display Pixels. *J. Disp. Technol.* **2015**, *12*, 200–205. [CrossRef]

18. Supekar, O.D.; Zohrabi, M.; Gopinath, J.T.; Bright, V.M. Enhanced Response Time of Electrowetting Lenses with Shaped Input Voltage Functions. *Langmuir* **2017**, *33*, 4863–4869. [CrossRef]

19. Wu, H.; Hayes, R.; Li, F.; Henzen, A.; Shui, L.; Zhou, G. Influence of fluoropolymer surface wettability on electrowetting display performance. *Displays* **2018**, *53*, 47–53. [CrossRef]

20. Choi, S.; Lee, J. Open-structure electrowetting display with capacitive sensing feedback system. In Proceedings of the 2015 28th IEEE International Conference on Micro Electro Mechanical Systems (MEMS), Estoril, Portugal, 18–22 January 2015; pp. 956–959. [CrossRef]

21. Zhou, R.; Fu, S.; Jiang, H.; Li, X.; Zhou, G. Thermal accelerated aging study of water/fluoropolymer/ITO contact in electrowetting display systems. *Results Phys.* **2019**, *15*, 102737. [CrossRef]

22. Dou, Y.; Wang, B.; Jin, M.; Yu, Y.; Zhou, G.; Shui, L. A review on self-assembly in microfluidic devices. *J. Micromech. Microeng.* **2017**, *27*, 113002. [CrossRef]

23. Dou, Y.; Tang, B.; Groenewold, J.; Li, F.; Yue, Q.; Zhou, R.; Li, H.; Shui, L.; Henzen, A.; Zhou, G. Oil Motion Control by an Extra Pinning Structure in Electro-Fluidic Display. *Sensors* **2018**, *18*, 1114. [CrossRef] [PubMed]

24. Wu, H.; Shui, L.; Li, F.; Hayes, R.; Henzen, A.; Mugele, F.; Zhou, G. Large-Area High-Contrast Hydrophobic/Hydrophilic Patterned Surface for Robust Electrowetting Devices. *ACS Appl. Nano Mater.* **2019**, *2*, 1018–1026. [CrossRef]

25. Chen, Z.; Lin, S.; Lin, Z.; Liao, Q.; Li, T.; Tang, B. Design of Video Display Driving System for Low-power Electrowetting Display. *Acta Photonica Sin.* **2020**, *49*, 222002. [CrossRef]

26. Chiang, H.-C.; Tsai, Y.-H.; Yan, Y.-J.; Huang, T.-W.; Mang, O.-Y. Oil defect detection of electrowetting display. *Opt. Eng. Appl.* **2015**, *9575*, 957514. [CrossRef]

27. Lin, S.; Zeng, S.; Qian, M.; Lin, Z.; Guo, T.; Tang, B. Improvement of display performance of electrowetting displays by optimized waveforms and error diffusion. *J. Soc. Inf. Disp.* **2019**, *27*, 619–629. [CrossRef]

28. Dou, Y.; Chen, L.; Li, H.; Tang, B.; Henzen, A.; Zhou, G. Photolithography Fabricated Spacer Arrays Offering Mechanical Strengthening and Oil Motion Control in Electrowetting Displays. *Sensors* **2020**, *20*, 494. [CrossRef]

29. Xie, Y.; Sun, M.; Jin, M.; Zhou, G.; Shui, L. Two-phase microfluidic flow modeling in an electrowetting display microwell. *Eur. Phys. J. E* **2016**, *39*, 16. [CrossRef]

30. Luo, Z.J.; Zhang, W.N.; Liu, L.W.; Xie, S.; Zhou, G. Portable multi-gray scale video playing scheme for high-performance electrowetting displays. *J. Soc. Inf. Disp.* **2016**, *24*, 345–354. [CrossRef]

31. Yi, Z.; Feng, W.; Wang, L.; Liu, L.; Lin, Y.; He, W.; Shui, L.; Zhang, W.; Zhang, Z.; Zhou, G. Aperture Ratio Improvement by Optimizing the Voltage Slope and Reverse Pulse in the Driving Waveform for Electrowetting Displays. *Micromachines* **2019**, *10*, 862. [CrossRef]

Article

Driving Waveform Design of Electrophoretic Display Based on Optimized Particle Activation for a Rapid Response Speed

Wenyao He [1,2], **Zichuan Yi** [1,*], **Shitao Shen** [2], **Zhenyu Huang** [2,3], **Linwei Liu** [2,3], **Taiyuan Zhang** [2,3], **Wei Li** [2,3], **Li Wang** [2,3], **Lingling Shui** [1,2], **Chongfu Zhang** [1] and **Guofu Zhou** [2,3]

[1]   College of Electron and Information, University of Electronic Science and Technology of China, Zhongshan Institute, Zhongshan 528402, China; wenyao.he@guohua-net.com (W.H.); Shuill@m.scnu.edu.cn (L.S.); cfzhang@uestc.edu.cn (C.Z.)

[2]   Institute of Electronic Paper Displays, South China Academy of Advanced Optoelectronics, South China Normal University, Guangzhou 510006, China; shenshitao@m.scnu.edu.cn (S.S.); 2018023170@m.scnu.edu.cn (Z.H.); yoeksome@sina.com (L.L.); taiyuan.zhang@guohua-oet.com (T.Z.); Wei.li@guohua-oet.com (W.L.); creekxi@163.com (L.W.); guofu.zhou@m.scnu.edu.cn (G.Z.)

[3]   Shenzhen Guohua Optoelectronics Tech. Co., Ltd., Shenzhen 518110, China

*   Correspondence: yizichuan@zsc.edu.cn; Tel.: +86-0755-2941-5855

Received: 3 April 2020; Accepted: 13 May 2020; Published: 14 May 2020

Abstract: Electrophoretic displays (EPDs) have excellent paper-like display features, but their response speed is as long as hundreds of milliseconds. This is particularly important when optimizing the driving waveform for improving the response speed. Hence, a driving waveform design based on the optimization of particle activation was proposed by analyzing the electrophoresis performance of particles in EPD pixels. The particle activation in the driving waveform was divided into two phases: the improving particle activity phase and the uniform reference grayscale phase. First, according to the motion characteristics of particles in improving the particle activity phase, the real-time EPD brightness value can be obtained by an optical testing device. Secondly, the derivative of the EPD brightness curve was used to obtain the inflection point, and the inflection point was used as the duration of improving particle activity phase. Thirdly, the brightness curve of the uniform reference grayscale phase was studied to set the driving duration for obtaining a white reference grayscale. Finally, a set of four-level grayscale driving waveform was designed and validated in a commercial E-ink EPD. The experimental results showed that the proposed driving waveform can cause a reduction by 180 ms in improving particle activity phase and 120 ms in uniform reference grayscale phase effectively, and a unified reference grayscale can be achieved in uniform reference grayscale phase at the same time.

Keywords: electrophoretic display; driving waveform; particle activation; response speed; reference grayscale

---

1. Introduction

Electrophoretic display (EPD) technology has been a research topic of interest for many years due to its wide market potential [1]. As a paper-like display technology, EPDs have the advantages of paper-like display effect, ultra-low power consumption, and being readable under bright light [2,3]. In recent years, EPDs have been widely used in e-books, electronic labels, and smart watches, amongst others [4,5]. The grayscale display of an EPD depends on the distribution of black and white particles in a microcapsule, which can be driven by voltage timing applied on the pixel, and the voltage timing is called the driving waveform. As a core part of EPDs, the length of the driving waveform can affect the response speed of EPDs directly. Therefore, it is of great significance to improve the response speed and reduce visibility of the ghost image of EPDs by optimizing the driving waveform design [6].

Currently, the driving waveform is divided into three phases: erase original image phase, particle activation phase, and new image phase [7]. In driving processes, the photoelectric properties of the microcapsule system are nonlinear, and it is difficult to form a grayscale display accurately [8]. A lot of time is required to unify the spatial position of particles for driving to a target grayscale, and this process takes hundreds of milliseconds, or even one second [9]. The erase original image phase can obtain a stable state which is in a white or black state. The new image phase is used for driving the EPD to a new grayscale [10]. When a particle has been kept at the same state for a long period, the mobility of charged particles becomes much lower than others. The particle activity can be improved by driving the panel several times between two optical extremes (black/white extremes states) in the particle activation phase [11,12]. In EPDs, the distribution of individual black or white particles is random, so it is necessary to find a relatively stable grayscale as a reference grayscale. The white grayscale is usually used as the reference grayscale, and the display of other grayscale are obtained by using the white grayscale as the driving start point. This driving mode has been used up to now, and plays an important role in EPD products [13]. However, the particle activation phase is longer than the other two phases and has a significant impact on the response speed of EPDs. In order to optimize the activation phase, the particle activation can be optimized by a high-frequency voltage mode, but it is difficult to obtain a stable reference grayscale and the driving duration can reach 960 ms [14]. Some scholars have removed the particle activation phase and added a response latency to eliminate the original image, but the reference grayscale uniformity cannot be obtained, resulting in ghosting after writing new images many times [15]. At the same time, some short waveforms have been proposed, which were divided into three phases: zero voltage bias phase, high-frequency activation phase, and driving phase [16,17], but with reduced display quality of EPDs [18,19].

By analyzing the electrophoresis performance of particles, a driving waveform design of EPDs based on optimized particle activation phase for a rapid response speed was proposed in this study. The activation phase of the driving waveform was divided into the improving particle activity phase and the uniform reference grayscale phase. In one phase, the derivative of the EPD brightness curve was used to obtain the inflection point, which was used as the duration of improving particle activity phase. In the other phase, the brightness curve of the uniform reference grayscale phase was studied to set the driving duration for obtaining a white reference grayscale. Currently, the designed waveforms have been validated in commercial E-ink EPDs. This can shorten the driving duration, improve the response speed, and reduce visibility of the ghost image effectively.

2. System Design Principle

2.1. Display Principle

The motion and distribution of electrophoretic particles in microcapsules is very complex [20]. The trajectories of charged particles interfere with each other, and the attraction and repulsion among particles exist at the same time [21]. During a grayscale display process in EPDs, the spatial position of the particles contained in the microcapsule determines the grayscale value in a pixel. The positively charged black particles can move toward the common electrode by the force of the electric field when a positive voltage is applied to the pixel electrode, and the negatively charged white particles can move toward the pixel electrode such that the display observed by the human eye is black, as shown in Figure 1 [22]. When all of the black particles are driven to the common electrode, the state is referred to as the black extreme state. The white particles move to the common electrode when a negative voltage is applied to the pixel electrode, and the display observed by the human eye is white [23]. When all of white particles have been driven to the common electrode, the state is referred to as the white extreme state. The voltage sequence applied on pixels is called the driving waveform, and its quality directly determines the display effect of EPDs. The internal electrophoretic particles cannot be displaced when the electric field is not applied to the pixel, which is referred to as bistability EPDs [24].

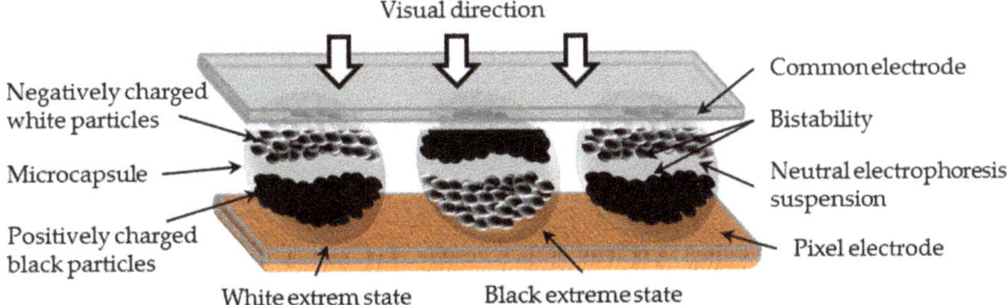

Figure 1. Schematic of an electrophoretic display (EPD) structure. The bottom is the pixel electrode, the middle is the microcapsule which includes a neutral electrophoresis suspension and white and black electrophoretic particles, while the top is the common electrode. The spatial position of particles cannot be changed after the removal of the pixel electrode voltage, which is referred to as bistability.

With the driving of the electric field, the charged particles are subjected to electric field forces as Equation (1):

$$F_c = q \times E \tag{1}$$

where F_c is an electric field force, q is the charge of a particle, and E is the electric field intensity. The particle motion is hampered by Stokes forces when particles are driven in the liquid, and the expression is shown in Equation (2):

$$Fd = 6\pi\mu vr \tag{2}$$

where Fd is Stokes force, μ is liquid viscosity coefficient, v is the motion relative rate between particles and fluids, and r is sphere particle diameter. In addition, the number of charges which are carried by a particle is 50–100 when the particle radius is within 1 μm. The materials in microcapsules include insulating oil, electrophoretic particles, and density balancing agent [25]. The parameters related to electrophoretic phenomena are particles mass and zeta potential, dielectric properties, viscosity, and electric field strength of the insulating oil. The particles can be driven and are in an accelerated motion state when the electric field force is greater than the Stokes force, and the combined force is shown in Equation (3):

$$F = Fc - Fd = m\frac{dv}{dt} \tag{3}$$

where m is the mass of a particle and $\frac{dv}{dt}$ is the acceleration. The differential equation of Equation (3) is shown as Equation (4) [26]:

$$vi = \frac{qE}{6\pi\mu V}[1 \pm \exp(-\frac{6\pi\mu V}{m}t)] \tag{4}$$

where vi is the velocity of a particle, which is proportional to the electric field strength. The particle's motion distance can be obtained by Equation (5):

$$s = vi \times t \tag{5}$$

where t is the particle's driving time and s is the particle's motion distance, which can be reflected by the change in EPD reflectivity. According to the CIELab (International Committee on Illuminate Lab color space) standard, the relationship between brightness and reflectivity can be calculated as Equation (6):

$$L^* = 116 \times (\sqrt[3]{\frac{R}{R0}}) - 16 \tag{6}$$

where R is the reflectivity of an EPD, $R0$ is the reference standard of 100% reflectivity, and L^* is the brightness of the EPD. The brightness value error of EPDs are mainly caused by the ghost image [27], as shown in Equation (7) [28]:

$$\Delta L^*_{GS(n)} = L^*_{GS(n),Max} - L^*_{GS(n),Min} \tag{7}$$

where $\Delta L^*_{GS(n)}$ is the ghost value, $L^*_{GS(n),Max}$ is the maximum brightness value of EPDs, $L^*_{GS(n),Min}$ is the minimum brightness value of EPDs, and n is the grayscale level. The uniformity of the grayscale needs to be detected when the image is refreshed. By using the gray differential statistics, a grayscale image is expressed as a function $H(x,y)$, and (x,y) is a point in the image. $(x + \Delta x, y + \Delta y)$ is the adjacent point of (x,y), and the gray difference between the two points is shown as Equation (8):

$$H_\Delta(x,y) = H(x,y) - H(x + \Delta x, y + \Delta y) \tag{8}$$

where H_Δ is the grayscale difference. The probability of getting each grayscale differential is $p(i)$. Equation (9) is used to calculate the texture feature value entropy e in a grayscale image, and texture is a phenomenon of nonuniform grayscale display in an EPD. The nonuniformity of the texture in the grayscale image is reflected, and the smaller the value, the more even the texture in the grayscale image, where the expression is shown in Equation (9):

$$e = -\sum_{i=0}^{n} p(i) \log_2 [p(i)] \tag{9}$$

2.2. Design Principles for Driving Waveforms

The characteristics of conventional driving waveform are shown in Figure 2. It is usually divided into three phases: erase the image, activation phase, and new image phase. The driving durations of three phases are 240, 480, and 240 ms, respectively.

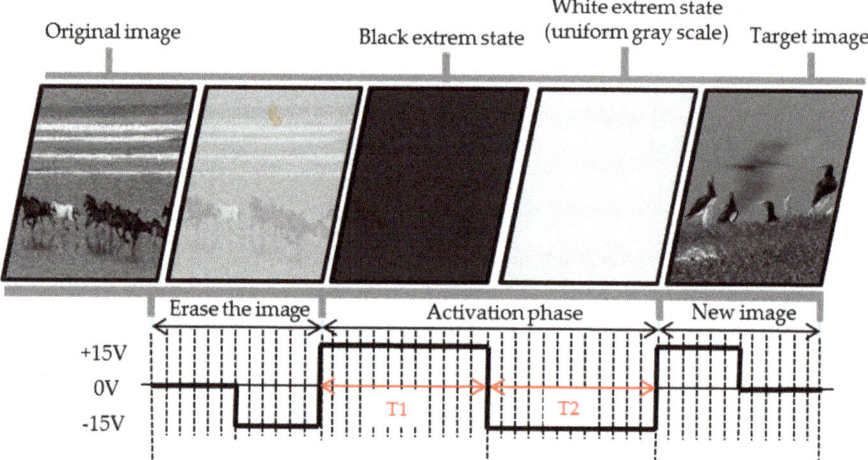

Figure 2. A typical EPD conventional driving waveform contains three phases for image updating. The T1 phase is used as improving particle activity phase and the T2 phase is used as uniform reference grayscale phase in the activation phase. The image in T1 is erased in T2.

In the original image, the particle spatial position in each pixel is different. Hence, there are different grayscale values in each pixel when the driving voltage is the same as each other. It's easy to form a ghost image, and the EPD can be driven to a uniform optical extremes states (black or white extremes states) for eliminating ghost images, which is called as the erase the image phase.

In addition, the longer a pixel remains in the same state, the lower the mobility of charged particles, which can hinder the driving of a new grayscale in EPDs. By driving the display several times between two optical extremes, it can improve the activity of particles and further eliminate ghosts, and this phase is called as particle activation phase. In this study, the particle activation phase was divided into two phases: improving particle activity phase and uniform reference grayscale phase (white extreme state). The main function of the improving particle activity phase is to increase particle activity, which can improve the particle driving speed. Then, the main function of the uniform reference grayscale phase is to unify a reference grayscale, so as to ensure that particles can drive to a same spatial position form different initial grayscale before new image phase.

In the new image phase, the target grayscale is produced in the end of pulse sequence. Different driving voltages with different driving durations are applied to different pixels for driving different grayscale. These three phases are combined into a complete driving waveform, and the n-level grayscale requires n^2 driving waveforms, then multilevel grayscale driving waveforms are combined into a lookup table [29]. In the display process, the driving waveform can be called by the driver system according to different initial grayscale and target grayscale.

3. Design of the Activation Phase in Driving Waveforms

The particle motion characteristics in the improving particle activity phase were studied by us according to the change of display brightness value. The inflection point of particles brightness was obtained by solving the first derivative of the brightness curve, which was the duration of the improving particle activity phase. The driving duration for the uniform reference grayscale phase was then studied to shorten the driving waveform.

3.1. Response Time Characteristics of the Improving Particle Activity Phase

The brightness changes in improving particle activity phase of driving waveforms for a four-level grayscale were measured, and black (B), dark gray (DG), light gray (LG), and white (W) were set as the initial grayscale, as shown in Figure 3.

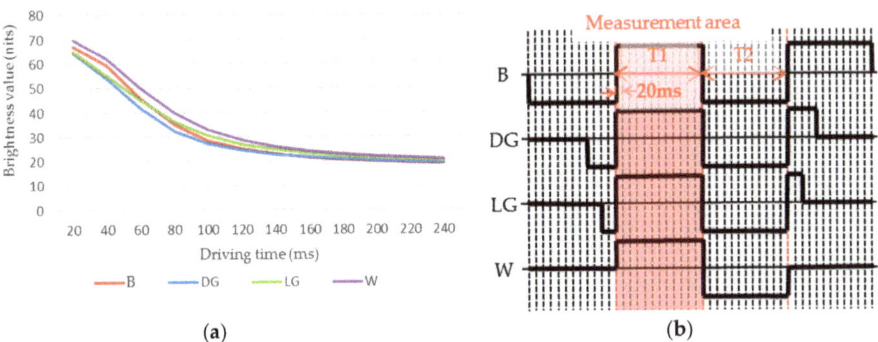

(a) (b)

Figure 3. Brightness change curve of the improving particle activity phase at different initial grayscale. (a) The brightness curve of the T1. (b) The driving waveform used in measurements.

The initial brightness values of the four different initial grayscale are not the same, but the brightness values can be closed together after the image erasure phase in the driving waveform (the phase before T1). Hence, when the measuring area is T1 phase, the brightness value is consistent. The motion speed of particles is directly proportional to the change speed of the display brightness value. The motion of particles is shown as the brightness change of EPDs. In Figure 3a, it can be seen that the brightness value decreases gradually, and the speed of the decrease rate increased and then decreased, so the motion speed of particles was increased and then decreased. In order to calculate the change of particles motion speed, the first order derivative was carried out for each curve, as shown in Figure 4.

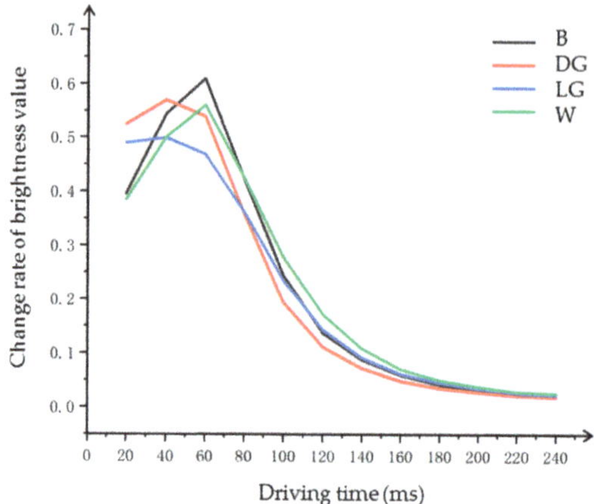

Figure 4. First order derivative curve of brightness value in improving particle activity phase. "B", "DG", "LG", and "W" indicate different initial grayscale in the driving waveform.

As can be seen, the change rate of the EPD brightness was the highest when the time was 40–60 ms, when the particle motion had a highest speed. Thus, the particle activity was highest when the driving duration of T1 was 40–60 ms, which was conducive for driving effect in the next phase.

3.2. Response Time Characteristics of the Uniform Reference Grayscale Phase

The brightness change value of T2 was measured when the white grayscale was set as the initial grayscale. As shown in Figure 5, the driving duration of T1 was 40, 45, 50, 55, and 60 ms, respectively.

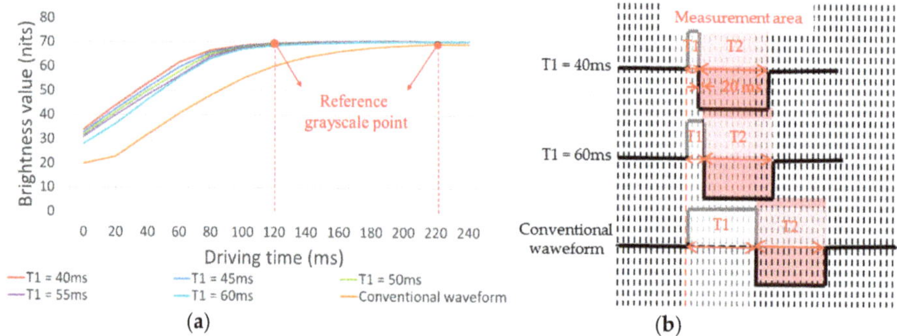

Figure 5. The influence of T1 timing on T2. (**a**) The brightness curve of T2. (**b**) The driving waveform used in the measurement.

As shown in Figure 5a, the time required for reaching the reference grayscale is 120 ms when T1 is 40, 45, 50, 55, and 60 ms. In the conventional driving waveform, the time required for reaching the reference grayscale is 220 ms. The change in brightness value of T2 is shown in Table 1.

Table 1. Change value of EPD brightness in T2.

T1 Durations	40 ms	45 ms	50 ms	55 ms	60 ms	Conventional Waveform
Time for reaching reference grayscale (ms)	120	120	120	120	120	220
Initial brightness values (nits)	33.93	32.81	31.79	30.66	28.13	19.85

The initial brightness of the EPD was more than 30 nits, when the T1 was 40, 45, 50, and 55 ms respectively, which showed that the EPD cannot reach optical extremes (black or white states) when the driving duration of 40, 45, 50, and 55 ms was applied continuously in T1, and it was not conducive to the particle driving in the next phase [30]. Therefore, 40, 45, 50, and 55 ms were not suitable for setting as the duration of T1. The driving duration to stable reference grayscale was 120 ms when T1 was 60 ms, and the driving duration to a stable reference grayscale was 220 ms when T1 was 240 ms in the conventional driving waveform, which showed that it was easier to obtain a stable reference grayscale with the uniform reference grayscale phase than the conventional driving waveform. The curve of brightness when T1 was 60 ms and the conventional driving waveform are shown in Figure 6.

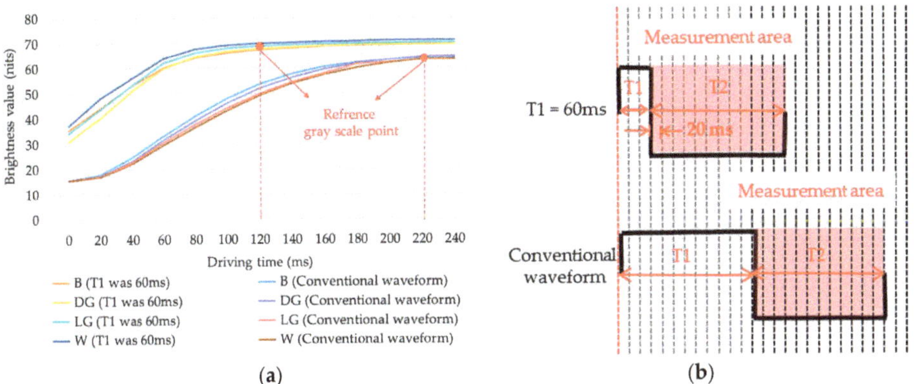

Figure 6. Brightness curves at different initial grayscale. (**a**) The brightness curve of T2. (**b**) The driving waveform used in this measurement.

3.3. Four-Level Grayscale Driving Waveform Design

In this paper, a four-level grayscale driving waveform was taken as an example to test the performance of the driving waveform. The proposed driving waveform was divided into three phases. The first phase was used for erasing the original image. The negative voltage duration of 240, 80, 40, and 0 ms were used to drive EPDs continuously when the initial grayscale was "B", "DG", "LG", and "W", respectively. The grayscale of the original image was driven to white. The second phase was the particle activation phase. A positive voltage of 60 ms was adopted to drive the particle for getting an activation state, and a negative voltage duration of 120 ms was adopted to drive the particle to form a unified reference grayscale according to the driving waveform design of the activation phase. The third phase was used to display a new grayscale. The positive voltage duration of 240, 80, 40, and 0 ms were used to obtain target grayscale of "B", "DG", "LG", and "W", respectively, and the uniform white reference grayscale was driven to target grayscale, as shown in Figure 7.

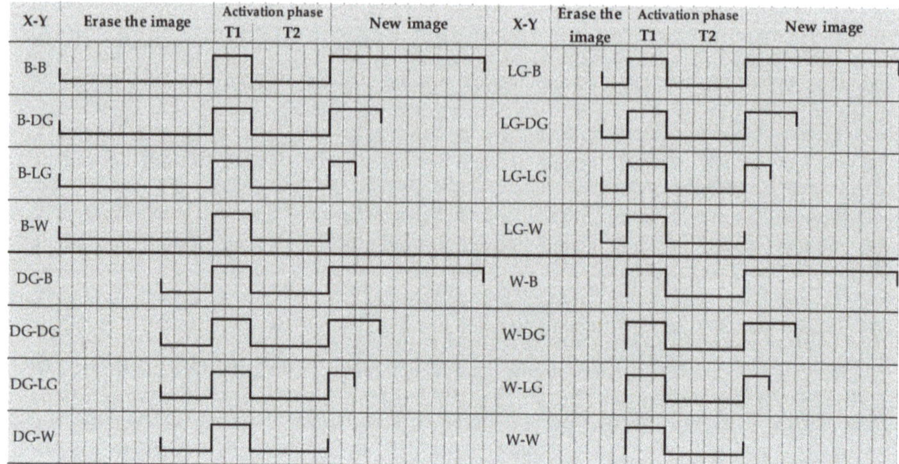

Figure 7. A four-level grayscale driving waveform set. X is displayed as an original grayscale, and Y can be displayed when the EPD has been driven by the driving waveform.

4. Experimental Results and Discussion

4.1. Testing System

Microcapsule EPDs (ED060SC7, 6.0 inch, screen resolution is 600×800) produced by E-ink (Holding Inc., Hsinchu, Taiwan) were introduced for the experiments in this study. The experimental equipment comprised mainly into two parts: the driving waveform editing system and the brightness data acquisition system. A driving waveform design can be completed by the computer software LabVIEW (LabVIEW10.0.1f3, National Instruments, Austin, TX, USA). The brightness data acquisition system was composed of a driving power supply, a closed lab case, two light sources, a camera, and a temperature controller, as shown in Figure 8. The details of the EPD and the temperature controller are shown in Figure 9.

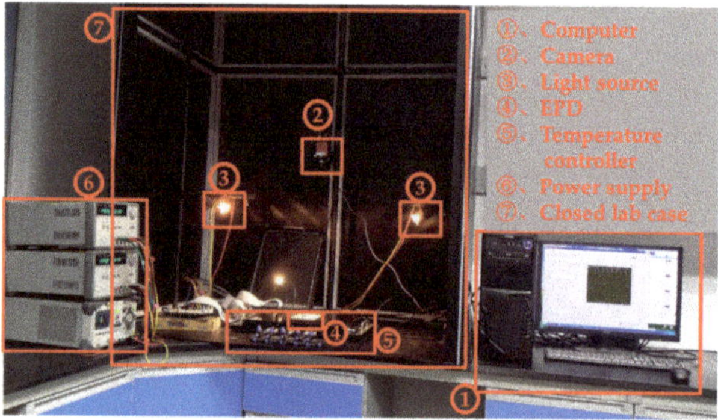

Figure 8. Optical testing device for EPDs. Camera was ADIMEC-1600m. Light sources were Brilliantline 12V 20W 36D Halogen. Temperature controller was Julabo FP25-ME. The power supply was an Agilent 3161A three-way output power supply.

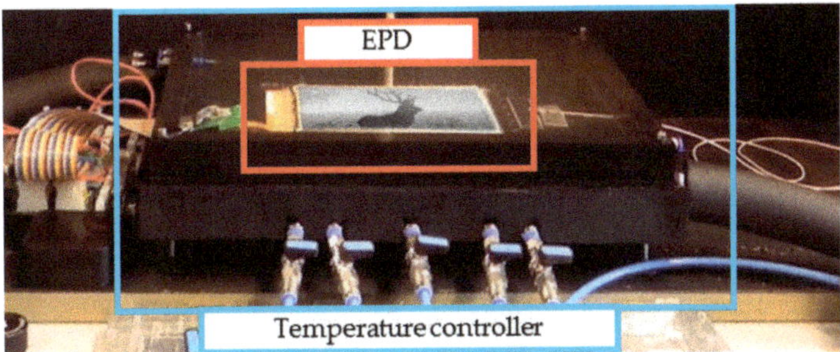

Figure 9. The details of the EPD and the temperature controller. Temperature controller was Julabo FP25-ME.

A closed lab case was designed to prevent the interference from external light and ensure uniform lighting across the EPD surface. Two light sources were used as the external light source for the data acquisition system. In a confined space, the light environment provided by internal light sources was relatively stable, which provided good comparability for testing the driving effect of driving waveforms. In addition, the electrophoretic motion of particles in an EPD system is easily affected by temperature [31,32], so the experiment temperature was set at 25 °C.

The data acquisition system can test the brightness value of EPDs. Firstly, the data acquisition system is initialized, camera parameters are set, and the data acquisition system is calibrated effectively. Then, the system is idled for awaiting calibration results, and the waveform binary file is loaded into the driving system. The system can measure the brightness automatically and export the measurement data. The flow chart of data acquisition is shown in Figure 10.

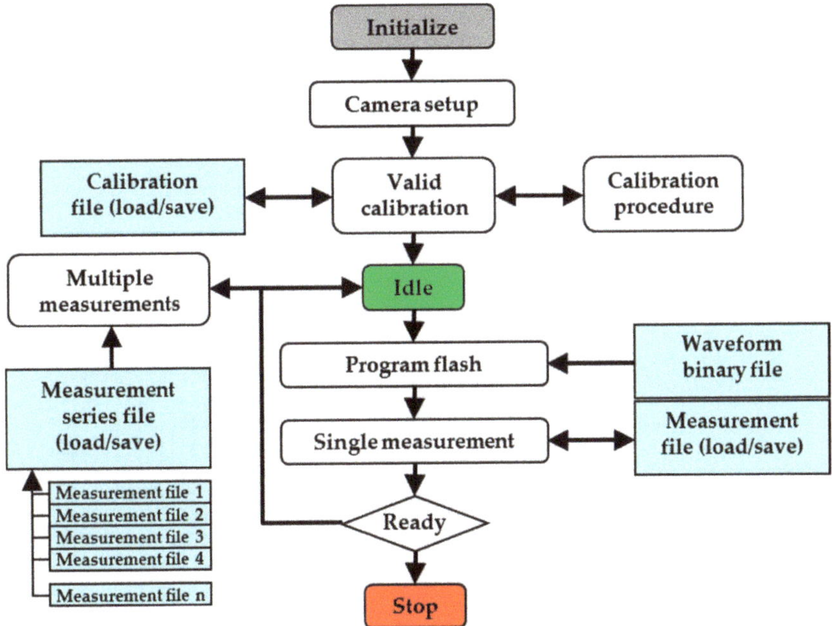

Figure 10. The flow chart of data acquisition. Calibration includes the effective number of measurement area blocks, the international white and black reference grayscale reflectance. The measurement mode includes single measurement and multiple measurement.

LabVIEW, a software development platform of NI, was applied as a waveform editing system, and the driving waveform can be edited by G language of graphical programming. The waveform editing interface is shown in Figure 11 The designed driving waveform can be converted into a binary file and downloaded directly to the driver circuit.

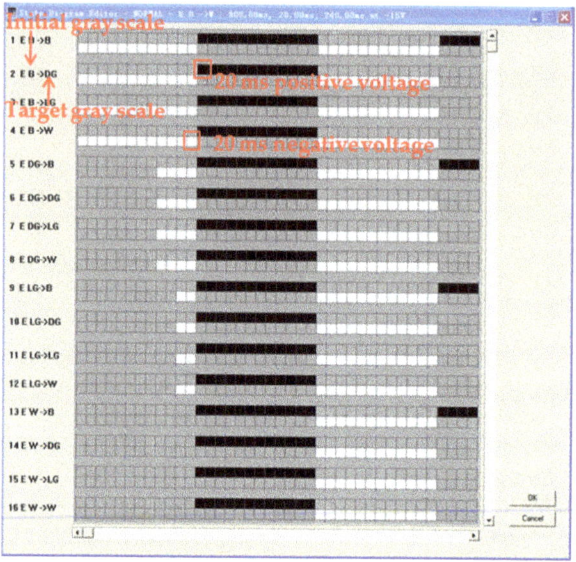

Figure 11. Waveform editing interface. A black grid indicates that a 20 ms positive voltage of +15 V is applied to the pixel electrode, a white grid indicates that a 20 ms negative voltage of −15 V is applied to the pixel electrode, and the rest states are 0 V.

4.2. Timing Comparison of Driving Waveforms

In different initial grayscale, it only takes 120 ms from the uniform reference grayscale to the white reference grayscale, which is 100 ms less than the conventional driving waveform, whose improving particle activity phase is 60 ms. The activation phase waveform of the proposed driving waveform and the conventional driving waveform is shown in Figure 12.

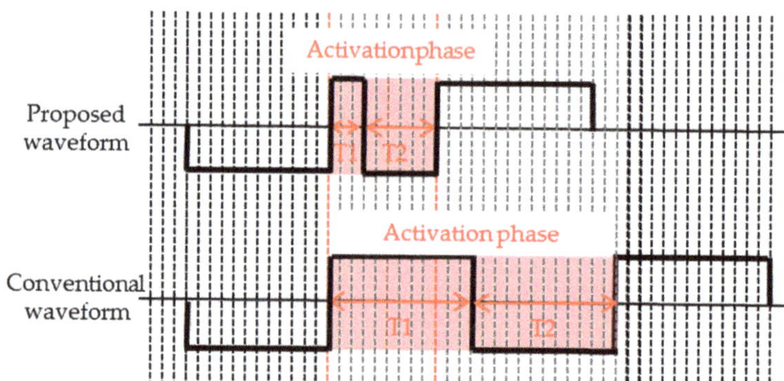

Figure 12. Driving timing comparison of activation phases in driving waveforms, and each vertical line represents 20 ms. In the proposed waveform, T1 is 60 ms and T2 is 120 ms. In the conventional waveform, T1 is 240 ms and T2 is 240 ms.

It can be seen that the activation phase in the proposed driving waveform was shortened by 180 ms, and the uniform reference grayscale phase was shortened by 120 ms. The overall proposed driving waveform was shortened by 300 ms compared to the conventional driving waveform.

4.3. Performance Testing

The conventional driving waveform and the proposed driving waveform were used to drive EPDs to the white state. The results show that the proposed driving waveform can increase the white grayscale by 2.7, as shown in Figure 13.

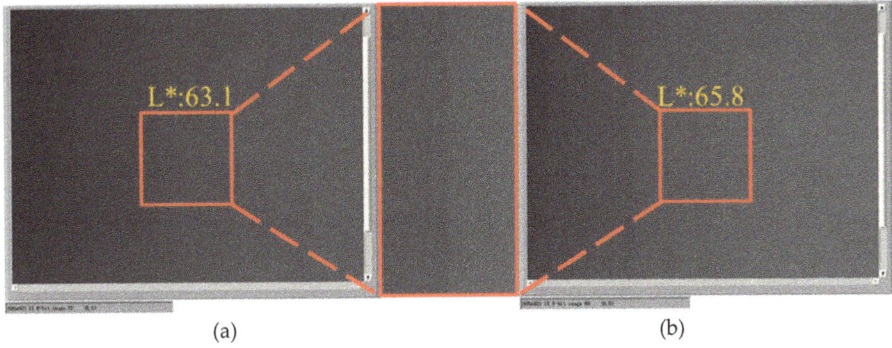

Figure 13. Comparison of the white reference grayscale. (**a**) An EPD was driven to white grayscale by the conventional driving waveform. (**b**) An EPD was driven to white grayscale with the proposed driving waveform.

Then, the proposed driving waveform was used to verify the image quality through several experiments. Firstly, a picture "E" was loaded in an EPD, and the proposed driving waveform and the conventional driving waveform were each used to drive it to the white grayscale, so as to explore visibility of the ghost image of each driving waveform, as shown in Figure 14.

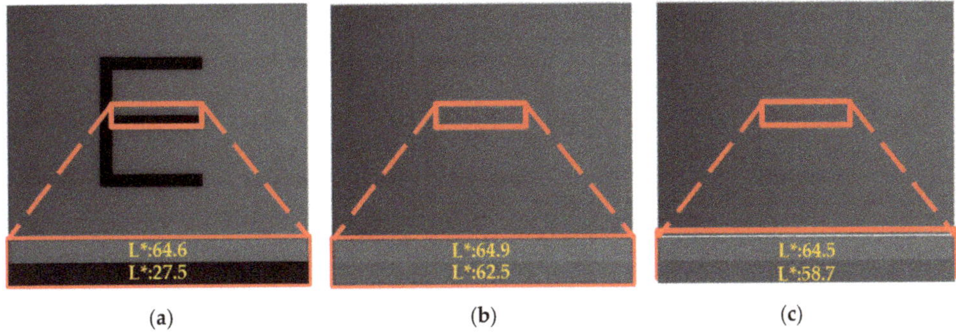

Figure 14. Visibility of the ghost image comparison between the proposed driving waveform and the conventional driving waveform. (**a**) Original image. (**b**) Driving effects of the proposed driving waveform. (**c**) Driving effects of the conventional waveform.

Figure 14a is an original image with a white background and a black font. When it was driven to full white by the proposed driving waveform, the background brightness value of the image was 64.9 nits, and the shadow brightness value was 62.5 nits, so the ghost image brightness was 2.4 nits, as shown in Figure 14b. When it was refreshed to full white by the conventional driving waveform, the background brightness value of the image was 64.5 nits and the shadow brightness value was

58.7 nits, so the ghost image brightness was 5.8 nits, as shown in Figure 14c. Hence, the proposed driving waveform can reduce visibility of the ghost image by 57%.

The uniformity of the grayscale was analyzed by calculating the change rate of entropy value and the gray value. As shown in Table 2, the proposed driving waveform was used to an EPD, and it can be driven the original image to full white, and the change rates of the entropy value and the gray value were −6.3% and −79.7%, respectively. Then, the conventional waveform was used to test related data, and the change rates of entropy value and gray value were −5.78% and −79.32% respectively. Hence, the proposed driving waveform can reduce the image grayscale texture by 0.52% and improve the image grayscale uniformity by 0.38%. Hence, the proposed driving waveform can optimize the activation phase without affecting the image grayscale quality and reduce the driving duration of 300 ms.

Table 2. The change of entropy and gray value.

Image Type	Entropy	Change Rate of Entropy (%)	Gray Value	Change Rate of Gray Value (%)
Original image	5.71	–	44.98×10^7	–
Proposed driving waveform	5.35	−6.30	9.13×10^7	−79.70
Conventional driving waveform	5.38	−5.78	9.30×10^7	−79.32

In addition, the conventional driving waveform has serious flicker when updating an image, especially in the refresh process of B-W. There are three phases in this process: In the erase the image phase, the original image is erased from black to full white. In the particle activation phase, the image needs to be driven to full black and then to full white. In the new image phase, the image is erased from white to full black. The human eye can feel a flicker due to the high contrast switch between black and white, as shown in Figure 15a. However, the proposed driving waveform cannot drive the EPD to full black in the activation phase, so the switching intensity of EPDs can be decreased, resulting in the decrease of the flicker, as shown in Figure 15b.

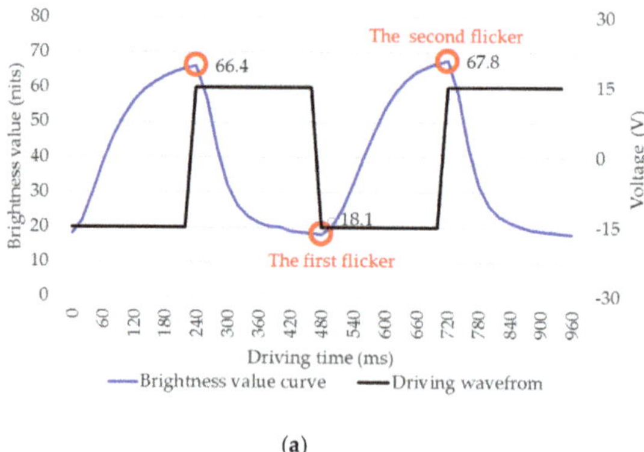

(**a**)

Figure 15. *Cont.*

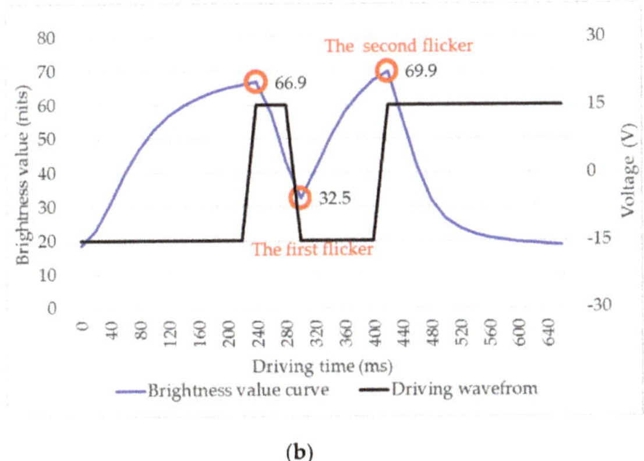

(**b**)

Figure 15. The change curve of EPD brightness. (**a**) When the EPD was driven by the conventional driving waveform. (**b**) When the EPD was driven by the proposed driving waveform.

The extreme values of the brightness curve in the activation phase were 66.4, 18.1, and 67.8 nits, respectively, when the EPD was driven by the conventional waveform, so flickers with the intensity of 48.3 and 49.6 nits were respectively formed. When the EPD was driven by the proposed driving waveform, the extreme values of the brightness curve in the activation phase was 66.9, 32.5, and 69.9 nits, respectively, so flickers with the intensity of 34.4 and 37.4 nits were respectively formed. Therefore, the proposed driving waveform can reduce the flicker intensity by 28.8% and 24.6%, and the overall flicker intensity by 26.7%, as shown in Table 3.

Table 3. The change in flicker intensity.

Waveform Type	Intensity of the First Flicker (nits)	Intensity of the Second Flicker (nits)	Total
Conventional driving waveform	48.3	49.6	97.3
Proposed driving waveform	34.4	37.4	41.8
Reduction rate of flicker intensity (%)	28.8	24.6	26.7

5. Conclusions

In this study, a fast response driving waveform for EPDs based on optimized activation phase was proposed by analyzing the characteristics of EPDs. The duration of the driving waveform can be effectively shortened by 300 ms compared to the conventional driving waveform, and the reference white grayscale can be increased by 2.7 at the same time. In addition, it can reduce visibility of the ghost image by 57%, which can improve display quality, and the flicker intensity can be reduced by 26.7% for enhancing the visual experience of human eyes.

Author Contributions: Z.Y. and W.H. designed this project. W.H. and Z.Y. carried out most of the experiments and data analysis. Z.H. and S.S. performed part of the experiments and helped with discussions during manuscript preparation. L.L. and T.Z. contributed to the data analysis and correction. L.W., C.Z., W.L., L.S. and G.Z. gave suggestions on project management and provided helpful discussions on the experimental results. All authors have read and agreed to the published version of the manuscript.

Funding: This research was funded by the Guangdong Basic and Applied Basic Research Foundation (no. 2020A1515010420), the Key Research Platforms and Research Projects in Universities and Colleges of Guangdong Provincial Department of Education (no. 2018KQNCX334), the Zhongshan Innovative Research Team Program (no. 180809162197886), the Zhongshan Institute high-level talent scientific research startup fund project (no. 416YKQ04), the Project for Innovation Team of Guangdong University (no. 2018KCXTD033), and the National Key R&D Program of China (no. 2018YFB0407100-02).

Conflicts of Interest: The authors declare no conflicts of interest.

References

1. Meng, X.; Wen, T.; Qiang, L.; Ren, J.; Tang, F. Luminescent electrophoretic particles via miniemulsion polymerization fornight-vision electrophoretic displays. *ACS Appl. Mater. Interfaces* **2013**, *5*, 3638–3642. [CrossRef] [PubMed]
2. Hertel, D.; Penczek, J. Predicting the viewing direction performance of e-paper displays with front light under ambient lighting conditions. *J. Soc. Inf. Disp.* **2015**, *23*, 510–522. [CrossRef]
3. Kao, W.C.; Wu, G.F.; Shih, Y.L. Design of real-time image processing engine for electrophoretic displays. In Proceedings of the IEEE International Conference on Consumer Electronics, Las Vegas, NV, USA, 9–12 January 2011. [CrossRef]
4. Heikenfeld, J.; Drzaic, P.; Yeo, J.S.; Koch, T. A critical review of the present and future prospects for electronic paper. *J. Soc. Inf. Disp.* **2011**, *19*, 129–156. [CrossRef]
5. Jablonski, C.; Grundler, G.; Pieles, U.; Stebler, S.; Oehrlein, R.; Szamel, Z. Synthesis and electrophoretic properties of novel Nanoparticles for coloredElectronic ink and e-paper applications. *Chimia* **2016**, *70*, 366–368. [CrossRef]
6. Kao, W.C. Electrophoretic display controller integrated with real-time halftoning and partial region update. *J. Disp. Technol.* **2010**, *6*, 36–44. [CrossRef]
7. Kao, W.C.; Chen, H.; Liu, Y.; Liou, S. Hardware engine for supporting gray-tone paintbrush function on electrophoretic papers. *J. Disp. Technol.* **2014**, *10*, 138–145. [CrossRef]
8. Li, W.C.; Keh, H.J. Electrophoretic mobility of charged porous shells or microcapsules and electric conductivity of their dilute suspensions. *Colloids Surf. A Physicochem. Eng. Asp.* **2016**, *497*, 154–166. [CrossRef]
9. Kao, W.C.; Liu, C.H. Real-time video signal processor for electrophoretic displays. In Proceedings of the 2014 IEEE International Conference on Consumer Electronics—Taiwan (ICCE-TW), Taipei, Taiwan, 26–28 May 2014. [CrossRef]
10. Wang, L.; Yi, Z.C.; Peng, B.; Zhou, G. An improved driving waveform reference grayscale of electrophoretic displays. In Proceedings of the AOPC 2015: Advanced Display Technology; and Micro/Nano Optical Imaging Technologies and Applications, Beijing, China, 15 October 2015. [CrossRef]
11. Yang, S.H.; Lin, F.C.; Huang, Y.P.; Al, E. Ghosting reduction driving method in electrophoretic displays. *Sid Symp. Dig. Tech. Pap.* **2012**, *43*, 1361–1364. [CrossRef]
12. Qian, Y.; Bai, P.; Zhou, G. Driving waveform optimization based on electrophoretic displays. *Appl. Electron. Tech.* **2016**, *42*, 33–35. [CrossRef]
13. Johnson, M.T.; Zhou, G.; Zehner, R.; Amundson, K.; Kamer, J.V.D. High-quality images on electrophoretic displays. *J. Soc. Inf. Disp.* **2006**, *14*, 175–180. [CrossRef]
14. Yi, Z.C.; Bai, P.F.; Wang, L. An electrophoretic display driving waveform based on improvement of activation pattern. *J. Cent. South Univ.* **2014**, *21*, 3133–3137. [CrossRef]
15. Kao, W.C.; Chang, W.P.; Ye, J.A. Driving waveform design based on response latency analysis of electrophoretic displays. *J. Disp. Technol.* **2012**, *8*, 596–601. [CrossRef]
16. Shen, S.T.; Gong, Y.X.; Jin, M.L.; Yan, Z.; Xu, C.; Yi, Z.; Zhou, G.; Shui, L. Improving electrophoretic particle motion control in electrophoretic displays by eliminating the fringing effect via driving waveform design. *Micromachines* **2018**, *9*, 143. [CrossRef]
17. Li, W.; Yi, Z.C.; Jin, M.L.; Shui, L.; Zhou, G. Improvement of video playback performance of electrophoretic displays by optimized waveforms with shortened refresh time. *Displays* **2017**, *49*, 95–100. [CrossRef]
18. Kao, W.C.; Liu, C.H.; Liou, S.C.; Tsai, J.C.; Hou, G.H. Towards video display on electronic papers. *J. Disp. Technol.* **2017**, *12*, 129–135. [CrossRef]
19. Lu, C.M.; Wey, C.L. A controller design for micro-capsule active matrix electrophoretic displays. *J. Disp. Technol.* **2011**, *7*, 434–442. [CrossRef]
20. Comiskey, B.; Albert, J.D.; Yoshizawa, H.; Jacobson, J. An electrophoretic ink for all-printed reflective electronic displays. *Nature* **1996**, *394*, 253–255. [CrossRef]
21. Pesce, G.; Rusciano, G.; Zito, G.; Sasso, A. Simultaneous measurements of electrophoretic and dielectrophoretic forces using optical tweezers. *Opt. Express* **2015**, *23*, 9363. [CrossRef]
22. Wang, Z.; Liu, Z. The key technology of ereader based on electrophoretic display. In Proceedings of the 2010 2nd International Conference on Software Technology and Engineering, San Juan, PR, USA, 3–5 October 2010. [CrossRef]

23. Kao, W.C.; Liu, S.C.; Chang, W.T. Signal processing for playing videos on electrophoretic displays. In Proceedings of the Circuits and Systems (MWSCAS), 2012 IEEE 55th International Midwest Symposium on, Boise, ID, USA, 5–8 August 2012. [CrossRef]

24. Yang, Z.; Hu, W.B. Developments of microcapsule electrophoretic display. *Vac. Electron.* **2012**, 54–59. [CrossRef]

25. Yu, D.; An, J.H.; Bae, J.Y.; Jung, D.; Kim, S.; Ahn, S.D.; Kang, S.; Suh, K.S. Preparation and characterization of acrylic-based electronic inks by in situ emulsifier-free emulsion polymerization for electrophoretic displays. *Chem. Mater.* **2004**, *16*, 4693–4698. [CrossRef]

26. Bert, T.; De Smet, H. Dielectrophoresis in electronic paper. *Displays* **2003**, *24*, 223–230. [CrossRef]

27. Yang, B.R. Overview of design considerations for electrophoretic e-paper and strategies for achieving full-color. In Proceedings of the 2016 23rd International Workshop on Active-Matrix Flatpanel Displays and Devices (AM-FPD), Kyoto, Japan, 6–8 July 2016. [CrossRef]

28. Feng, G.; Gormish, M.J. Ghosting reduction using digital halftoning for electrophoretic displays. *Sid Symp. Dig. Tech. Pap.* **2012**, *39*, 697–700. [CrossRef]

29. Kao, W.C.; Ye, J.A.; Lin, F.S.; Lin, C.; Sprague, R. Configurable timing controller design for active matrix electrophoretic display with 16 gray levels. *IEEE Trans. Consum. Electron.* **2009**, *55*, 1–5. [CrossRef]

30. Kao, W.C.; Tsai, J.C. Driving method of three-particle electrophoretic displays. *IEEE Trans. Electron Devices* **2018**, *65*, 1023–1028. [CrossRef]

31. Kao, W.C.; Liu, J.J.; Chu, M. Integrating photometric calibration with adaptive image halftoning for electrophoretic displays. *J. Disp. Technol.* **2010**, *6*, 625–632. [CrossRef]

32. Kao, W.C.; Liu, J.J.; Chu, M.I.; Wang, Y.K.; Yang, T.H. Photometric calibration for image enhancement of electrophoretic displays. In Proceedings of the IEEE International Symposium on Consumer Electronics (ISCE 2010), Braunschweig, Germany, 26 July 2010. [CrossRef]

 *micromachines*

Review

High-Sensitivity in Dielectrophoresis Separations

Benjamin G. Hawkins * , **Nelson Lai and David S. Clague**

Biomedical Engineering Department, California Polytechnic State University, San Luis Obispo, CA 93407, USA;
nelai@calpoly.edu (N.L.); dclague@calpoly.edu (D.S.C.)
* Correspondence: bghawkin@calpoly.edu; Tel.: +1-805-756-6203

Received: 19 February 2020; Accepted: 26 March 2020; Published: 9 April 2020

Abstract: The applications of dielectrophoretic (DEP) techniques for the manipulation of cells in a label-free fashion within microfluidic systems continue to grow. However, a limited number of methods exist for making highly sensitive separations that can isolate subtle phenotypic differences within a population of cells. This paper explores efforts to leverage that most compelling aspect of DEP—an actuation force that depends on particle electrical properties—in the background of phenotypic variations in cell size. Several promising approaches, centering around the application of multiple electric fields with spatially mapped magnitude and/or frequencies, are expanding the capability of DEP cell separation.

Keywords: dielectrophoresis; microfluidics; cell separation

---

1. Introduction

Dielectrophoresis (DEP) is in the process of growing from a developmental technology to an integrated research tool. A testament to this is the number of research articles that continue to demonstrate novel device designs and separations. There are number of excellent recent review articles that establish a clear foundation in the theory [1], technologies [2,3], and applications [4–6]. With many preeminent researchers, an active research portfolio, and strong surveys of the field, there is little need at this time for a broad survey of DEP theory and applications; instead, this review focuses specifically on high-performance techniques that aim to achieve highly sensitive separations and overcome some of the traditional challenges found in DEP applications. For completeness and context, however, a short summary of the foundational physics and assumptions that are integrated in the oft-cited governing equations for DEP are presented here.

DEP is the transport of polarizable particles in response to an externally applied electric field. Actuation is achieved by application of a non-uniform electric field which simultaneously induces polarization and exerts force on the interface between two electrically dissimilar media. In general, the DEP force can be written in terms of Maxwell's Stress Tensor, $\vec{\vec{T}}$:

$$\vec{F}_{DEP} = \iiint_V (\nabla \cdot \vec{\vec{T}}) dV \tag{1}$$

$$\vec{\vec{T}} = \epsilon \vec{E}\vec{E} + \frac{1}{\mu}\vec{B}\vec{B} - \frac{1}{2}\left(\epsilon \vec{E} \cdot \vec{E} + \frac{1}{\mu}\vec{B} \cdot \vec{B}\right)\vec{\vec{I}}, \tag{2}$$

where $\vec{E}$ is the electric field, $\vec{B}$ is the magnetic field, $\vec{n}$ is the unit normal to a surface over which we are integrating to calculate the total force, ϵ is the material permittivity, μ is the material permeability, and $\vec{\vec{I}}$ is the identity tensor. Equation (2) is an expression of the volumetric electromagnetic force on an object with non-uniform electrical properties. Though the DEP force can be generated by static or time-varying fields, the applications of interest here—those that achieve some form of separation that is sensitive to particle electrical properties—typically employ harmonic (sinusoidal) electric fields with frequencies that allow a quasi-static approximation. The force observed is the result of time-averaging the DEP force equation [7]:

$$\langle \vec{F}_{DEP} \rangle = \iiint_V (\nabla \cdot \langle \overleftrightarrow{T} \rangle) dV \tag{3}$$

$$\langle \overleftrightarrow{T} \rangle = \frac{1}{4} \Re(\tilde{\epsilon}) \left(\vec{E}\vec{E}^* + \vec{E}^*\vec{E} - |\vec{E}|^2 \overleftrightarrow{I} \right), \tag{4}$$

where $\vec{E}^*$ is the complex conjugate of the electric field. An accurate model for the force on a particle would be based on the volume integral of divergences of the MST within the particle. Application of the divergence theorem and assumption of locally homogeneous electrical properties allows replacement of Equation (4) with:

$$\langle \vec{F}_{DEP} \rangle = \iint_A (\langle \overleftrightarrow{T} \rangle \cdot \vec{n}) dA, \tag{5}$$

where $\vec{n}$ is the unit vector normal to a surface at the interface between two media with homogeneous, but dissimilar, electrical properties and indicates that DEP force is generated at teh interfaces between materials. This simplification provides the basis for the multi-shell models employed by most researchers in the field.

Approximations

With the application of several approximations, the expression for $\vec{F}_{DEP}$ can be simplified to the form presented in most applications in literature:

$$\langle \vec{F}_{DEP} \rangle = \pi \epsilon_m a^3 \Re(\tilde{f}_{CM}) \nabla \left(\vec{E} \cdot \vec{E} \right), \tag{6}$$

where $\tilde{f}_{CM}$ is the complex Clausius-Mossotti factor and a is the particle radius. $\tilde{f}_{CM}$ depends on the complex permittivity, $\tilde{\epsilon}$ of material on either side of the interface under consideration:

$$\tilde{f}_{CM} = \frac{\tilde{\epsilon}_p - \tilde{\epsilon}_m}{\tilde{\epsilon}_p + 2\tilde{\epsilon}_m} \tag{7}$$

$$\tilde{\epsilon} = \epsilon - j\frac{\sigma}{\omega}, \tag{8}$$

where ϵ is the electrical permittivity, σ is the electrical conductivity, and ω is the frequency of the applied electric field in radians/second. $\Re(\tilde{f}_{CM})$ varies from -0.5 to 1.0; the point where $\Re(\tilde{f}_{CM}) = 0$ is termed the "cross-over frequency" and here notated as $f_{CM,0}$. The approximations that lead to Equation (6) should be carefully examined for a given application:

- **Isotropic Media** The material on either side of the interface is assumed to have electrical properties that are independent of the orientation of the electric field.
- **Homogeneous Media** The material on either side of the interface is assumed to have spatially uniform electrical properties.
- **Spherical Particle** The most common assumption addressed is the spherical particle assumption. Good approximations exist for spheroidal particles and are often employed when particle shape deviates significantly from spherical.
- **Semi-infinite Domain** The domain is assumed to be large relative to the size of the particle; this also assumes that the particle is not in close proximity to other particles. Other particles perturb the local electric field solution and alter the DEP force.
- **Dipole Field** Equation (6) assumes that the perturbation to the externally applied, non-linear electric field is well-approximated by an equivalent dipole, that is, that the variation of the field over the particle is approximately linear. Multipole terms exist and can be used for suitably non-linear electric field gradients [8].

2. Sensitivity

Here, we consider the effects of variability among the major components of a multi-shell model for a subjected to a DEP force. To examine the sensitivity of the magnitude of the DEP force to changes in cellular components, we consider the magnitude of the DEP force normalized by the squared field gradient, which we define as Γ_{DEP}:

$$\Gamma_{DEP} \equiv \left(\frac{\vec{F}_{DEP}}{\nabla(\vec{E} \cdot \vec{E})} \right) = \pi a^3 \varepsilon_m \Re \left(\tilde{f}_{CM} \right). \tag{9}$$

We consider a 4-shell, spherical cell with parameters reported by Rohani, et al. [9] and indicated in Table 1. This work is further discussed in Section 3.5. This model includes permittivity and conductivity for the cell membrane, cytoplasm, nuclear envelope, and nucleoplasm. While no definitive library exists for such parameters, a survey of available literature indicates that permittivity values vary slightly compared to conductivity values for each of these cellular compartments.

Table 1. Model parameters for DEP force sensitivity evaluation. These baseline values were used as a starting point and increased by 10% for permittivities and 100% for conductivities. Variation of cell radius increased the overall cell size, but did not change the thickness of any layer.

Variable	Value	Units	Definition
a	10.01	μm	Outer cell radius
a_m	10	μm	Inner membrane radius
a_{nm}	5.02	μm	Nuclear outer radius
a_{np}	20	nm	Nucleoplasm (inner) radius
ε_{mem}	14	(rel)	Membrane permittivity
σ_{mem}	0.11	μS/m	Membrane conductivity
ε_{cyt}	60	(rel)	Cytoplasm permittivity
σ_{cyt}	0.52	S/m	Cytoplasm conductivity
ε_{nm}	25	(rel)	Nucleus membrane permittivity
σ_{nm}	3	mS/m	Nucleus membrane conductivity
ε_{np}	60	(rel)	Nucleoplasm permittivity
σ_{np}	1.4	S/m	Nucleoplasm conductivity
ε_m	80	(rel)	Media permittivity
σ_m	varies	S/m	Media conductivity

To determine the sensitivity, a MATLAB model was developed to calculate Γ_{DEP} between 1 Hz and 10^{12} Hz for a given set of electrical model cell parameters. Values for Γ_{DEP} were calculated for a 10% increase of permittivity and a 100% increase of conductivity for each compartment over 10 equally spaced intervals, for example, $\Delta\varepsilon$ or $\Delta\sigma$. The average of these changes at each frequency was calculated to determine the sensitivity as a function of frequency for each parameter.

$$\Delta F_{DEP} = \frac{1}{N-1} \sum_{j=1}^{N} |\Gamma_{DEP,j} - \Gamma_{DEP,j+1}|, \tag{10}$$

with summation over j representing the interval. The multishell model equation is a monotonic function for the values considered, so refinement of the intervals would yield no additional information. As shown by Rohani, et al. and others [10], media conductivity can be chosen to enhance the variation of the DEP force in response to changes in cell properties. Accordingly, MATLAB's optimization package was used to vary media conductivity between 1 μS/m and 0.2 S/m to maximize the variation of ΔF_{DEP} in response to changes in a given compartment electrical parameter, independent of frequency. The optimized conductivities for each parameter are indicated in Table 2 and Figure 1.

The average sensitivity of the DEP force, ΔF_{DEP}, is plotted as a function of frequency in Figure 1.

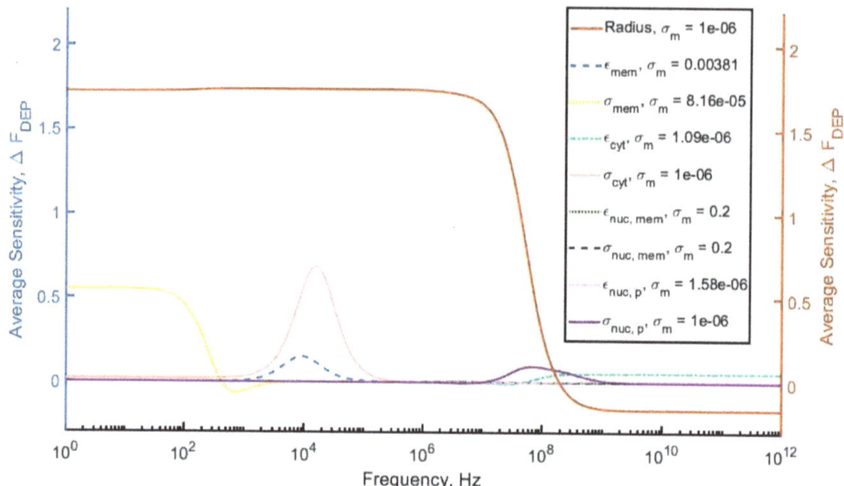

Figure 1. The average variation in the DEP force, ΔF_{DEP}, that arises from variations in radius and other electrical parameters in a 4-shell model for HEK cells. Model parameters are given in Table 1.

Table 2. Optimized media conductivity values to maximize the variation in ΔF_{DEP} in response to changes in cell model electrical properties.

Variable	Optim. σ Value	Units	Definition
a	1	µS/m	Outer cell radius
ε_{mem}	3.81	mS/m	Membrane permittivity
σ_{mem}	81.6	µS/m	Membrane conductivity
ε_{cyt}	1.09	µS/m	Cytoplasm permittivity
σ_{cyt}	1	µS/m	Cytoplasm conductivity
ε_{nm}	0.2	S/m	Nucleus membrane permittivity
σ_{nm}	0.2	S/m	Nucleus membrane conductivity
ε_{np}	1.58	µS/m	Nucleoplasm permittivity
σ_{np}	1	µS/m	Nucleoplasm conductivity

The largest change in the DEP force is due to variation in cell radius, with other parameters contributing significantly less (by approximately a factor of 2) to the overall variation. The maximum average variation across all frequencies was considered as well (Figure 2) and shows similar results.

In order to achieve the optimized sensitivity, solutions of quite high or low conductivity are required, potentially harming or altering cell populations of interest, so the realistic limits for the relative contributions of cellular electrical properties to the overall DEP force remains significantly lower than that of cell size. This underscores the need for DEP techniques that aim to reduce the contribution of overall size variation.

Beyond the need for sensitive techniques that can reduce the influence of biological size variability, DEP techniques are needed that can distinguish between subtle variations in the magnitude of the force, rather than the cross-over frequency. For example, a model for viral vesicle inclusions in Hepatitis C infected hepatocytes is developed using a combination of multi-shell models and Maxwell's mixture theorem. As Hepatitis-C virus infection continues, vesicles containing viral bodies build up within the cytoplasm of hepatocytes. Isolation of these cells from a background of healthy hepatocytes could have significant impact on diagnosis and treatment. The results indicate that increasing viral vesicle load within the cytoplasm causes subtle variation in $\Re(\tilde{f}_{CM})$ in the pDEP frequency regime and relatively small variation in the cross-over frequency, $f_{CM,0}$ (Figure 3) [11].

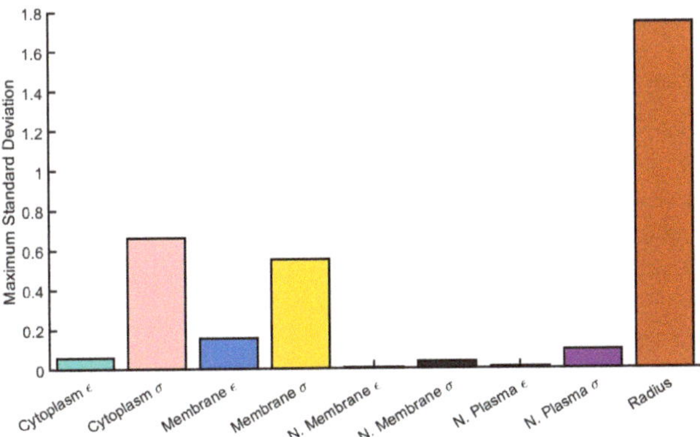

Figure 2. The maximum average variation in ΔF_{DEP} across all frequencies that arises from variations in radius and other electrical parameters in a 4-shell model for HEK cells. Model parameters are given in Table 1.

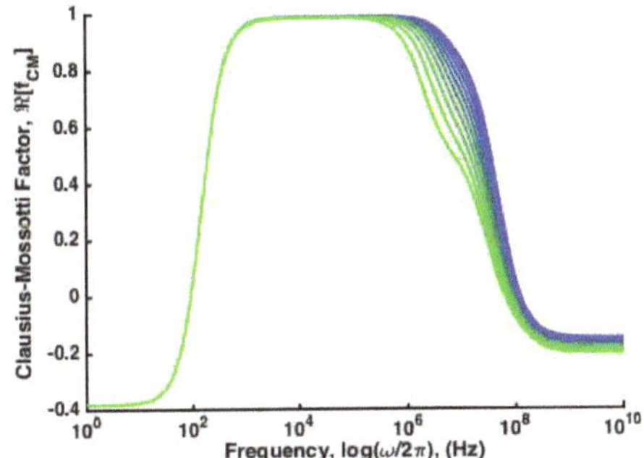

Figure 3. A multi-shell model that incorporates the presence of increasing volume fraction of Hepatitus-C viral vesicles in an example ellipsoidal hepatocyte model. The primary variation that results is a change in the magnitude of pDEP at high frequencies.

In order to realize the potential of DEP to perform truly robust, repeatable, label-free separations, a collection of techniques and approaches must be developed that can successfully separate cells with subtle variations like those identified in Figure 3 independent of biological size variability.

3. Approaches to Increasing Sensitivity

In general, methods to improve the sensitivity of DEP separations introduce another force that acts in opposition to the primary actuating DEP force and also carries with it a size-dependence. In this way, if the opposing force is volume-dependent, the effects of particle size can largely be omitted. Here we consider the primary methods for generating an opposing force and characterize each method by whether it is a batch process or continuous flow, whether it is an equilibrium method or not, and whether it is a dynamic process or static. For clarity, batch processes are those that separate a population of particles once, and must be cleaned and reloaded to perform additional separation; continuous flow separations operate continuously and do not require "resetting". Equilibrium methods separate particles to a particular position where the DEP force is cancelled out by the opposing force; non-equilibrium methods have a net force on actuated particles, so they will continue to move in response. Dynamic separations are those that change the separation or actuation force during the separation process, static separations yield a single result for the separation of particles.

The highest performance DEP separation will be one that achieves continuous, equilibrium separation in a dynamically adjustable manner.

3.1. Gravity-Based Systems: DEP-FFF

The first practical method to counteract the size-dependence of DEP was demonstrated by Wang, et al. [12], and used gravity to perform a combination of DEP and field flow fractionation (FFF). Subsequently, the process has been refined and separations of differentiated stem cells, [13], circulating tumor cells [14,15] and bacteria [16] have been demonstrated.

In a recent 'tour-de-force' in the application of DEP-FFF, Gascoyne and co-workers [14,17,18] demonstrated isolation of circulating tumor cells (CTCs) from blood, specifically preferentially capturing CTCs over peripheral blood mononuclear cells. This demonstration, overcoming one of the main confounding factors that reduce the purity of other separation methods, highlights the need for further development of DEP technologies. The DEP-FFF method used for capture and characterization in this suite of studies was based on the elution time in a DEP-FFF system; operating as a batch, non-equilibrium, dynamic method. In addition to CTC isolation and characterization, Waheed, et al. identified a "dielectric phenotype" factor $(1/R\phi)$ that depends on the particle radius (R) and the degree of membrane folding (ϕ). This factor was shown to be the primary determining factor of the cell cross-over frequency, $f_{CM,0}$ [14].

DEP-FFF suffers from a number of confounding factors, however, that limit its size-independent nature. The reliance on a pressure-driven flow field that varies significantly over the size of the typical particle leads to a drag induced torque and subsequent rotational lift force in certain flow regimes [19]. In addition to this lift effect, DEP-FFF methods lead to a equilibrium separation position that is a function of device depth, so orientation of the device to gravity is a factor and recovering separate streams from such a separation poses very challenging fabrication [20]. Separating multiple streams vertically within a microfabricated system requires multilayer device alignment, increasing design and manufacturing cost. For this reason, DEP-FFF is typically operated as a batch process, with elution time being the primary measurement.

3.2. 2D Electrode Systems: Single-Field/Single-Frequency

There are several methods that utilize DEP trapping effects to reduce the size-dependent nature of the force. In contrast to DEP-FFF, where the DEP force and fluid drag forces are orthogonal, these techniques typically involves a balance between nDEP forces and fluid drag in the same direction which reduces—but does not eliminate—the radius term. When these forces are balanced, the resulting DEP velocity is dependent on a^2,

$$\vec{u}_{DEP} = \frac{\varepsilon_m a^2 \Re(\tilde{f}_{CM})}{6\eta} \nabla(\vec{E} \cdot \vec{E}), \tag{11}$$

where η is the fluid viscosity. As a result, the approach to achieving size-independent separations using a single DEP force relies on leveraging the so-called "cross-over frequency" of DEP force. In this manner, a single field with a single frequency can be applied to a device, and particles experiencing a non-zero DEP force will be actuated, with those at the cross over frequency being unaffected. This approach has been used to separate cells from latex beads [21], bacteria from blood [22], and fluorescently-labeled cancer cells from blood [4]. Electrode configurations that are employed consist of interdigitated [23,24], castellated [25], ratchet [26] and trapezoidal [27] configurations.

There are a few limitations to this approach—(i) the dependence on the cross over frequency, (ii) the near-zero value of the DEP force near the cross over frequency, and (iii) the binary nature of the separation mechanism. In consideration of the former, many alterations to the electrical properties of particles of interest do lead to changes in the cross-over frequency, however there are many examples of more subtle changes that could be actuated via DEP that do not lead to significant changes in cross-over frequency (see Figure 3). While the force at cross over is zero, and therefore the lack of actuation from DEP will be independent of size, the magnitude of the DEP force near the cross over frequency is small, and other size-dependent actuation forces will be potentially dominant for particles with $\Re(\tilde{f}_{CM})$ near the cross over frequency, again leading to a reduction in sensitivity and an increase in size dependence. Finally, the binary nature of this approach leads to a separation for particles that have a cross over frequency near the field frequency, and those that do not. There is no ability to separate a spectrum of particles as a function of cross-over frequency. These factors combined mean that single-field, single-frequency approaches lack sufficient resolving power to truly engage the potential of DEP for highly sensitive separations.

The concept of "iso-motive" DEP also rests within this category, with electrode configurations that generate a spatially uniform DEP force (hence "iso-motive") [28,29]. The approach counters one of the challenges to DEP—the non-uniform nature of the force—but all the above approaches lead to batch-mode separations, or continuous-flow separations that remain size-dependent.

Of particular interest for this review is the application of CMOS memory fabrication techniques to realize an addressable electrode grid in a microchannel [30]. The authors created an array of 32,768 individually addressable 11 μm-square electrodes. Bordering on hyperresolution digital microfluidics, the ability to apply an RF-frequency field to each local electrode leads to exceptionally promising platform for highly flexible separation design and testing. In contrast to many purpose-built DEP separation systems, this programmable array integrated circuit approach could be dynamically reconfigured to a wide range of separations, or tuned for a specific separation in-situ.

3.3. 2D Electrode Systems: Multiple-Fields/Multiple-Frequencies

The combination of multiple electric fields operating at the same or different frequencies holds the greatest promise for highly sensitive DEP separations. Multiple field configurations can be split in to two major types: combined field DEP and travelling-wave DEP. Multiple, overlapping fields have been generated in planar electrode configurations and leveraged by many to accomplish various cell separations. In the combined-field separation area, work by Urdaneta and Smela showed separation of live and dead yeast cells, using different frequencies to preferentially attract each cell type to a different set of electrodes [31]. Similar techniques have been recently applied, taking advantage of multi-frequency signals (amplitude, frequency, or phase modulated) on a single electrode array to generate dissimilar DEP actuation forces on particles of interest to separate polystyrene microspheres based on size [32], algae cells based on lipid content [33], and MCF7 cancer cells from diluted blood [34]. Planar multifield configurations have also been extended to concentrate viruses, proteins, and bacteria [35].

In a thoughtful extension of DEP-FFF, Gascoyne and coworkers developed the "electrosmear" assay by spatially mapping the magnitude of the DEP force along the channel [36]. nDEP forces are generated on the bottom of a wide channel, and variations in the magnitude of the DEP force are accomplished by changing the local electric field magnitude or by changing the local electric field frequency. This, combined with the gravitational sedimentation force leads to cell adhesion to the interdigitated electrode array at locations where the nDEP force is overcome by gravity. This allows a spatial mapping of cellular DEP response to physical location on the electrode array. The electrosmear technique effectively minimizes size dependence, but because it is a temporal (batch) separation technique, it is relatively low throughput. Also, as with DEP-FFF, dependence on gravity limits the flexibility and versatility of the system.

Travelling-wave approaches leverage an electric field that is typically applied across 4 electrodes, each 90° out of phase with one another, leading to a field force that is based on the out of phase component of $\tilde{f}_{CM}$ [37,38]:

$$\langle \vec{F}_{twDEP} \rangle = -\frac{2\pi^2 a^2 \varepsilon_m}{\lambda} \Im\left(\tilde{f}_{CM}\right) \nabla\left(\vec{E} \cdot \vec{E}\right),\tag{12}$$

where $\Im(\cdots)$ refers to the imaginary part and λ defines the wavelength of the twDEP field (i.e., distance between electrodes at each phase). twDEP systems have the advantage of reduced size dependence and can be applied perpendicular to the drag force and the electrode array, meaning that the twDEP field simultaneoulsy suspends and translates cells. The electrodes are actuated at a frequency corresponding to nDEP, suspending them above the array, and the difference in phase drives them transverse to the array. One of the initial applications was developed in theory by the Mezic group to demonstrate a multifield appraoch using twDEP to achieve separation [39]. twDEP methods have also been used by many and show promise as a method for size-independent cell separation [40–43]. In each case, the twDEP field is used to simultaneously levitate cells via nDEP and transport them transverse to the direction of a fluid flow field via the phase-dependent component of the twDEP force (proportional to $\Im(\tilde{f}_{CM})$). However, twDEP remains a non-equilibrium technique. In order to achieve predictable separation, careful tuning of force magnitude and flow velocity is required, or binary separation can be achieved as the equilibrium condition.

3.4. 3D Electrode Systems

Three-dimensional electrode systems refers to the use of electrodes on multiple sides of a microfluidic channel, rather than confining the electrode pattern to one side. This can mean an energized channel top with an interdigitated array on the bottom, interdigitated arrays embedded in the side of a microchannel, conductive posts, or electrodes that encompass the entire channel. Some 3D electrode systems also have incorporated multiple electric fields at different frequencies.

In a straightforward extension of the planar interdigitated electrode array, Lai, et al. used a solid conductive ITO sheet to close the top of the channel and energized an interdigitated array formed using a Ti/Al thin film on the bottom. The device successfully isolated RBCs from plasma in a serpentine channel [25]. Another example of top/bottom electrode configuration for continuous flow was demonstrated by Tada, et al. using multiple fields applied to an interdigitated array on the channel bottom with an indium-tin-oxide (ITO) coated glass top layer to generate electric field gradients spanning channel depth. The result was a device that trapped dead cells via pDEP on the channel bottom and concentrated live cells via opposing nDEP forces in the center of the channel [44]. Top-bottom electrode configurations generate gradients throughout channel depth, overcoming the challenge faced by planar configurations whose gradients do not reach sufficiently far in to the device to affect all particles. The distribution of field gradients allows for better control over their location as well, but at the cost of lower gradient magnitudes, meaning such devices require larger applied fields to achieve actuation. While top/bottom configurations offer better uniformity across channel depth, they require precise alignment if electrodes are patterned on both sides of the device, or suffer slightly non-uniform gradients across channel depth if one surface is a single electrode. Separations that leverage the top/bottom electrode configuration either rely on the cumulative effects of successive particle-gradient interactions or repulsion type interactions to manipulate particles. Applying a single field between electrodes leads to separations that depend on the equilibrium between nDEP and fluid drag forces, as shown in Equation (11). While features can be patterned to achieve better control over particle trajectories, making for more sensitive separations that can be performed in continuous flow, they remain strongly size-dependent.

Spanning the field from top to bottom of the channel has been accomplished with physical structures as well. Energized post array systems have been developed by Voldman and co-workers to achieve broad field distributions that were able to achieve single-cell trapping [45]. These traps were extruded metal posts that generated a quadrupolar DEP force field. Martinez-Duarte and co-workers have demonstrated a technique to easily generate arrays of conductive pillars by pyrolysis of SU-8 and used the resulting structures to filter bacterial cells [46,47]. This approach is highly efficient for fabrication and allows distribution of the electric field throughout the depth of the microchannel. These structures overcome the challenge of varying field magnitude across channel depth encountered in the top/bottom configurations. Conductive pillars allow control of field gradients uniformly across channel depth without requiring precise alignment of top and bottom structures, and require lower field magnitudes. Like top/bottom electrode configurations, applying a single field leads to continuous-flow separations that rely on successive interactions or exclusion, and remain size dependent.

Electrodes have also been incorporated in to the sidewall of a channel, fashioned out of conductive PDMS [48], solder [49], and various metals [50,51]. The advantage of 3D- and sidewall electrode systems is that they distribute the electric field (and its gradient) throughout the channel. Rather than limited by proximity to a planar interdigitated electrode array on a single channel surface, placing potential and ground on opposite sides of the channel allows the DEP force to be applied further in to the fluid domain. Extending the application, electrodes on either sidewall have been developed to achieve lateral separation of a number of analytes [14,52,53]. By changing the distance between electrodes the field magnitude can be spatially mapped, and cell separation can be tuned even further, improving sensitivity and allowing for either batch or continuous separations. Without a countering, volume dependent force, however, the resulting separation is size-dependent.

A few researchers have demonstrated successful actuation of cells using 3D electrode structures and multiple field frequencies, successfully sorting particles transverse to the direction of flow and achieving a continuous flow, size-independent separation [54]. In this work, opposing nDEP and pDEP fields were applied to side-wall electrodes to separate HEK293 from N115 mouse neuroblastoma cells. The benefits of such a system are clear. The total DEP force exerted on a particle due to overlapping fields at different frequencies. Under ideal circumstances, the field applied at the sidewall exerts a force transverse to the direction of flow such that the equilibrium position of a single particle can be determined according to:

$$\sum \vec{F}_{DEP} = 2\pi\varepsilon_m a^3 \Re\left(\tilde{f}_{CM,1}\right) \nabla \left(\vec{E}_{rms,1} \cdot \vec{E}_{rms,1}\right) +$$

$$2\pi\varepsilon_m a^3 \Re\left(\tilde{f}_{CM,2}\right) \nabla \left(\vec{E}_{rms,2} \cdot \vec{E}_{rms,2}\right) = 0 \tag{13}$$

$$\frac{\nabla \left(\vec{E}_{rms,1} \cdot \vec{E}_{rms,1}\right)}{\nabla \left(\vec{E}_{rms,2} \cdot \vec{E}_{rms,2}\right)} = \frac{\Re\left(\tilde{f}_{CM,2}\right)}{\Re\left(\tilde{f}_{CM,1}\right)}, \tag{14}$$

where the equilibrium lateral position of a particle subjected to fields with different frequency and magnitude occurs at the point where the ratio of squared field gradients equals the ratio of Clausius-Mossotti factors. This approach indicates four possible experimental tuning parameters—field magnitudes and field frequencies. While this approach achieves size-independence, application of the field across the channel width requires high field magnitudes or narrow channel dimensions. Furthermore, gradients between electrodes decay rapidly, meaning that a component of the applied DEP forces oppose fluid drag and potentially reintroduce a measure of size-dependence.

The furthest extension of the 3D electrode based dielectrophoresis concept was demonstrated by Kung, et al. and termed "tunnel" DEP. Researchers fabricated independent electrodes at the corners of a microchannel parallel to the direction of flow. Selective actuation of the electrodes at each corner allowed positioning of a stream of particles to an equilibrium position within the channel cross-section [55,56]. Like Wang, et al. this size-independent separation offers advantages in the form of tunability and continuous flow separation; additionally, this approach offers a second dimension of spatial control, in theory allowing the positioning of particles within the channel cross section based on DEP responses to multiple field frequencies and magnitudes. In order to achieve this level of control, however, complicated fabrication procedures must be applied and channel dimension must remain small to maintain field gradient magnitudes.

3.5. Media Conductivity

In addition to cell electrical properties, the DEP force depends on the properties of the surrounding media. Throughout it development, researchers have chosen media conductivity to allow for larger

DEP force magnitudes and potential separations. Swami and co-workers have demonstrated a number of sensitive separations, isolating cells with mitochondrial structure variations [9], extracellular vesicles from pancreatic tumor cells based on invasiveness [57], and bacterial (Clostridium difficile) cells with altered envelope structure [58]. While the approach does not eliminate size dependence, as we show in Section 2, it can be leveraged to increase the relative contribution of particle electrical properties to variations in the resultant DEP force.

In another approach to minimize the dependence on size in cell discrimination and increase sensitivity, Voldman and coworkers developed "isodielectric focusing" which leverages the media dependence of $\tilde{f}_{CM}$ eliminating the majority of size dependence in a continuous flow separation [59,60]. Vahey, et al. established the isodielectric focusing technique by first creating a spatial gradient in the electrical conductivity of the media. Suspended cells then have a varying $\tilde{f}_{CM}$ across the gradient. Using an angled array of interdigitated electrodes to induce an nDEP force, they were able to deflect polystyrene beads and cells to a location where the nDEP force goes to 0. Particles introduced by a pressure driven flow are deflected transverse to the direction of flow—and transverse to the media conductivity gradient—by nDEP. As the local media conductivity changes, $\tilde{f}_{CM}$ changes and approaches 0. When $\tilde{f}_{CM} = 0$, $\langle \vec{F}_{DEP} \rangle = 0$ and particles are transported by fluid flow. While this technique also minimizes size dependence and achieves continuous flow separation, it requires a conductivity gradient that is both stable and properly varies $\Re\left(\tilde{f}_{CM}\right)$ to achieve separation. Such a stable flow and proper design of the conductivity gradient may not be achievable in all applications.

4. Conclusions

In this article, we have identified a number of approaches that variously combine 2D and 3D electrode structures, single or multiple electric field magnitudes, single or multiple electric field frequencies, and media variation strategies aimed at improving the sensitivity of label-free DEP separations to small variations in cell electrical properties. An analysis of the sensitivity of the DEP force indicates that particle size variation is potentially the largest factor, confounding efforts to isolate other variations. Therefore, we focus on efforts to increase sensitivity while also reducing or eliminating size dependence. The first counteracting force was gravity. DEP-FFF encompasses these techniques and there is a deep body of work in the area. The technique has the potential to be sensitive and size-independent, but either operates in batch mode, limiting throughput, or requires significant manufacturing investment to develop a continuous flow implementation. 2D electrodes are often used to actuate DEP forces. For the most part, these techniques can be sensitive, and are easy to fabricate, but without an opposing force, are largely size dependent. 2D electrode structures where multiple field magnitudes and/or multiple field frequencies are applied overcome the size-dependent nature with a second DEP force. While eliminating the size-dependence, 2D structures offer limited control of the spatial distribution of electric field gradients. 3D electrode structures, on the other hand, distribute electric field gradients more uniformly, at the expense of slightly more involved fabrication, but without multiple fields, remain size-dependent. Combining 3D structures with multiple fields overcomes both the size-dependent nature of the DEP force as well as the challenge of controlling the spatial distribution of the electric field.

As we develop techniques that get closer and closer to isolating minute particle electrical properties, it is highly likely that our assumptions of homogeneity and isotropy will break down. The multi-shell model, in these cases, will likely not completely reflect or predict particle response. In order to more accurately model the DEP force, some have considered multipolar approximations, but the MST approach (Equation (2)), should be considered when attempting to model particle behavior and characterizing these highly sensitive separations.

5. Future Directions

The field of DEP cell separation continues to iterate on new device designs and configurations that take advantage of various chemical, hydrodynamic, and electrical properties of devices and analytes.

The field is bustling with new proof of concept demonstrations that achieve challenging separations in useful ways. As microfabrication techniques become more and more accessible and see application to the microfluidic field in more useful ways, dielectrophoresis and electrokinetics will play a crucial role in new and novel separations that answer valuable research questions.

The biggest challenge to the field will be successfully bridging the gap between increasingly niche and narrowly focused demonstrations of a particular separation to robust and flexible separation tools that can be routinely applied in the biomedical and basic science labs. To date, DEP has not successfully bridged that gap, with DEP-FFF and the DEPTech 3DEP instrument being notable exceptions. The potential for truly label-free separations based solely on subtle variations in electrical phenotype in a robust, repeatable, and flexible platform remains a persistent challenge. Novel approaches that leverage spatially tunable electric field magnitudes and frequencies will be at the forefront of the next generation of DEP separation devices; tools and instruments that will enable groundbreaking work in the hands of a broader community of biomedical, biological, and basic science researchers.

Author Contributions: N.L. surveyed the literature and summarized relevant articles. D.S.C. provided valuable perspective and assisted with writing. B.G.H. completed the modeling analysis, compiled additional literature sources, and conceptualized and wrote the article. All authors have read and agreed to the published version of the manuscript.

Funding: This research received no external funding.

Conflicts of Interest: The authors declare no conflict of interest.

Abbreviations

The following abbreviations are used in this manuscript:

MDPI	Multidisciplinary Digital Publishing Institute
DEP	Dielectrophoresis
nDEP	Negative Dielectrophoresis
pDEP	Positive Dielectrophoresis
DEP-FFF	Dielectrophoretic Field-Flow Fractionation
CTC	Circulating Tumor Cell

References

1. Pethig, R. Review Article-Dielectrophoresis: Status of the theory, technology, and applications. *Biomicrofluidics* **2010**, *4*. [CrossRef] [PubMed]
2. Yao, J.; Zhu, G.; Zhao, T.; Takei, M. Microfluidic device embedding electrodes for dielectrophoretic manipulation of cells-A review. *Electrophoresis* **2019**, *40*, 1166–1177. [CrossRef] [PubMed]
3. Hughes, M.P. Fifty years of dielectrophoretic cell separation technology. *Biomicrofluidics* **2016**, *10*, 032801. [CrossRef] [PubMed]
4. Fernandez, R.E.; Rohani, A.; Farmehini, V.; Swami, N.S. Review: Microbial analysis in dielectrophoretic microfluidic systems. *Anal. Chim. Acta* **2017**, *966*, 11–33. [CrossRef]
5. Zhang, J.; Song, Z.; Liu, Q.; Song, Y. Recent advances in dielectrophoresis-based cell viability assessment. *Electrophoresis* **2020**. [CrossRef]
6. Zhang, H.; Chang, H.; Neuzil, P. DEP-on-a-chip: Dielectrophoresis applied to microfluidic platforms. *Micromachines* **2019**, *10*, 423. [CrossRef]
7. Xujing, W.; Wang, X.; Gascoyne, P. General expressions for dielectrophoretic force and electrorotational torque derived using the Maxwell stress tensor method. *J. Electrost.* **1997**, *39*, 277–295. [CrossRef]
8. Washizu, M.; Jones, T. Multipolar dielectrophoretic force calculation. *J. Electrost.* **1994**, *33*, 187–198. [CrossRef]
9. Rohani, A.; Moore, J.H.; Kashatus, J.A.; Sesaki, H.; Kashatus, D.F.; Swami, N.S. Label-Free Quantification of Intracellular Mitochondrial Dynamics Using Dielectrophoresis. *Anal. Chem.* **2017**, *89*, 5757–5764. [CrossRef]
10. Gascoyne, P.R.C.; Shim, S.; Noshari, J.; Becker, F.F.; Stemke-Hale, K. Correlations between the dielectric properties and exterior morphology of cells revealed by dielectrophoretic field-flow fractionation. *Electrophoresis* **2013**, *34*, 1042–1050. [CrossRef]

11. Asami, K. Characterization of biological cells by dielectric spectroscopy. *J. Non-Cryst. Solids* **2002**, *305*, 268–277. [CrossRef]
12. Wang, X.B.; Vykoukal, J.; Becker, F.F.; Gascoyne, P.R. Separation of polystyrene microbeads using dielectrophoretic/gravitational field-flow-fractionation. *Biophys. J.* **1998**, *74*, 2689–2701. [CrossRef]
13. Wang, X.B.; Yang, J.; Huang, Y.; Vykoukal, J.; Becker, F.F.; Gascoyne, P.R.C. Cell Separation by Dielectrophoretic Field-flow-fractionation. *Anal. Chem.* **2000**, *72*, 832–839. [CrossRef] [PubMed]
14. Waheed, W.; Alazzam, A.; Mathew, B.; Christoforou, N.; Abu-Nada, E. Lateral fluid flow fractionation using dielectrophoresis (LFFF-DEP) for size-independent, label-free isolation of circulating tumor cells. *J. Chromatogr. B* **2018**, *1087–1088*, 133–137. [CrossRef] [PubMed]
15. Gascoyne, P.R.; Shim, S. Isolation of circulating tumor cells by dielectrophoresis. *Cancers* **2014**, *6*, 545–579. [CrossRef] [PubMed]
16. Yang, J.; Huang, Y.; Wang, X.B.; Becker, F.F.; Gascoyne, P.R.C. Cell Separation on Microfabricated Electrodes Using Dielectrophoretic/Gravitational Field-Flow Fractionation. *Anal. Chem.* **1999**, *71*, 911–918. [CrossRef]
17. Shim, S.; Gascoyne, P.; Noshari, J.; Stemke Hale, K. Dynamic physical properties of dissociated tumor cells revealed by dielectrophoretic field-flow fractionation. *Integr. Biol.* **2011**, *3*, 850. [CrossRef]
18. Shim, S.; Stemke-Hale, K.; Noshari, J.; Becker, F.F.; Gascoyne, P.R. Dielectrophoresis has broad applicability to marker-free isolation of tumor cells from blood by microfluidic systems. *Biomicrofluidics* **2013**, *7*, 011808. [CrossRef]
19. Gascoyne, P.R.C. Dielectrophoretic-field flow fractionation analysis of dielectric, density, and deformability characteristics of cells and particles. *Anal. Chem.* **2009**, *81*, 8878–85. [CrossRef]
20. Vykoukal, J.; Vykoukal, D.M.; Freyberg, S.; Alt, E.U.; Gascoyne, P.R. Enrichment of putative stem cells from adipose tissue using dielectrophoretic field-flow fractionation. *Lab Chip* **2008**, *8*, 1386–1393. [CrossRef]
21. Schnelle, T. Paired microelectrode system: Dielectrophoretic particle sorting and force calibration. *J. Electrost.* **1999**, *47*, 121–132. [CrossRef]
22. D'Amico, L.; Ajami, N.J.; Adachi, J.A.; Gascoyne, P.R.C.; Petrosino, J.F. Isolation and concentration of bacteria from blood using microfluidic membraneless dialysis and dielectrophoresis. *Lab Chip* **2017**, *17*, 1340–1348. [CrossRef] [PubMed]
23. Sadeghian, H.; Hojjat, Y.; Soleimani, M. Interdigitated electrode design and optimization for dielectrophoresis cell separation actuators. *J. Electrost.* **2017**, *86*, 41–49. [CrossRef]
24. Jiang, A.Y.L.; Yale, A.R.; Aghaamoo, M.; Lee, D.H.; Lee, A.P.; Adams, T.N.G.; Flanagan, L.A. High-throughput continuous dielectrophoretic separation of neural stem cells. *Biomicrofluidics* **2019**, *13*, 064111. [CrossRef]
25. Lai, C.C.; Yeh, Y.T.; Chung, C.K. Design and fabrication of a Ti/Al thin-film electrode in the meander-shaped microchannel and its application for promoting capillary-driven dielectrophoresis blood separation. *J. Micromech. Microeng.* **2020**, *30*, 025002. [CrossRef]
26. Gonzalez, C.F.; Remcho, V.T. Fabrication and evaluation of a ratchet type dielectrophoretic device for particle analysis. *J. Chromatogr. A* **2009**, *1216*, 9063–9070. [CrossRef]
27. Choi, S.; Park, J.K. Microfluidic system for dielectrophoretic separation based on a trapezoidal electrode array. *Lab Chip* **2005**, *5*, 1161–1167. [CrossRef]
28. Shkolnikov, V.; Xin, D.; Chen, C. Continuous dielectrophoretic particle separation via isomotive dielectrophoresis with bifurcating stagnation flow. *Electrophoresis* **2019**, *40*, 2988–2995. [CrossRef]
29. Tada, S.; Omi, Y.; Eguchi, M. Analysis of the dielectrophoretic properties of cells using the isomotive AC electric field. *Biomicrofluidics* **2018**, *12*, 044103. [CrossRef]
30. Hunt, T.P.; Issadore, D.; Westervelt, R.M. Integrated circuit/microfluidic chip to programmably trap and move cells and droplets with dielectrophoresis. *Lab Chip* **2007**, *8*, 81–87. [CrossRef]
31. Urdaneta, M.; Smela, E. Multiple frequency dielectrophoresis. *Electrophoresis* **2007**, *28*, 3145–3155. [CrossRef] [PubMed]
32. Giesler, J.; Pesch, G.R.; Weirauch, L.; Schmidt, M.P.; Thöming, J.; Baune, M. Polarizability-Dependent Sorting of Microparticles Using Continuous-Flow Dielectrophoretic Chromatography with a Frequency Modulation Method. *Micromachines* **2019**, *11*, 38. [CrossRef] [PubMed]
33. Hadady, H.; Redelman, D.; R. Hiibel, S.; J. Geiger, E. Continuous-flow sorting of microalgae cells based on lipid content by high frequency dielectrophoresis. *AIMS Biophys.* **2016**, *3*, 398–414. [CrossRef]
34. Modarres, P.; Tabrizian, M. Frequency hopping dielectrophoresis as a new approach for microscale particle and cell enrichment. *Sens. Actuators B Chem.* **2019**, *286*, 493–500. [CrossRef]

35. Han, C.H.; Woo, S.Y.; Bhardwaj, J.; Sharma, A.; Jang, J. Rapid and selective concentration of bacteria, viruses, and proteins using alternating current signal superimposition on two coplanar electrodes. *Sci. Rep.* **2018**, *8*, 1–10. [CrossRef] [PubMed]

36. Das, C.M.; Becker, F.; Vernon, S.; Noshari, J.; Joyce, C.; Gascoyne, P.R.C. Dielectrophoretic segregation of different human cell types on microscope slides. *Anal. Chem.* **2005**, *77*, 2708–19. [CrossRef]

37. Van Den Driesche, S.; Rao, V.; Puchberger-Enengl, D.; Witarski, W.; Vellekoop, M.J. Continuous separation of viable cells by travelling wave dielectrophoresis. *Procedia Eng.* **2010**, *5*, 41–44. [CrossRef]

38. Hagedorn, R.; Fuhr, G.; Muller, T.; Gimsa, J. Traveling-wave dielectrophoresis of microparticles. *Electrophoresis* **1992**, *13*, 49–54. [CrossRef]

39. Loire, S.; Mezic, I. Separation of bioparticles using the travelling wave dielectrophoresis with multiple frequencies. In Proceedings of the 42nd IEEE International Conference on Decision and Control (IEEE Cat. No.03CH37475), Maui, HI, USA, 9–12 December 2003; Volume 6, pp. 6448–6453. [CrossRef]

40. Wu, Y.; Ren, Y.; Tao, Y.; Jiang, H. Fluid pumping and cells separation by DC-biased traveling wave electroosmosis and dielectrophoresis. *Microfluid. Nanofluidics* **2017**, *21*. [CrossRef]

41. van den Driesche, S.; Bunge, F.; Tepner, S.; Kotitschke, M.; Vellekoop, M.J. Travelling-wave dielectrophoresis allowing flexible microchannel design for suspended cell handling. In *Bio-MEMS and Medical Microdevices III*; van den Driesche, S., Giouroudi, I., Delgado-Restituto, M., Eds.; SPIE: Barcelona, Spain, 2017; Volume 10247, p. 102470H. [CrossRef]

42. De Gasperis, G.; Yang, J.; Becker, F.F.; Gascoyne, P.R.C.; Wang, X.B. Microfluidic Cell Separation by 2-dimensional Dielectrophoresis. *Biomed. Microdevices* **1999**, *2*, 41–49. [CrossRef]

43. Sim, K.; Shi, L.; He, G.; Chen, S.; Liu, D.; Yu, C. Mechanically flexible microfluidics for microparticle dispensing based on traveling wave dielectrophoresis. *J. Micromech. Microeng.* **2020**, *30*, 024001. [CrossRef]

44. Tada, S.; Hayashi, M.; Eguchi, M.; Tsukamoto, A. High-throughput separation of cells by dielectrophoresis enhanced with 3D gradient AC electric field. *Biomicrofluidics* **2017**, *11*. [PubMed]

45. Voldman, J. BioMEMS: Building with cells. *Nat. Mater.* **2003**, *2*, 433–434. [CrossRef] [PubMed]

46. Jaramillo, M.d.C.; Torrents, E.; Martínez-Duarte, R.; Madou, M.J.; Juárez, A. On-line separation of bacterial cells by carbon-electrode dielectrophoresis. *Electrophoresis* **2010**, *31*, 2921–2928. [CrossRef] [PubMed]

47. Martinez-Duarte, R.; Gorkin, R.A., III; Abi-Samra, K.; Madou, M.J. The integration of 3D carbon-electrode dielectrophoresis on a CD-like centrifugal microfluidic platform. *Lab Chip* **2010**, *10*, 1030. [CrossRef]

48. Lewpiriyawong, N.; Yang, C.; Lam, Y.C. Continuous sorting and separation of microparticles by size using AC dielectrophoresis in a PDMS microfluidic device with 3-D conducting PDMS composite electrodes. *Electrophoresis* **2010**, *31*, 2622–2631. [CrossRef]

49. Shafiee, H.; Caldwell, J.L.; Sano, M.B.; Davalos, R.V. Contactless dielectrophoresis: A new technique for cell manipulation. *Biomed. Microdevices* **2009**, *11*, 997–1006. [CrossRef]

50. Wang, Y.; Wang, J.; Wu, X.; Jiang, Z.; Wang, W. Dielectrophoretic separation of microalgae cells in ballast water in a microfluidic chip. *Electrophoresis* **2019**, *40*, 969–978. [CrossRef]

51. Piacentini, N.; Mernier, G.; Tornay, R.; Renaud, P. Separation of platelets from other blood cells in continuous-flow by dielectrophoresis field-flow-fractionation. *Biomicrofluidics* **2011**, *5*, 034122. [CrossRef]

52. Zhao, K.; Larasati.; Duncker, B.P.; Li, D. Continuous Cell Characterization and Separation by Microfluidic Alternating Current Dielectrophoresis. *Anal. Chem.* **2019**, *91*, 6304–6314. [CrossRef]

53. Demierre, N.; Braschler, T.; Muller, R.; Renaud, P. Focusing and continuous separation of cells in a microfluidic device using lateral dielectrophoresis. *Sens. Actuators B Chem.* **2008**, *132*, 388–396. [CrossRef]

54. Wang, L.; Lu, J.; Marchenko, S.A.; Monuki, E.S.; Flanagan, L.A.; Lee, A.P. Dual frequency dielectrophoresis with interdigitated sidewall electrodes for microfluidic flow-through separation of beads and cells. *Electrophoresis* **2009**, *30*, 782–791. [CrossRef] [PubMed]

55. Kung, Y.C.; Huang, K.W.; Chong, W.; Chiou, P.Y. Tunnel Dielectrophoresis for Tunable, Single-Stream Cell Focusing in Physiological Buffers in High-Speed Microfluidic Flows. *Small* **2016**, *12*, 4343–4348. [CrossRef] [PubMed]

56. Kung, Y.C.; Huang, K.W.; Chong, W.; Chiou, P.Y. Microfluidics: Tunnel Dielectrophoresis for Tunable, Single-Stream Cell Focusing in Physiological Buffers in High-Speed Microfluidic Flows (Small 32/2016). *Small* **2016**, *12*, 4302. [CrossRef]

57. Moore, J.H.; Varhue, W.B.; Su, Y.H.; Linton, S.S.; Farmehini, V.; Fox, T.E.; Matters, G.L.; Kester, M.; Swami, N.S. Conductance-Based Biophysical Distinction and Microfluidic Enrichment of Nanovesicles Derived from Pancreatic Tumor Cells of Varying Invasiveness. *Anal. Chem.* **2019**, *91*, 10424–10431. [CrossRef]
58. Su, Y.H.; Rohani, A.; Warren, C.A.; Swami, N.S. Tracking Inhibitory Alterations during Interstrain Clostridium difficile Interactions by Monitoring Cell Envelope Capacitance. *ACS Infect. Dis.* **2016**, *2*, 544–551. [CrossRef]
59. Vahey, M.D.; Voldman, J. Iso-dielectric Separation of Cells and Particles. In *Microsystems Technology Laboratories Annual Research Report*; Massachusetts Institute of Technology: Cambridge, MA, USA, 2009.
60. Vahey, M.D.; Voldman, J. High-throughput cell and particle characterization using isodielectric separation. *Anal. Chem.* **2009**, *81*, 2446–2455. [CrossRef]

 *micromachines*

Article

Polarizability-Dependent Sorting of Microparticles Using Continuous-Flow Dielectrophoretic Chromatography with a Frequency Modulation Method

Jasper Giesler [1], Georg R. Pesch [1,2,*], Laura Weirauch [1], Marc-Peter Schmidt [3], Jorg Thöming [1,2] and Michael Baune [1]

[1] Chemical Process Engineering, Faculty of Production Engineering, University of Bremen, Leobener Straße 6, 28359 Bremen, Germany; j.giesler@uni-bremen.de (J.G.); lweirauch@uni-bremen.de (L.W.); thoeming@uni-bremen.de (J.T.); mbaune@uni-bremen.de (M.B.)

[2] MAPEX Center for Materials and Processes, University of Bremen, 28359 Bremen, Germany

[3] Department of Engineering, Brandenburg University of Applied Sciences, Magdeburger Straße 50, 14770 Brandenburg an der Havel, Germany; schmmarc@th-brandenburg.de

* Correspondence: gpesch@uni-bremen.de

Received: 7 November 2019; Accepted: 23 December 2019; Published: 28 December 2019

Abstract: The separation of microparticles with respect to different properties such as size and material is a research field of great interest. Dielectrophoresis, a phenomenon that is capable of addressing multiple particle properties at once, can be used to perform a chromatographic separation. However, the selectivity of current dielectrophoretic particle chromatography (DPC) techniques is limited. Here, we show a new approach for DPC based on differences in the dielectrophoretic mobilities and the crossover frequencies of polystyrene particles. Both differences are addressed by modulating the frequency of the electric field to generate positive and negative dielectrophoretic movement to achieve multiple trap-and-release cycles of the particles. A chromatographic separation of different particle sizes revealed the voltage dependency of this method. Additionally, we showed the frequency bandwidth influence on separation using one example. The DPC method developed was tested with model particles, but offers possibilities to separate a broad range of plastic and metal microparticles or cells and to overcome currently existing limitations in selectivity.

Keywords: dielectrophoresis (DEP); microparticles; polystyrene; chromatography; interdigitated electrodes; microfluidic; separation

1. Introduction

Separating microparticles according to specific properties such as size, material, and shape is a research area of great interest for instance in cell or biomolecule manipulation [1–5] and waste recovery [6,7]. To separate microparticles, field-flow fractionation [8], gel electrophoresis [9], and size-exclusion chromatography [10] are state-of-the-art approaches. A major drawback of these approaches is their low throughput or low selectivity for particle mixtures with similar separation properties (e.g., shape, density) below a particle size of 10 μm [11–13]. Dielectrophoresis (DEP), which is referred to as the movement of polarizable particles in an inhomogeneous electric field, offers an alternative tool to address a wide range of particles and at the same time is able to achieve relevant throughputs [14,15]. The dielectrophoretic force not only depends on one specific property of a particle, but on a variety of particle properties, such as size [16,17], permittivity, and electrical conductivity [1], allowing for multi-dimensional particle fractionation. Apart from established DEP concepts such as field-flow fractionation [17,18], filtration [19], selective trapping (e.g., insulator-based dielectrophoresis) [20], dielectrophoretic particle chromatography (DPC) is a promising concept to achieve high throughput separation of particles. Since DPC was introduced by Washizu et al. [5], different approaches were done using selective trapping of particles [21,22], packed bed columns [23], or stepwise change of the frequency [24]. DEP chromatography proved to be very successful in isolating circulating breast tumor cells (CTCs) from blood [25] at a very low concentration. Such studies later led to the development of a clinical high throughput device to separate CTCs from blood samples [26,27]. Aldaeus et al. [28] developed an analytical model for a DPC device that was based on multiple trap and release cycles for fractionation. A related technique to manipulate micrometer sized particles is using traveling wave dielectrophoretic separators [29,30]. In these microfluidic devices, a 90° phase angle is present between adjacent electrodes, which changes the dielectrophoretic movement a particle experiences [31,32]. Such traveling wave systems offer versatile particle separation techniques, but are usually complex to fabricate and operate [30,33]. The other presented dielectrophoretic chromatography techniques have in common that they depend on strongly diverging polarizabilities (e.g., one type of particle showing positive dielectrophoresis, whereas the other particles show negative dielectrophoresis or exhibit no dielectrophoretic movement). This requirement limits the applicability when addressing particle mixtures with less pronounced differences in polarizability. Addressing binary (or more) mixtures in which there is heterogeneity in the two (or more) classes is even more complex, especially when the cross-over frequencies of the classes are so close that the heterogeneity causes an overlap (an example is the separation of cells according to only small differences in their expression).

Here, we introduce the novel concept of frequency modulated dielectrophoretic particle chromatography. The frequency of the applied field changes constantly to exploit small differences in the dielectrophoretic mobilities of target particles. In this technique, by switching the frequency, we switch between positive and negative dielectrophoretic movement of target particles to generate multiple trap-and-release cycles, which leads to a polarizability dependent chromatographic separation. In principle, this allows separating particles that even show only minute differences in their polarizability and to separate mixtures with heterogeneity in the classes. The simplicity of our approach allows for a simple fabrication and operation and could be easily scaled up by using different ways to introduce the electric field gradient (for example using a porous medium as demonstrated in our recent work [14]).

2. Method

2.1. Theory

In classic chromatographic processes (e.g., gas chromatography), mixtures are separated due to different interactions of the sample and stationary phase, leading to characteristic retention times for each class in the sample. In dielectrophoretic particle chromatography, the stationary phase is represented by the inhomogeneous electric field rising over interdigitated electrodes. The electrode chip forms the bottom of a microfluidic device, where a polydimethylsiloxane (PDMS) channel is used as the separation column. The microparticle suspension is injected into the flow chamber and further transported by a carrier flow. The electrodes are connected to an AC voltage source to generate a highly inhomogeneous electric field. This gives rise to a dielectrophoretic force on the particle caused by the action of the inhomogeneous field on the induced dipole (or multipole) of the particle. In the simple point-dipole approximation, the dielectrophoretic force $\mathbf{F}_{\text{DEP}}$ can be expressed as:

$$\mathbf{F}_{\text{DEP}} = \pi r_{\text{p}}^3 \varepsilon_{\text{m}} \text{Re}\left(\frac{\tilde{\varepsilon}_{\text{p}} - \tilde{\varepsilon}_{\text{m}}}{\tilde{\varepsilon}_{\text{p}} + 2\tilde{\varepsilon}_{\text{m}}}\right) \nabla |\mathbf{E}|^2, \tag{1}$$

with r_{p} representing the particles radius, $\nabla |\mathbf{E}|^2$ the electric field gradient squared, and $\tilde{\varepsilon}_{\text{p}}$ the complex permittivity of the particles and the medium ($\tilde{\varepsilon}_{\text{m}}$), respectively. The velocity due to dielectrophoresis, $\mathbf{v}_{DEP}$, in a stationary fluid can be calculated by dividing the dielectrophoretic force by the friction factor f^*:

$$\mathbf{v}_{DEP} = \mu_{DEP} \nabla |\mathbf{E}|^2 = \frac{\pi r_p^3 \varepsilon_m \mathrm{Re}\left(\frac{\tilde{\varepsilon}_p - \tilde{\varepsilon}_m}{\tilde{\varepsilon}_p + 2\tilde{\varepsilon}_m}\right) \nabla |\mathbf{E}|^2}{f^*}. \tag{2}$$

Here, μ_{DEP} is the dielectrophoretic mobility, which not only provides the direction of the movement of the microparticles, but incorporates the radius of the particles and fluid properties additionally. The direction of the DEP force can be determined by calculating the real part of the Clausius–Mossotti factor $\mathrm{Re}(CM)$:

$$CM = \frac{\tilde{\varepsilon}_p - \tilde{\varepsilon}_m}{\tilde{\varepsilon}_p + 2\tilde{\varepsilon}_m}. \tag{3}$$

The complex permittivity expands the permittivity ε of a material and incorporates the material's conductivity σ and the angular frequency ω of the electric field:

$$\tilde{\varepsilon} = \varepsilon_0 \varepsilon_r - i\frac{\sigma}{\omega}. \tag{4}$$

For low frequencies, $\mathrm{Re}(CM)$ is dominated by the conductivity of the material. With increasing frequency, the permittivity becomes more important. When particles are less polarizable than the surrounding medium ($\mathrm{Re}(CM) < 0$, negative DEP), they move against the electric field gradient and towards low field regions. On the contrary, more polarizable particles ($\mathrm{Re}(CM) > 0$, positive DEP) are directed with the gradient towards field maxima. In the current setup, field maxima are located close to the edges of the interdigitated electrodes at the bottom, and local field minima can be found at the top of the channel. Depending on the polarization, particles are either attracted to the edges of the electrode (positive, pDEP) or to the top (negative, nDEP). Therefore, the movement direction will be strongly affected by the applied field's frequency due to the frequency dependence of $\mathrm{Re}(CM)$ (Equation (4)). Particles can become trapped in potential wells (field extrema) due to DEP and can adhere to the walls of the device when they reach them.

The conductivity of small insulating particles (such as the polystyrene particles that are used in this study as a model) is dominated by their surface conductance K_S [34]:

$$\sigma_p = \frac{2K_S}{r_p}. \tag{5}$$

Usually, K_S is assumed to be around $1\,\mathrm{nS}$ for polystyrene particles [35]. Equation (5) leads to a (with increasing particle diameter decreasing) net conductivity of polystyrene particles ($1\,\mu\mathrm{m} < d_P < 10\,\mu\mathrm{m}$) of around $4\,\mu\mathrm{S}\,\mathrm{cm}^{-1}$ to $40\,\mu\mathrm{S}\,\mathrm{cm}^{-1}$, which is higher than some low conductive DEP buffers. This allows for positive DEP manipulation at low frequencies of even electrically insulating particles, when they are smaller than a certain threshold diameter.

To evaluate the resolution of a chromatographic separation, R_S can be calculated [28],

$$R_s = \frac{\Delta t}{\frac{1}{2}(w_1 + w_2)}, \tag{6}$$

with Δt as the separation time between the maximum values ($I_{\max}$) of two peaks and w_x, the width of the two residence time distributions. The width is defined as the distance in time between the half maximum values (FWHM).

2.2. Device Operation

The device proposed here, a microfluidic channel with interdigitated electrodes at the bottom of the channel (Figure 1b), uses periodic changes from pDEP to nDEP or vice versa to separate particles with respect to their polarizability. Since the polarizability of a particle directly depends on the frequency of the electric field, constant frequency changes (Equation (7)) can be used to manipulate the particles' position in the separator. To achieve a retardation, due to either nDEP or pDEP, particles are dragged out of the fast streamlines in the center of the channel to streamlines with low fluid velocity at the bottom or top. Then, when the frequency changes, the pDEP or nDEP effect is reversed, and particles are pushed back into the faster streamlines in the center of the channel. Depending on the strength of the interaction of a particle with the field (i.e., the absolute value of Equation (3)), particles with different polarizabilities experience different retardation. Unlike DEP field-flow fractionation, there is no particle equilibrium position. Here, the periodic change of frequency leads to a constant change of the particles position and, therefore, depending on the polarizability of a particle, to a different average velocity.

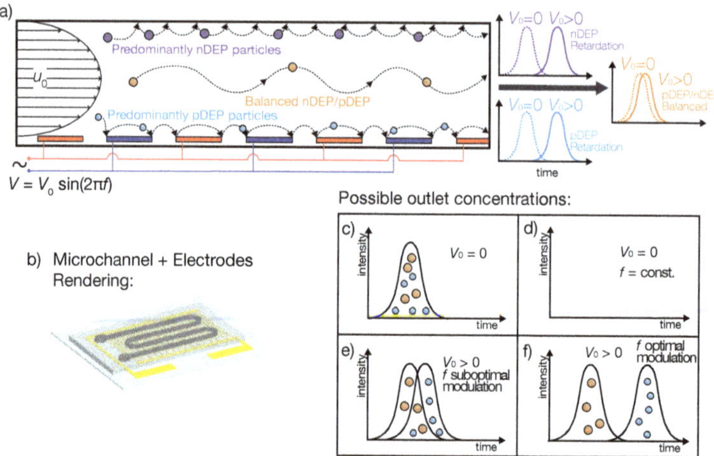

Figure 1. (**a**) Sketch of the DPC separation experiments. (**b**) Sketch of the DPC separation column. Meandering PDMS microchannel sealed by interdigitated electrodes on a glass chip. (**c–f**) Different possible outlet concentrations for DPC. (**c**) Without voltage, no retardation of the particles occurs, and both fractions elute at the same time. (**d**) When a voltage is applied and the frequency is fixed, the particles are trapped in the column due to DEP and will not exit the channel. If the frequency is modulated, a chromatographic separation occurs (**e**), which can be optimized by changing the frequencies and voltage (**f**).

In case a particle gets trapped in potential wells because of dielectrophoresis or adheres to the surface of the channel due to non-specific adsorption, a particle resuspension requires a force pointing away from the wall, which is in our case again DEP. Naturally, to reverse the trapping movement, particles trapped by pDEP now have to experience nDEP and vice versa (Figure 1a). Especially for particles trapped at the bottom of the channel, a resuspension via an external force becomes important, since no gravitational force contributes to their remobilization. Further, as the particles' diameter decreases, the gravitation force becomes less important and therefore may not be sufficient to resuspend small particles close to the ceiling of the channel. To achieve a retardation of the particles and consequently a chromatographic separation, it is in general not necessary to fixate particles at the bottom or ceiling. To generate an increase in retention time, particles are just required to be transported into regions of low fluid velocity, which are present at the bottom (transport via pDEP) or the ceiling (nDEP) of the channel. Apart from the approach taken here, which is to modulate the frequency to reverse the particle polarization and the DEP force vector's direction, in principle, it would also be possible to change the polarization by changing the medium's conductivity.

Depending on the particle's Clausius–Mossotti factor as a function of frequency (Equation (3)), three different scenarios can be distinguished (Figure 1a): (I) A particle shows substantial more pDEP than nDEP during the modulation spectrum and therefore predominantly moves towards the bottom of the channel, where the electrodes induce a high electric field strength. Since the fluids' velocity close to the bottom is low, particles are slowed down by the lower fluid velocity or by getting reversibly trapped at the electric field maxima. The particles are then pushed away from the electrodes by nDEP when the frequency changes. This scenario effectively increases the particle's residence time. (II) When a particle exhibits a balanced pDEP and nDEP movement, the retardation is less pronounced. These microparticles travel towards high field regions when the CM factor is positive and away from them when it is negative. Due to their constant movement orthogonal to the fluid-flow direction, they spend less time in regions with low fluid velocities and therefore are eluted fast. (III) If nDEP outweighs pDEP, particles are predominantly pushed towards low field regions, which here are present at the channel's ceiling. Like in Scenario I, only low fluid-flow is present at the field minima, and the particle's residence time is going to be enlarged. Although the polarizability of particles from Scenarios I and III is different, retention times can be the same. Nevertheless, since the extent of retardation depends on the chosen process parameters (e.g., frequency, voltage), a separation can be possible with a different set of parameters (Figure 1e,f).

Here, the frequency of the applied sinusoidal voltage was modulated using a triangle-shaped function. This allows changing the frequency of the electric field constantly between two values in a controllable time. Consequently, the frequency f can be described as a function of time t:

$$f(t) = f_{\mathrm{A}} \mathrm{tri}(2t f_{\mathrm{mod}}) + f_0 \tag{7}$$

with f_{A} as the amplitude of frequency modification, $\mathrm{tri}(x)$ as the triangle function, f_{mod} representing the modulation frequency, and f_0 for the offset of the frequency modulation. As an example, for achieving frequencies between 30 and 270 kHz. the following set of parameters was used: $f_{\mathrm{A}} = 120\,\mathrm{kHz}$, $f_{\mathrm{mod}} = 300\,\mathrm{mHz}$, and $f_0 = 150\,\mathrm{kHz}$. Other modulation functions may also be suitable for achieving a separation.

In this study, we used polystyrene (PS) particles to demonstrate the functionality of the proposed technique. Due to their surface conductance (Equation (5)), PS particles show pDEP at low frequencies and nDEP at high frequencies. With the usually assumed $K_{\mathrm{S}} = 1\,\mathrm{nS}$ [34–36] and a medium conductivity of $\sigma_{\mathrm{M}} = 1.2\,\mathrm{\mu S\,cm}^{-1}$, the cross-over frequency from negative to positive DEP ($\mathrm{Re}(CM) = 0$ in Equation (3)) depends only on particle size (see Figure S1). The frequency dependent polarizability of the particles forms the fundamental aspect of this separation technique and can be used by varying the frequency over time periodically, as shown in Figure 2. These periodical changes from pDEP to nDEP generate multiple trapping and release cycles. The separation technique can also be used for other particle types that show frequency dependent polarizability.

Larger polystyrene particles showed pDEP in a smaller frequency bandwidth and, consequently, when varying the frequency as shown, for a shorter duration. Four different polystyrene particle sizes were chosen to demonstrate the separation effect. With our chosen frequency modulation from 30 kHz to 270 kHz, 3 μm particles showed predominantly positive DEP, 6 μm particles a balanced pDEP/nDEP behavior, and 10 μm particles predominantly negative DEP. Further, we used 2 μm particles to assess the possibility to separate two particle types that both experienced predominantly pDEP.

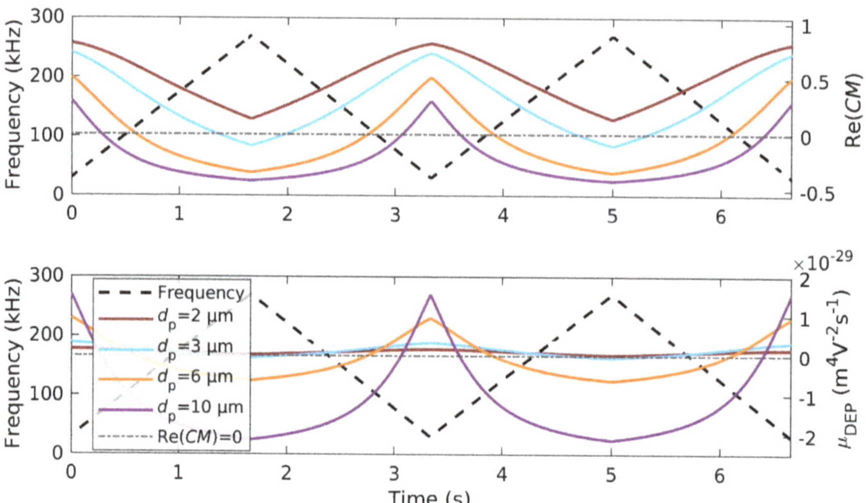

Figure 2. Real part of the Clausius–Mossotti factor Re(*CM*) (top) and dielectrophoretic mobility μ_{DEP} (bottom) of four different polystyrene particles over time for two full cycles (right ordinate axis of diagram). The modulated frequency is shown as well (left ordinate axis). Particles suspended in DI water with $\sigma_m = 1.2\,\mu\text{S cm}^{-1}$, $K_S = 1\,\text{nS}$, and $\varepsilon_m = 78.5$, calculated with Equations (3) and (4).

2.3. Device Fabrication

The microfluidic device consisted of two main parts. The column was formed by a 2 mm wide meandering PDMS channel (height 80 µm, length 17 cm), which provided the walls and the top of the channel (Figure 1b). The bottom was formed by the electric field generating electrode chip. Both parts were bonded using an intermediate layer as described later. The PDMS channel were produced using an SU8 master mold (soft lithography). The interdigitated electrodes (electrode arm width and gap width 100 µm) were fabricated using standard cleanroom techniques. Full details of the fabrication method can be found in Section S2.

The electrode covered glass slide was bonded to the PDMS channel using liquid PDMS (10:3, base:curing agent). PDMS was selected as the intermediate layer, because of its well known spinning curves, low toxic potential, and easy accessibility [37–39]. The PDMS mixture was spin coated at 6000 rpm for 330 s on the electrodes. Using these parameters, the thickness of the uncured PDMS layer should be below 3 µm [38]. Subsequently, the cleaned PDMS channel was manually aligned over the electrodes and placed onto them. The bonding was finalized by curing the intermediate layer at 80 °C for an additional two hours. The PDMS did not only allow bonding the electrodes to the channel, which proved to be unsuccessful in our lab using corona bonding; it also reduced the unspecific adhesion of the particles to the electrodes [40]. Since using PDMS as the intermediate layer creates a reversible bonding and PDMS channels are inexpensive to replace, several channels were used during the experiments, and no significant changes between them could be observed.

2.4. Experimental Setup

Two syringe pumps were connected to a manually actuated 4 way valve (H&S V-101D, IDEX Health & Science, LLC, Oak Harbor, WA, USA). One syringe pump (KDS-100-CE, KD Scientific Inc., Holliston, MA, USA) controlled the volume flow of the carrier fluid; the other pump (LEGATO 270, KD Scientific Inc., Holliston, MA, USA) provided the flow of the particle suspension (both 5 mL h^{-1}). In the normal position, the carrier flow was connected to the inlet of the separation column. To initiate the experiment, the valve was manually turned to allow a 2 s pulse of particle suspension to flow into the separator (Figure 1a). The injection in all experiments happened at $t = 10$ s. The carrier fluid was pure water containing 0.02 vol % Tween20 (Sigma-Aldrich, Steinheim, Germany) to reduce particle–wall interactions, 0.003 vol % 0.01 mol L^{-1} potassium hydroxide in deionized water to adjust pH, and potassium chloride to adjust the electrical conductivity to the desired value (1.2 µS cm^{-1}). The particles were suspended in the same suspension as the carrier flow, but without adding potassium chloride.

Monodisperse fluorescent polystyrene particles (Fluoresbrite, Polysciences Europe GmbH, Hirschberg, Germany) of different sizes and colors (2 µm polychromatic red, 3 µm yellow-green, 6 µm polychromatic red, and 10 µm yellow-green plain particles) were mixed and diluted in the described solution.

The inlet of the channel was connected to the manually actuated 4 way valve via a capillary (inner diameter: 100 µm) with a length of about 17 mm. To allow a controlled injection of the particles (i.e., to avoid dispersion of the peak), the internal volume of the connection from valve to channel inlet should be kept as small as possible. The chosen (short and with small diameter) inlet capillary resulted in a volume of 135 nL, resulting in an average residence time of less than 100 ms in this capillary.

The electrodes were connected to a voltage amplifier (PZD2000A, TREK, Lockport, New York, NY, USA) controlled by a signal generator (Rigol DG4062, Rigol Technologies EU GmbH, Puchheim, Germany). The signal generator provided the functionality of frequency modulation inherently. The amplifier's output signal was monitored using an oscilloscope (RIGOL DS2072A, Rigol Technologies EU GmbH, Puchheim, Germany). The amplification factor of the amplifier was not constant, but decreased with increasing frequency. The output decreased by 4.3 % per 10 kHz, which resulted in exponential decay in the applied voltage. All stated voltages were measured at 30 kHz. This circumstance may be overcome by using a different amplifier in future experiments.

The different fluorescent stains of the particles allowed to easily distinguish between them. To observe the particles, an inverted microscope (ECLIPSE Ts2R-FL, Nikon Instruments Europe BV, Amsterdam, The Netherlands) was used. For observation, a 40,6-diamidino-2-phenylindole/fluorescein isothiocyanate/tetramethylrhodamine isothiocyanate (DAPI/FITC/TRITC, excitation: 387/478/555 nm, emission: 433/517/613 nm) triple bandpass was selected, which allowed observing at least three different types of particles at once. However, only two particle colors could be observed simultaneously, since the current optics inhibited the DAPI excitation. Videos of the fluorescence were recorded at the outlet of the channel using a color CMOS camera (GS3-U3-51S5C-C, FLIR Systems Inc., Wilsonville, OR, USA), which were further processed using MATLAB (see Section S2, for further information). In MATLAB, the frames were segmented, resulting in different pictures for each particle and background. Finally, the intensity of each picture was counted and plotted over time.

3. Results and Discussion

Three different main experiments were conducted to demonstrate the different capabilities of the proposed separator: We firstly demonstrate the possibility to separate particles experiencing predominantly pDEP from particles with a balanced pDEP/nDEP behavior. This was done by separating 3 µm particles from 6 µm particles. We further show the separation of predominantly nDEP experiencing particles (10 µm) from particles experiencing balanced pDEP/nDEP (again, 6 µm particles). Finally, we show that even particles that both experience mostly pDEP in the modulated frequency spectrum can be separated by separating 3 µm particles from 2 µm particles. Figure S9 provides Particle Image Velocimetry (PIV) data of 10 µm particles at 100 V$_{pp}$ and 0 V to demonstrate the fluctuation of the velocity due to the nDEP effect. Further, Videos S2 and S3 visualize the separation of 3 µm and 6 µm particles at 80 V$_{pp}$ and 0 V. For such small particles, it was not possible to extract the velocity reliably from the video using PIV. Nevertheless, the velocity fluctuations due to the action of DEP were clearly visible for the 3 µm particles. Unfortunately, from the observation perspective and with the experimental methods at hand, it was not possible to infer if particles were slowed down because they were attracted to or pushed away from the electrode array.

For the 3 and 6 µm particles, without an electric field, both particles showed typical retention time distributions for a laminar flow without observing separation (as expected; Figure 3a). When applying a voltage with frequency modulation, we could observe a clear chromatographic separation for all investigated voltages, i.e., 60 V_{pp}, 80 V_{pp}, 100 V_{pp}, and 120 V_{pp} (see Figure S4 for the full dataset). To achieve separation, the frequency was varied between 30 kHz and 270 kHz in 3.33 s (full cycle length, 300 mHz). Various parameters for frequency modulation were tested in advance, but this set of parameters worked best. However, the influence of each parameter is not fully understood and needs to be investigated further. While we could observe separation at all voltages, the best resolution for the separation of 3 µm and 6 µm particles could be achieved at a voltage of 80 V_{pp}, resulting in an average resolution of $R_s = 3.60 \pm 0.31$ (number of experiments $N = 4$) (Figure 3b). To provide a visual impression of the separation of 3 µm and 6 µm at 0 V_{pp} and 80 V_{pp}, see Videos S2 and S3.

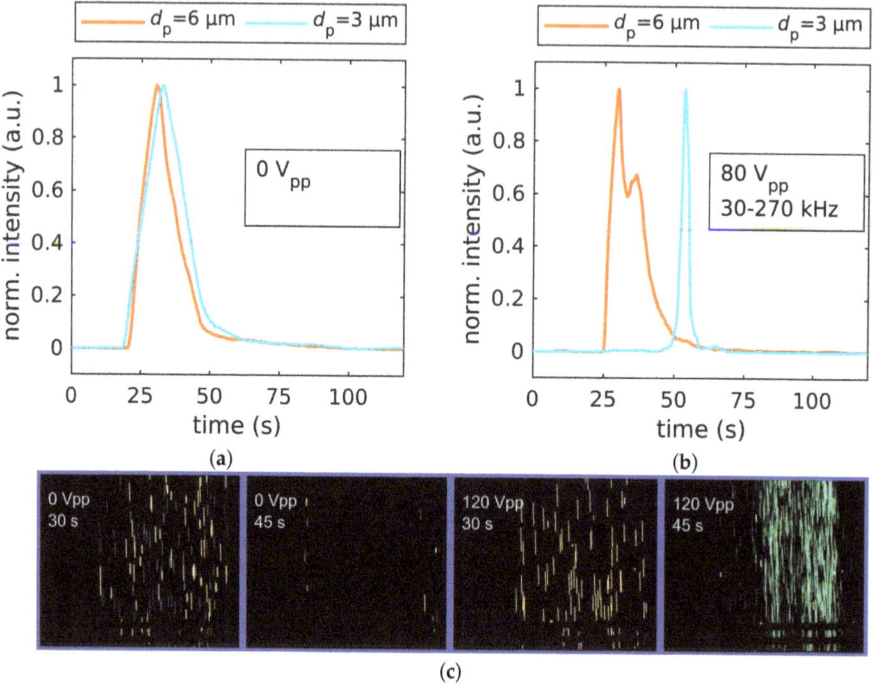

Figure 3. (**a**) Separation of 3 and 6 µm polystyrene particles: fluorescence intensity over time for no applied voltage, $R_s = 0.17 \pm 0.06$ ($N = 4$), and (**b**) when applying 80 V_{pp} at 30 kHz–270 kHz with a modulation frequency of 300 mHz, $R_s = 3.60 \pm 0.31$ ($N = 4$). (**c**) Single frames of different times of 3 µm (yellow-green) and 6 µm (orange/red) fluorescent polystyrene particles (brightness and contrast are adjusted for better visibility).

At all investigated voltages (see Figure S4 and Figure 3b), the 6 µm particles eluted earlier than the smaller particles, which showed a substantial delay with respect to measurements without the electric field. This was because the 3 µm particles showed predominately pDEP in the frequency modulation range and thus were substantially retarded due to the DEP interaction (Figure 2, blue line). Interestingly, the peak size of the 3 µm particles decreased significantly, which suggested that their retention time was dominated by DEP and not by their initial height in the channel, as was visible in the experiments without applied voltage. In contrast, the peak size and position of the 6 μm stayed almost the same, which was due to the balanced nDEP/pDEP ratio (Figure 2, orange line). This balanced nDEP/pDEP led to a negligible movement orthogonal to the fluid-flow direction over one cycle. Consequently, at low to moderate voltages, particles were only slightly retarded in the channel caused by moving along the different streamlines of the parabolic flow profile. Since the dielectrophoretic velocity increased with increasing electric field strength (Equation (2)) and therefore with increasing applied potential, we assumed that particles traveled greater distances orthogonal to the flow, eventually hitting either the electrode array or channel ceiling, as the voltage increased. We thus assumed that with increasing voltage, also 6 μm particles would experience retardation.

The calculations suggested (Figure 2) that the retardation of the 3 μm PS particles was based on their movement towards the interdigitated electrodes (pDEP). To investigate the effect of nDEP on the retention time, we separated 10 μm particles, which showed predominantly nDEP in the chosen frequency modulation spectrum (Figure 2), from the balanced 6 μm particles (Figure 4). This switch from pDEP dominated behavior, to an nDEP/pDEP -balanced behavior, to an nDEP dominated behavior with increasing particle size was due to the decreasing conductivity of polystyrene particles with increasing diameter (Equation (5)). We observed a chromatographic separation of 10 μm from 6 μm particles (see Figure S5 for the full dataset) for 80 V_{pp}, 100 V_{pp}, and 120 V_{pp} at 30 kHz–270 kHz. As before, the 6 μm particles showed almost no change in their retention time, whereas the larger and less polarizable 10 μm particles showed substantial delay, which indicated a retardation due to nDEP. Figure S9 shows PIV data of the 10 μm particles at 0 V and at 100 V_{pp} to demonstrate how their velocity periodically decreased and increased due to the nDEP effect. This periodic velocity fluctuation corresponded exactly to the applied frequency modulation.

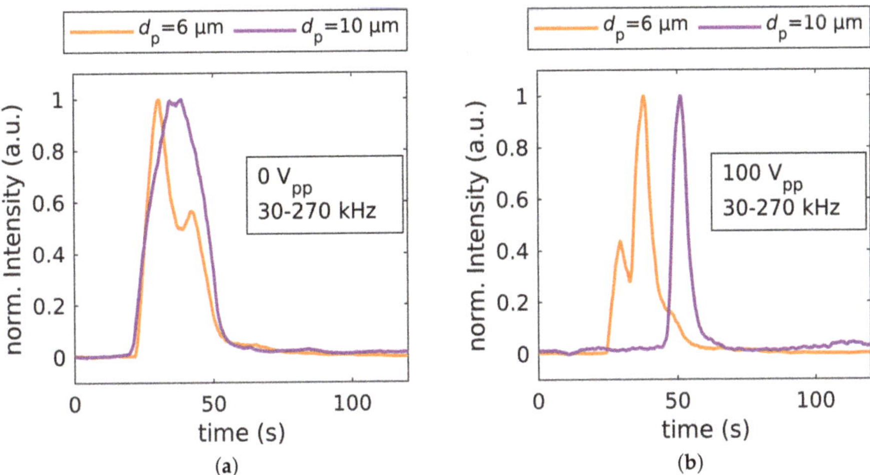

Figure 4. (**a**) Separation of 6 and 10 μm particles: fluorescence intensity over time of without applied voltage, $R_s = 0.21 \pm 0.19$ ($N = 4$), and (**b**) with application of 80 V_{pp} at 30 kHz–270 kHz with a modulation frequency of 300 mHz, $R_s = 1.95 \pm 0.33$ ($N = 4$).

Before we address the more challenging task of separating 2 μm and 3 μm particles that both experience pDEP in the modulation spectrum, we discuss the resolution for the separation of 6 μm from 3 μm and 6 μm from 10 μm (Figure 5). The resolution of the separation of 3 and 6 μm (Figure 5a, green) particles increased with voltage in all conducted experiments until a maximum at 80 V_{pp} was reached, after which the resolution decreased. This was because the retention time of bigger particles increased further with voltage (80 V_{pp}: 27.92 s ± 1.74 s to 160 V_{pp}: 40.8 s ± 3.33 s, both $N = 4$), while the time of the maximum fluorescence intensity for the smaller particles was constant for all voltages investigated, as long as a voltage was applied. We suspect the increase of the retention time of the 6 μm particles was because of the increased covered distances orthogonal to the fluid-flow. As previously discussed, the higher field strength caused the particles to reach the walls or at least enter regions close to a wall with low fluid velocity (at the top and bottom of the channel) and to be retarded as a consequence.

This decrease in resolution was not observed for the separation of 6 μm and 10 μm particles (Figure 5a, turquoise). Although the retention times of the 6 μm increased monotonically with voltage, the resolution simultaneously increased with applied voltage. The even stronger increase in retention time of the 10 μm particles (80 V_{pp}: 45.61 s $\pm$ 5.24 s to 120 V_{pp}: 57.71 s $\pm$ 1.6 s, both $N = 4$) compensated the increase from the 6 μm particles. We assumed that the 10 μm particles spent even more time in areas with low fluid velocity, and therefore, the retention time increased. Our PIV measurements (see Figure S9) indicated a periodic interaction of the particles with the electric field. However, the change in velocity was below 20 %, which showed additional potential for increasing the retention time of the 10 μm particles.

To investigate the effect of the voltage on the resolution of the separation process further, particles with diameters of 2 μm and 3 μm were selected. As the mobility of both particles was close to each other, this posed a more ambitious separation problem. Using the same set of parameters as before, we could again observe a voltage dependence of the peak time (see Figure 5b and also Figures S6–S8, for intensity profiles as a function of time). In contrast to the 3 vs. 6 μm and 6 vs. 10 μm experiments, the separation was low for all investigated voltages. This was because μ_{DEP} for both particle types was low and very close to each other.

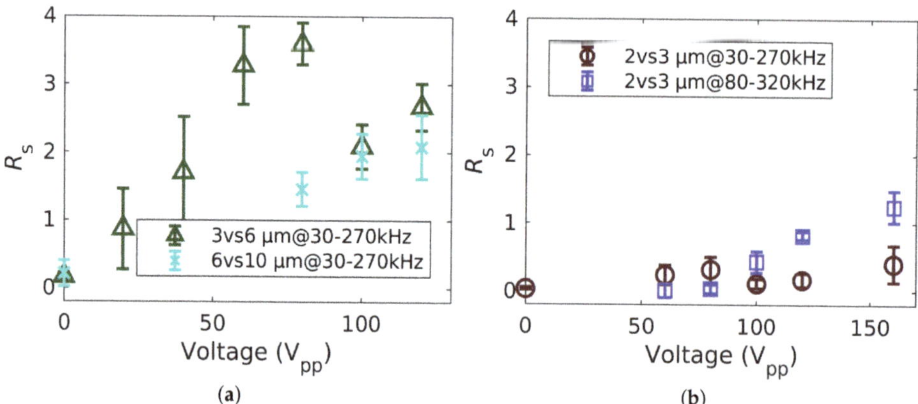

Figure 5. Resolution R_s of DPC over applied voltage for different particle suspensions and frequencies. (**a**) 3 μm vs. 6 μm and 6 μm vs. 10 μm PS particles at 30 kHz to 270 kHz. (**b**) 2 μm vs. 3 μm PS particles at 30 kHz to 270 kHz and 80 kHz to 320 kHz.

The separation of the 2 and 3 μm particles was improved, concerning peak width and peak distance, by changing the frequency between which was varied. Applying an offset of 50 kHz (now: 80 kHz to 320 kHz), the retention times of each particle type became more homogeneous (FWHM decreased) and the distance between the peaks increased (Figure 5b). Interestingly, the resolution was similarly low for both sets of frequencies for all voltages except for 120 V_{pp} and 160 V_{pp}, but the retention times were significantly different. At 120 V_{pp}, where the highest resolution using 30 kHz–270 kHz was achieved, the particles eluted almost 15 s later than when using 80 kHz–320 kHz at the same voltage (2 μm: 55.22 s $\pm$ 6.94 s vs. 40.73 s $\pm$ 0.75 s, 3 μm: 49.06 s $\pm$ 5.32 s vs. 36.86 s $\pm$ 0.84 s, both $N = 4$).

We propose that at the lower frequency (30 kHz to 270 kHz), both particles were dominated by pDEP (Figure 2) and thus showed a significant increase in retention time with increasing voltage.

When the frequency set switched to 80 kHz–320 kHz, the pDEP/nDEP behavior was more balanced, i.e., both particles exhibited less pDEP and more nDEP in the modulation spectrum, causing them to interact less with the field and thus to elute earlier. As expected, the subtle differences in polarizability of $2\,\mu m$ and $3\,\mu m$ particles in this frequency bandwidth were more pronounced, resulting in a better resolution. Although the residence time of the particles was much shorter when applying 80 kHz-320 kHz, the resolution stayed the same at $100\,V_{pp}$ and increased even further to $R_s = 1.25 \pm 0.23, (N = 4)$, when the voltage was set to $160\,V_{pp}$, which did not occur using the lower frequency set. This highlights one of the potentials of our separation technique, i.e., the possibility to separate particles with very equal polarizabilities by tuning the frequency modulation according to the target particle's polarizabilities. Again, the particles larger in diameter eluted earlier, and in contrast to the $6\,\mu m$ particles, both particles showed an increase in retention time with respect to the measurements without the electric field and therefore without superimposed dielectrophoretic movement. No maximum in retention time was found for the 2 vs. $3\,\mu m$ mixture at higher frequency, indicating that a further increase in voltage led to a further increase in resolution.

Since the dielectrophoretic velocity depended quadratically on the particle's radius and the applied electric field, the size and voltage dependency was not surprising. Due to this, smaller particles accelerated less due to DEP. Additionally, the cross-over frequency at which the force switched from pDEP to nDEP was higher, i.e., small particles experienced pDEP for a longer duration per cycle. Consequently, small particles, once they came close to the interdigitated electrodes, remained there and thus in regions of low fluid velocity. The latter point should become more important as the residence time in the separation column increases (i.e., at a longer column length).

Both nDEP and pDEP can be utilized to induce a retardation of the suspended particles. As a consequence, particles with polarizability (e.g., one showing more pDEP, another one dominated by nDEP) can elute at the same time. However, as the frequencies and the modulation frequencies can be adjusted, the nDEP/pDEP ratio can be tuned, which should result in different retention times and lead to a chromatographic separation.

Despite the fact that the parameters were chosen by evaluating the mobility of the particles over the frequency and only model particles were evaluated, the technique can become a tool for chromatographic separation of arbitrary particles that show a frequency dependent polarizability. One major advantage of this technique is that the columns' parameters were adjustable without actually changing the column itself. As shown, the electric field strength and the frequency bandwidth had an impact on the retention times and the peak width. Joule heating could disturb the separation when mediums with higher conductivities are used (e.g., cell buffer) [41]. To reduce the required voltage, by maintaining a similar electric field strength, the thickness of the isolating layer on the electrodes could be reduced. Promising alternatives to PDMS to achieve thinner coatings are polymers with a lower viscosity (e.g., SU-8). Additionally, since the electric field decreases with the height of the channel significantly, channels with a reduced height could be used.

4. Conclusions

We experimentally showed the separation of three binary mixtures of suspended particles using dielectrophoretic particle chromatography with a modulated electric field frequency. The current data further suggested that a separation of three different particle types (for example 2, 3 and $6\,\mu m$) in a single experiment should be possible. Unfortunately, it was not possible to observe all three different kinds of particles at the same time with the current hardware. We believe that an increasing column length led to a better separation. In addition to this, when the injection valve was operated automatically (in contrast to the current manual operation), standard deviations should decrease significantly. The influence of other parameters such as the modulating frequency, the medium's electrical conductivity, the linearity of amplification, and the carrier fluids' volume flow are complex and not yet understood in detail. Comprehensive studies regarding their impact using experiments and simulations are under way. Nevertheless, we demonstrated the principle and discussed the effect of the applied voltage. We further showed how adapting the modulation frequency to the target particle's polarizabilities further increased the resolution. In the proposed chromatography column, no single trap-and-release mechanism was used to achieve a chromatographic separation, but the particles showed different interactions with the permanently present and adjustable stationary phase. Although we only studied model particles in this study, the presented method allowed chromatographically separating arbitrary particles with frequency dependent polarizabilities. We believe that the presented technique can potentially separate particle mixtures that are traditionally difficult to separate, for instance cell separation in liquid biopsy or the recovery of precious materials from waste streams.

Supplementary Materials: The following are available online at http://www.mdpi.com/2072-666X/11/1/38/s1, Section S1: Method; Section S2: Device Fabrication; Section S3: Post-Processing; Section S4: Further Experimental Results; Figure S1: Real part of Clausius-Mossotti factor and DEP mobility; Figure S2: Gold electrodes and photomask; Figure S3: Result of segmentation process; Figure S4: Fluorescence intensity over time of 3 μm and 6 μm particles; Figure S5: Fluorescence intensity over time of 6 μm and 10 μm particles; Figure S6: Fluorescence intensity over time of 2 μm and 3 μm particles at 120 V_{pp}; Figure S7: Fluorescence intensity over time of 2 μm and 3 μm particles at 160 V_{pp}; Figure S8: Fluorescence intensity over time of 2 μm and 3 μm particles at different voltages; Figure S9: PIV data of 10 μm PS particles; Video S1: 10um_100Vpp_PIVdata.mp4, Video of periodic velocity changes of 10 μm particles; Video S2: DPC_0Vpp_3vs6umPS.mp4, Video of 3 μm and 6 μm particles without applied voltage; Video S3: DPC_80Vpp_3vs6umPS.mp4, Video of 3 μm and 6 μm particles at 80 V_{pp}.

Author Contributions: J.G., G.R.P., M.B., and J.T. conceived of the experiments. J.G. conducted the experiments. M.-P.S. fabricated the electrodes and SU8 master and contributed their layout. J.G., G.R.P., L.W., M.B., and J.T. analyzed the results. G.R.P. and M.B. supervised the project. J.G. wrote the manuscript with input from all other authors. All authors have read and agreed to the published version of the manuscript.

Funding: This work was funded by the German Research Foundation (DFG) within the priority program, "MehrDimPart—Highly specific and multidimensional fractionation of fine particle systems with technical relevance" (SPP2045, Grant Numbers BA 1893/2-1, TH 893/20-1, and project number 382064650).

Acknowledgments: The authors thank Jonathan Kottmeier (IMT, TUBraunschweig) for the help in manufacturing the components of the separation column and Fei Du (CVT, Universität Bremen) for fruitful discussions.

Conflicts of Interest: The authors declare no conflict of interest.

Abbreviations

The following abbreviations are used in this manuscript:

AC	alternating current
CM	Clausius–Mossotti factor
DEP	dielectrophoresis
DPC	dielectrophoretic particle chromatography
FWHM	full width at half maximum
nDEP	negative dielectrophoresis
pDEP	positive dielectrophoresis
PDMS	polydimethylsiloxane
PS	polystyrene

References

1. Vahey, M.D.; Voldman, J. An equilibrium method for continuous-flow cell sorting using dielectrophoresis. *Anal. Chem.* **2008**, *80*, 3135–3143, doi:10.1021/ac7020568. [CrossRef] [PubMed]
2. Moon, H.S.; Kwon, K.; Kim, S.I.; Han, H.; Sohn, J.; Lee, S.; Jung, H.I. Continuous separation of breast cancer cells from blood samples using multi-orifice flow fractionation (MOFF) and dielectrophoresis (DEP). *Lab Chip* **2011**, *11*, 1118, doi:10.1039/c0lc00345j. [CrossRef] [PubMed]
3. Henslee, E.A.; Sano, M.B.; Rojas, A.D.; Schmelz, E.M.; Davalos, R.V. Selective concentration of human cancer cells using contactless dielectrophoresis. *Electrophoresis* **2011**, *32*, 2523–2529, doi:10.1002/elps.201100081. [CrossRef] [PubMed]

4. Wang, X.B.; Yang, J.; Huang, Y.; Vykoukal, J.; Becker, F.F.; Gascoyne, P.R.C. Cell separation by dielectrophoretic field-flow-fractionation. *Anal. Chem.* **2000**, *72*, 832–839, doi:10.1021/ac990922o. [CrossRef] [PubMed]
5. Washizu, M.; Suzuki, S.; Kurosawa, O.; Nishizaka, T.; Shinohara, T. Molecular dielectrophoresis of bio-polymers. In Proceedings of the Conference Record of the 1992 IEEE Industry Applications Society Annual Meeting, Houston, TX, USA; pp.1446–1452, doi:10.1109/IAS.1992.244397. [CrossRef]
6. Spengler, T.; Ploog, M.; Schröter, M. Integrated planning of acquisition, disassembly and bulk recycling: A case study on electronic scrap recovery. *OR Spectr.* **2003**, *25*, 413–442. [CrossRef]
7. Du, F.; Baune, M.; Kück, A.; Thöming, J. Dielectrophoretic gold particle separation. *Sep. Sci. Technol.* **2008**, *43*, 3842–3855, doi:10.1080/01496390802365779. [CrossRef]
8. Williams, S.K.R.; Runyon, J.R.; Ashames, A.A. Field-flow fractionation: Addressing the nano challenge. *Anal. Chem.* **2011**, *83*, 634–642, doi:10.1021/ac101759z. [CrossRef]
9. Hanauer, M.; Pierrat, S.; Zins, I.; Lotz, A.; Sönnichsen, C. Separation of nanoparticles by gel electrophoresis according to size and shape. *Nano Lett.* **2007**, *7*, 2881–2885, doi:10.1021/nl071615y. [CrossRef]
10. Wei, G.T. Shape separation of nanometer gold particles by size-exclusion chromatography. *Anal. Chem.* **1999**, *71*, 2085–2091, doi:10.1021/ac990044u. [CrossRef]
11. Wills, B.A.; Finch, J.A. Dewatering. In *Wills' Mineral Processing Technology*; Elsevier: Amsterdam, The Netherlands, 2016; Chapter 15, pp. 417–438, doi:10.1016/b978-0-08-097053-0.00015-7. [CrossRef]
12. Wills, B.A.; Finch, J.A. Classification. In *Wills' Mineral Processing Technology*; Number 1993; Elsevier: Amsterdam, The Netherlands, 2016; Chapter 9, pp. 199–221, doi:10.1016/B978-0-08-097053-0.00009-1. [CrossRef]
13. Contado, C. Field flow fractionation techniques to explore the "nano-world". *Anal. Bioanal. Chem.* **2017**, *409*, 2501–2518, doi:10.1007/s00216-017-0180-6. [CrossRef]
14. Pesch, G.R.; Lorenz, M.; Sachdev, S.; Salameh, S.; Du, F.; Baune, M.; Boukany, P.E.; Thöming, J. Bridging the scales in high-throughput dielectrophoretic (bio-)particle separation in porous media. *Sci. Rep.* **2018**, *8*, 10480, doi:10.1038/s41598-018-28735-w. [CrossRef] [PubMed]
15. Suehiro, J.; Zhou, G.; Imamura, M.; Hara, M. Dielectrophoretic filter for separation and recovery of biological cells in water. *IEEE Trans. Ind. Appl.* **2003**, *39*, 1514–1521, doi:10.1109/TIA.2003.816535. [CrossRef]
16. Modarres, P.; Tabrizian, M. Frequency hopping dielectrophoresis as a new approach for microscale particle and cell enrichment. *Sens. Actuators B Chem.* **2019**, *286*, 493–500, doi:10.1016/j.snb.2019.01.157. [CrossRef]
17. Wang, Y.; Du, F.; Pesch, G.R.; Köser, J.; Baune, M.; Thöming, J. Microparticle trajectories in a high-throughput channel for contact-free fractionation by dielectrophoresis. *Chem. Eng. Sci.* **2016**, *153*, 34–44, doi:10.1016/j.ces.2016.07.020. [CrossRef]
18. Park, S.; Zhang, Y.; Wang, T.H.; Yang, S. Continuous dielectrophoretic bacterial separation and concentration from physiological media of high conductivity. *Lab Chip* **2011**, *11*, 2893, doi:10.1039/c1lc20307j. [CrossRef] [PubMed]
19. Pesch, G.R.; Du, F.; Schwientek, U.; Gehrmeyer, C.; Maurer, A.; Thöming, J.; Baune, M. Recovery of submicron particles using high-throughput dielectrophoretically switchable filtration. *Sep. Purif. Technol.* **2014**, *132*, 728–735, doi:10.1016/j.seppur.2014.06.028. [CrossRef]
20. Weirauch, L.; Lorenz, M.; Hill, N.; Lapizco-Encinas, B.H.; Baune, M.; Pesch, G.R.; Thöming, J. Material-selective separation of mixed microparticles via insulator-based dielectrophoresis. *Biomicrofluidics* **2019**, *13*, 064112, doi:10.1063/1.5124110. [CrossRef]
21. Sano, H.; Kabata, H.; Kurosawa, O.; Washizu, M. Dielectrophoretic chromatography with cross-flow injection. In Proceedings of the Technical Digest. MEMS 2002 IEEE International Conference. Fifteenth IEEE International Conference on Micro Electro Mechanical Systems (Cat. No.02CH37266), Las Vegas, NV, USA, 24 January 2002; pp. 2–5, doi:10.1109/MEMSYS.2002.984043. [CrossRef]
22. Kikkeri, K.; Ngu, B.; Agah, M.; Engineering, C.; Tech, V. Submicron Dielectrophoretic Chromatography. In Proceedings of the 2018 IEEE Micro Electro Mechanical Systems (MEMS), Belfast, UK, 21–25 January 2018; pp. 1177–1180.
23. Umezawa, Y.; Kobayashi, O.; Kanai, S.; Hakoda, M. Development of Particle Packed Bed Type Chromatography Using Dielectrophoresis. *Key Eng. Mater.* **2013**, *534*, 88–92, doi:10.4028/www.scientific.net/KEM.534.88. [CrossRef]

24. Hakoda, M.; Otaki, T. Analytical Characteristic of Chromatography Device Using Dielectrophoresis Phenomenon. *Key Eng. Mater.* **2012**, *497*, 87–92, doi:10.4028/www.scientific.net/KEM.497.87. [CrossRef]

25. Gascoyne, P.R.; Noshari, J.; Anderson, T.J.; Becker, F.F. Isolation of rare cells from cell mixtures by dielectrophoresis. *Electrophoresis* **2009**, *30*, 1388–1398, doi:10.1002/elps.200800373. [CrossRef]

26. Gupta, V.; Jafferji, I.; Garza, M.; Melnikova, V.O.; Hasegawa, D.K.; Pethig, R.; Davis, D.W. ApoStream™, a new dielectrophoretic device for antibody independent isolation and recovery of viable cancer cells from blood. *Biomicrofluidics* **2012**, *6*, 024133, doi:10.1063/1.4731647. [CrossRef]

27. Shim, S.; Stemke-Hale, K.; Tsimberidou, A.M.; Noshari, J.; Anderson, T.E.; Gascoyne, P.R. Antibody-independent isolation of circulating tumor cells by continuous-flow dielectrophoresis. *Biomicrofluidics* **2013**, *7*, 011807, doi:10.1063/1.4774304. [CrossRef]

28. Aldaeus, F.; Lin, Y.; Amberg, G.; Roeraade, J. Multi-step dielectrophoresis for separation of particles. *J. Chromatogr. A* **2006**, *1131*, 261–266, doi:10.1016/j.chroma.2006.07.022. [CrossRef]

29. Green, N.G.; Hughes, M.P.; Monaghan, W.; Morgan, H. Large area multilayered electrode arrays for dielectrophoretic fractionation. *Microelectron. Eng.* **1997**, *35*, 421–424, doi:10.1016/S0167-9317(96)00122-0. [CrossRef]

30. Cheng, I.F.; Froude, V.E.; Zhu, Y.; Chang, H.C.; Chang, H.C. A continuous high-throughput bioparticle sorter based on 3D traveling-wave dielectrophoresis. *Lab Chip* **2009**, *9*, 3193–3201, doi:10.1039/b910587e. [CrossRef]

31. Sun, T.; Morgan, H.; Green, N.G. Analytical solutions of ac electrokinetics in interdigitated electrode arrays: Electric field, dielectrophoretic and traveling-wave dielectrophoretic forces. *Phys. Rev. E Stat. Nonlinear Soft Matter Phys.* **2007**, *76*, 046610, doi:10.1103/PhysRevE.76.046610. [CrossRef]

32. García-Sánchez, P.; Ramos, A.; González, A.; Green, N.G.; Morgan, H. Flow reversal in traveling-wave electrokinetics: An analysis of forces due to ionic concentration gradients. *Langmuir* **2009**, *25*, 4988–4997, doi:10.1021/la803651e. [CrossRef]

33. Hughes, M.P. Fifty years of dielectrophoretic cell separation technology. *Biomicrofluidics* **2016**, *10*, 032801, doi:10.1063/1.4954841. [CrossRef]

34. Pethig, R. Dielectrophoresis: Status of the theory, technology, and applications. *Biomicrofluidics* **2010**, *4*, 022811, doi:10.1063/1.3456626. [CrossRef]

35. Ermolina, I.; Morgan, H. The electrokinetic properties of latex particles: Comparison of electrophoresis and dielectrophoresis. *J. Colloid Interface Sci.* **2005**, *285*, 419–428, doi:10.1016/j.jcis.2004.11.003. [CrossRef]

36. Arnold, W.M.; Schwan, H.P.; Zimmermann, U. Surface conductance and other properties of latex particles measured by electrorotation. *J. Phys. Chem.* **1987**, *91*, 5093–5098, doi:10.1021/j100303a043. [CrossRef]

37. Satyanarayana, S.; Karnik, R.N.; Majumdar, A. Stamp-and-stick room-temperature bonding technique for microdevices. *J. Microelectromech. Syst.* **2005**, *14*, 392–399, doi:10.1109/JMEMS.2004.839334. [CrossRef]

38. Gajasinghe, R.W.; Senveli, S.U.; Rawal, S.; Williams, A.; Zheng, A.; Datar, R.H.; Cote, R.J.; Tigli, O. Experimental study of PDMS bonding to various substrates for monolithic microfluidic applications. *J. Micromech. Microeng.* **2014**, *24*, 075010, doi:10.1088/0960-1317/24/7/075010. [CrossRef]

39. Koschwanez, J.H.; Carlson, R.H.; Meldrum, D.R. Thin PDMS films using long spin times or tert-butyl alcohol as a solvent. *PLoS ONE* **2009**, *4*, e4572, doi:10.1371/journal.pone.0004572. [CrossRef]

40. Wang, X.B.; Vykoukal, J.; Becker, F.F.; Gascoyne, P.R. Separation of polystyrene microbeads using dielectrophoretic/gravitational field-flow-fractionation. *Biophys. J.* **1998**, *74*, 2689–2701, doi:10.1016/S0006-3495(98)77975-5. [CrossRef]

41. Castellanos, A.; Ramos, A.; González, A.; Green, N.G.; Morgan, H. Electrohydrodynamics and dielectrophoresis in microsystems: Scaling laws. *J. Phys. D Appl. Phys.* **2003**, *36*, 2584–2597, doi:10.1088/0022-3727/36/20/023. [CrossRef]

MDPI

St. Alban-Anlage 66

4052 Basel

Switzerland

Tel. +41 61 683 77 34

Fax +41 61 302 89 18

www.mdpi.com

Micromachines Editorial Office

E-mail: micromachines@mdpi.com

www.mdpi.com/journal/micromachines